AF516434

ANANDA K. COOMARASWAMY
WRITINGS ON GEOLOGY AND MINERALOGY

Ananda K. Coomaraswamy
Writings on

GEOLOGY and MINERALOGY

Scientific Papers and Comments

Edited by

A. Ranganathan

K. Srinivasa Rao

INDIRA GANDHI NATIONAL CENTRE FOR THE ARTS
NEW DELHI

MANOHAR
2001

First published 2001

ISBN 81-7304-373-6

Published by
Indira Gandhi National Centre for the Arts
Janpath, New Delhi 110001
in association with
Ajay Kumar Jain for
Manohar Publishers & Distributors
4753/23 Ansari Road, Daryaganj
New Delhi 110002

Rajkamal Electric Press
B 35/9 GT Karnal Road Indl. Area
Delhi 110033

Contents

Preface

Airavatham Ranganthan (Rangu to his close friends), distinguished freelance journalist and writer – well known as an art critic, science writer, book reviewer and political commentator – made the life and works of Dr. Ananda Kentish Coomaraswamy (22 Aug. 1877 – 9 Sep. 1947), a comprehensive subject for a life-long study. His well directed efforts to draw the attention of Mrs. Indira Gandhi resulted in the issuance of a stamp by the Government of India in honour of the great scholar, Dr. Ananda Coomaraswamy, in 1977, on the occasion of his birth Centenary. Rangu lectured on Dr. Coomaraswamy at several centers, including the India International Center, New Delhi; the National Museum, Colombo; the Max Meuller Bhavan, New Delhi and the Autonomous Postgraduate Center, Tiruchirapalli. His seven All India Radio talks broadcast from Guwahati, Madras and New Delhi stations have been included in a book entitled: *A Coomaraswamy Spectrum* (published by The Christian Literature Society of Madras, 1981).

Mr. Ranganathan was a regular visitor to the Anniversary Symposia and special lectures by distinguished Visiting Professors to the Institute of Mathematical Sciences (IMSc), since the inception of the Institute in 1962, as a special invitee of Prof. Alladi Ramakrishnan, the Founder Director of IMSc. This association brought Rangu into close contact first with my seniors, Prof. R. Vasudevan, Prof. N.R. Ranganathan and Prof. V. Radhakrishnan and later with me. I introduced Rangu to the subsequent Directors of IMSc, Prof. E.C.G. Sudarshan and Prof. R. Ramachandran, and he continued to visit the Institute periodically. This contact of decades led Rangu to reveal to me, in 1996, that he had an invitation from Dr. Kapila Vatsyayan, of the Indira Gandhi National Center for Arts (IGNCA), to compile the writings of Dr. Ananda Coomaraswamy on Geology and Mineralogy Eventually, in July 1996, he showed me the old, fading 'xerox' copies of the works of Coomaraswamy, in his possession for several years and asked me to get them into a Camera Ready Copy (CRC) format for publication – an impossible task! I offered to get him modern xerox copies of the papers from Europe, since I was planning a visit to Germany and Belgium that year. Armed with the list of Dr. Coomaraswamy's publications in International Journals (of 1903 – 1906), I left for the Arnold Sommerfeld Institute of the Technical University of Clausthal, in Germany, in Aug. 1996.

As luck would have it, the Arnold Sommerfeld Institute and the Institute of Geology and Paleontology of the Technical University of Clausthal were housed in the same building. I requested Prof. Dr. H.-D. Doebner, the Director of the Arnold Sommerfeld Institute, and my host, to introduce me to Prof. Dr. Horst Quade, the Director of the Institute of Geology and Paleontology. Prof. Horst Quade in turn introduced me to his assistant, Dr. Doris Schmidt, and with her help I was able to gather all the required papers from the Central Documentation Library in Hannover by the end of 1996, before I returned to my Institute. A few of the required papers were available from the library of the Technical University itself and the helpful staff of the TU-library made xerox copies of these papers which I promptly posted in Sep. 96 itself to Rangu telling him that with such copies it would be easy to prepare, a Camera Ready Copy of the book on the writings of Ananda Coomaraswamy on Geology and Mineralogy.

Meanwhile, on his part, Rangu obtained copies of the four annual Ceylon Reports of Coomaraswamy from Sri Lanka's Mineralogical Director. He also had written his introduction to the book. At his instance, several bound volumes containing the newly acquired xerox copies of the papers of Dr. Ananda Coomaraswamy and Ranganathan's introduction were made. He took them with him on one of his visits to Delhi and discussed the matter of how the CRC can be prepared with Drs. Kapila Vatsyayan and Lalit M. Gujral, on the basis of his discussions with me.

On Rangu's return, he asked me to get the work on the CRC of the book started, with the help of professionals in LaTeX in April 1997. Over a period of three months, a series of files, containing the rough draft version of all the 28 papers of Coomaraswamy on Geology and Mineralogy, was prepared. This was when the tragedy struck Rangu, in the form of a renal failure. With the help of his good friends, Mr. T.C.A. Ramanujam (Retired Income Tax Commissioner) and Ms. Seeta Badrinath (Advocate), and my reliable Doctor friends, Dr. Radha Rajagopal and Dr. S. Rajagopal, I got Rangu admitted in the Wellingdon Nursing Home for periodic dialysis.

When I left on a four month trip to Europe in Aug. 1997, I was under the impression that Rangu, on his road to recovery, would be back on rails when I returned. Unfortunately this was not to be. Two weeks after my return, on Dec. 8, 1997, Rangu went with his Housekeeper, Lakshmi Ammal, to keep his appointment for dialysis at the Apollo Hospital and in the early morning hours he breathed his last in the hospital. Soon after, I was shocked to hear from Lakshmi Ammal, on the telephone, that Rangu was no more. Rangu was a bachelor and had lived with his mother for most of his life, since his father, Airavatham, ICS, drowned in a boat tragedy on the river Hoogly when Rangu was barely three years old. Rangu died, as he lived, sans publicity, unsurrounded by members of the family and relatives, waited upon by his faithful and loyal Housekeeper, Lakshmi Ammal. I informed his close friends and his cousin, Mr. S.K. Sarma, in Delhi about the death. Thanks to the good offices of Mr. C.R. Ramaswami (formerly of AIR), Rangu's death was reported in the Madras edition of the *Indian Express* (of Dec. 9, 1997):

> **Veteran Journalist Passes away**
> Chennai, Dec. 8: A. Ranganathan (60), freelance journalist, passed away on Monday following renal failure ... He was a close friend of Nobel laureate Dr. S. Chandrasekhar and ECG Sudarshan. His letters to S. Chandrasekhar on Ramanujan have been published in *Ramanujan: Letters and Commentary*, by Bruce C. Berndt and Robert A. Rankin.

Since the half finished manuscript, on Ananda Coomaraswamy's writings on Geology & Mineralogy, was in my possession and I had by then got all the figures in the xerox copies of the papers photographed (by Mr. Kannappan of Lakshmi Studio), I informed Dr. Kapila Vatsyayan about the status of Rangu's manuscript and also about the unfortunate, untimely demise of Rangu.

In due course, by Feb. 1998, Dr. Kapila Vatsyayan and Dr. Lalit M. Gujral suggested that the CRC work should be continued and completed by me. The main difficulty was in resolving the problem of pagination. The journal articles and the Ceylon Reports were not only using different fonts and fontsizes, but the latter were on foolscap pages. This necessitated the dual pagination for the book – a running pagination for the present LaTeX version of the book and another pagination wherein the page numbering was that of the source: viz. the Journal / Report. To make this obvious, it was decided to use *boxes* for the pages of the book and the pagination of the source retained with the headings of the Journal / Report inside the *boxes.*

It is fitting that the cover design depicts the postal stamp of Ananda Coomaraswamy, and Rangu considered its commissioning and release as a successful attempt of his in convincing the powers that be about the meritorious life and works of Ananda Coomaraswamy.

The last work of Rangu, started just before he was struck by end stage renal failure, is now completed and this is his greatest tribute to the prolific Dr. Ananda Coomaraswamy, who, in the words of the French savant, Romain Rolland*:

> *is one of those great Hindus who nourished like Tagore on the cultures of Europe and Asia, and justifiably proud of their splendid civilization, have conceived the task of working for the Union of Eastern and Western thought for the good of humanity.*

Dr. K. Srinivasa Rao
Senior Professor
The Institute of Mathematical Sciences.

Chennai - 600 113
May 2, 2000

**Ananda K. Coomaraswamy (1877 - 1947): A Critical Appreciation*, S. Chandrasekhar, Blackie & Son Publishers Pvt. Ltd., Bombay (1977) p.32.

Acknowledgements

It is a pleasure to acknowledge and record our thanks to Dr. Kapila Vatsyayan, Indira Gandhi National Center for Arts (IGNCA), who made the suggestion for this book to Mr. A. Ranganthan, and later, in 1998, wanted me to complete the work to produce a Camera Ready Copy (CRC) of the manuscript.

I am thankful to Dr. R. Ramachandran, Director, Institute of Mathematical Sciences (IMSc), for granting me the necessary permission to complete the task of preparing the CRC and for allowing the use of the excellent computer facilities of the Institute in the final stages of the work and for the laser print out of the CRC.

Dr. Lalit M. Gujral, Advisor, IGNCA, offered all possible help and was willing to wait patiently until I could find time to coordinate and complete the work. I am thankful to Dr. Gujral, in particular, for sending me a copy of A. Ranganathan's:
A Coomaraswamy Spectrum (The Christian Literature Society, Madras (1981)).

The typesetting work in LaTeX was done by professionals. The dual pagination idea of mine, necessitated by the nature of the originals, was solved for me by Mr. P. Karthikeyan of the pagination section of Scientific Publishing Services Pvt. Ltd. (SPS). I am thankful to Mr. R. Balasubramanian (Technical Director), Mr. V. Narendra Kumar (associate Director) and Mr. M.S. Viswanathan (Personnel Manager) of SPS for allowing this.

My thanks are due to Prof. Dr. H.-D. Doebner, Director of the Arnold Sommerfeld Institute, for introducing me to Prof. Dr. Horst Quade, Director, Institute for Geology and Paleontology, Technical University of Clausthal, Germany, in Aug. 1996. Prof. Quade in turn intoduced me to Dr. Doris Schmidt who obtained for me xerox copies of most of the articles from the Hannover documentation center in Germany. Mr. Manfred Steinwedel and Frau Ines Jobman of the Technical University of Clausthal Library helped me with the xerox copies of some of the articles from the Journal of Geological Society, London and the Quarterly Journal of Geological Society, London.

My thanks are due to Mrs. and Mr. S. Jayaraman (Post Master General at Chennai) for directing me to Mrs. Radha Dwarakanath – daughter of the Philatelists, Mr. A.C. Thrilokanath and Mrs. A.C. Sukanyanath – who was kind to search, find and offer me from her parents' collection a stamp of Ananda Kentish Coomarswamy used in the cover design; and to Dr. G. Subramoni, Systems Analyst, IMSc and Ms. R. Indra for their help in the realization of my cover design.

Last but not least, I am grateful to my wife Dr. Geetha S. Rao and my sons, K.S. Aravind and K.S. Anand who were sympathetic and helpful to all my efforts to complete a job I had undertaken for Mr. A. Ranganathan, who was a good friend of the family.

Dr. K. Srinivasa Rao, F.N.A.Sc.,
Senior Professor
The Institute of Mathematical Sciences.

Chennai - 600 113
May 2, 2000

Ananda Coomaraswamy's Contributions to Geology and Mineralogy

INTRODUCTION by A. RANGANATHAN

Dr. Ananda Kentish Coomaraswamy – familiarly known as Ananda Coomaraswamy – comes close to deserving the title of a universal genius of the twentieth century. Indeed Ananda Coomaraswamy was one of the most remarkable examples of the *homo universalis* in an age of specialization. His universal approach stemmed undoubtedly from his unlimited curiosity and brilliant powers of observation. *Sapere Videre* (knowing how to see) was the theme behind all his work, whether it was a geological discovery or a treatise on Buddhist Iconography. In fact his ability to switch from science to art history, and from art history through religion on to metaphysics, revealed an impressive many-sidedness that was possible only to one whose mind was cast in the mould of the great classical philosophers like Plato, Meister Eckhart, St. Thomas Aquinas and Sankara. For, they were deeply attached to the *Adhyatmic* view of *the unity of all life – one source, one essence and one goal.*

Ananda Coomaraswamy is important in the history of the twentieth century for two reasons: he was a brilliant scientist-writer, and he was a seminal influence. As a scientist-writer, Coomaraswamy was a skilled practioner in the art of scientific communication. Indeed Madame Curie's account of the discovery of Radium, Albert Einstein's formulation of the Special Theory of Relativity in his paper **On the Electro-Dynamics of Moving Bodies**, Professor C.V. Raman's paper **On The Molecular scattering of Light in Water and the Colour of the Sea** and Ananda Coomaraswamy's **Report on the discovery of Thorianite** are pieces of great science writing. Again Coomaraswamy broke new ground in metaphysical subject matter, form and style. Furthermore, since nothing sacred was alien to Coomaraswamy, nothing sacred was omitted from his later writings. And one of the most important Coomaraswamy achievements was to extend the range of the English language in order to encompass the complexity, depth and grandeur of traditional thought.

In his admirably clear study of Coomaraswamy's theory of poet's creation entitled **The Traditional Theory of Literature**, Professor Ray Livingston has stressed the importance of a great critic like Coomaraswamy who makes our mind receptive to the depths of meaning and profound life-giving perennial works such as the **Vedas, Brahmanas, Upanishads**, as well as the writings of **Plato, Platinus, St. Augustine, Rumi, Dante, Eckhart** and **Blake**. Professor Livingston has also observed that in the process of winning back his birthright as an Oriental, Coomaraswamy responded as few men ever have to the **Faustian** challenge:

"What from your fathers' heritage is lost,
Earn it anew to really possess it."

However, Coomaraswamy's response to the **Faustian** challenge was fundamentally aesthetic. Indeed Coomaraswamycould focus his image, whether of the **Divine Comedy**, or a bronze figure of a dancing Shiva in a lightning flash of gnosis, revealing a new wave-length of aesthetic perception. For his horizon encompassed not merely the sublime figure of the Nataraja but also the majesty of the great Gothic cathedrals, the great poetry of **Dante**, the manifold meanings of the **Vedic** incarnations, **Paradise Lost** and the transcendental thought of **Eckhart** which reflected the "Spiritual being of Europe and its highest tension".

Juat as the Quantum Theory, which postulates the relationship between different aspects of one and the same physical process, led Niels Bohr to arrive at the epistemological truth of the Principle of Complementarity, so did Ananda Coomaraswamy forge the traditional symbols and figures of thought into a sensitive idiom of modern English prose. And though Coomaraswamy (born on the 22nd August 1877 and died on the 9th September 1947) will be considered here as a scientist, it is important to remember that his artisitic and scholarly interests led him to pioneer the disciplines of comparative aesthetics and comparative iconography, as well as make significant contributions to the comparative literature and the Perennial Philosophy.

Ananda Coomaraswamy was educated at Wycliffe College and University College, London. At a very early age, he developed an enthusiasm for geological studies. In fact, he gave a series of three lectures during the summer of 1894 – **The Rhaetic Beds**, **Darwin's Theory of Evolution** and **Fossils – Their Formation and Significance**. Coomaraswamy contributed an article on **Doverow, once more, Geological** to his School magazine, *The Wycliffe Star*, in 1895. Subsequently, he entered University College, London, in 1897 and obtained a first class B.Sc. (Honours) degree in Geology and Botany, in 1900.

Coomaraswamy's first formal scientific paper: **On Ceylon Rocks and Graphite**, was published in the August 1900 issue of the *Quarterly Journal of the Geological Survey* of London. Here it would be relevant to quote a sentence from this Coomaraswamy paper: *"Previous to 1887, remarks on the geology of Ceylon which have been made in books dealing with the island have been of a general character. The articles on the geology of Ceylon in the works of Sir Emerson Tennet (6), 'And an Officer of the Ceylon Rifles' (9) are worthy of note. Since then, important contributions to our knowledge of Ceylon Geology have been made by Sandberger (14), Lacriox (17), Walther (18), Melzi (29) and Diersche (30). No Geological Survey is in progress in Ceylon and it is much to be hoped that the Government will soon realize the importance of intiating one."* The bracketed numerals indicate bibliographical references, and the diagrams, the maps as well as the microphotographs that illustrate the paper give us an idea of his scientifically rigorous and meticulously documented work. And not surprisingly, Coomaraswamy was elected a

Fellow of University College, London, just before he left England for Ceylon in 1903.

Ananda Coomaraswamy was appointed Director of the Mineralogical Survey of Ceylon in 1903. During the next four years, Coomaraswamy's scientific work concerned in one way or another with what may be termed as the Geology of Ceylon (currently known as Sri Lanka). Coomaraswamy published not only geological maps (including the well-known map of Kandy district in 1906) but also several papers on the new minerals and the crystaline limestones and rocks of Ceylon. He also visualized the economic importance of the island in the work of the finds of such occurrences and moonstone, rare-earth minerals, iron ores, mica, graphite and corundum (ruby and saphire). Furthermore, his masterly **Administration Reports of the Mineralogical Survey of Ceylon** – for the years 1903, 1904, 1905 and 1906 – cover a wide range of geological, mineralogical and petrological subjects.

In this Introduction, we shall content ourselves with making a few general comments appearing in this volume. The first section of this volume entitled Annanda Coomaraswamy's **Contributions to the Geology of Ceylon** contains most of the scientific papers which earned for Coomaraswamy, as an external student of the University of London, the degree of Doctor of Science in 1906; as well as those scientific papers, geological and mineralogical, which complement his D.Sc. thesis. The second section includes the classic Ceylon **Administration Reports** which were written in 1903, 1904, 1905 and 1906. The final section which is mainly concerned with several **Comments** on Coomaraswamy's work – Letters to the Editor as well as scientific papers – highlight Ananda Coomaraswamy's reputation as a mainstream geologist, quite different from Coomaraswamy's undoubted greatness as an exponent of the Perennial philosophy.

Ananda Coomaraswamy is relevant to history of Geology and Mineralogy in Sri Lanka on five counts: First, his discoveries of the occurrence of the Radiolaria in Gondawana Beds near Madras, Serendibite (along with G.T. Prior) and Thorianite, helped to set a scientific standard for the new developing disciplines in Geology in Ceylon. Second, he had compiled a **Glossary of Sinahalese Mining Terms**, and a **Glossary of Sinhalese Terms used for Gems and Gemming**. Third, he brought out a comprehensive **Bibliography of Ceylon Geology** (1906) which has since been revised and published seperately by the eminent Indian geologist Prof. D.N. Wadia, F.R.S. Fourth, Coomaraswamy's brilliant papers on (a) **The Crystalline Limestones of Ceylon**, (b) **On Ceylon Rocks and Graphites**, (c) **The Wollastonite Scapolite Gneisses of Ceylon** and (d) the four-part essay entitled **Contributions to the Geology of Ceylon** have not only raised some interesting problems of petrology and mineralogy, but also reveal the inter-disciplinary nature of Coomaraswamy's scientific work. Finally, his discoveries relating to the ancient crystalline rocks of Ceylon soon entered into almost every branch of geology, so that they became an essential background in several geological works.

One of his important findings was that of the remains of Radiolaria (Protozoa possessing a siliceous skeleton of elements constituting a lattice-like structure) in the Gondawana Shales of Sriperamatur near Madras. Ananda Coomaraswamy wrote: *"Last year, while staying at Madras, I paid a short visit to Sriperamatur, accompanied by Mr. Rangachari, M.A., of Madras University. I did not obtain many plant-remains but was struck by the very porcellanic character of the shales containing them. These are thin-bedded, white to cream or pale coffee-coloured, and often have a smooth conchoidal fracture. A few specimens were collected from a wayside exposure, on the trace between Sriperamatur and Vellum. In the thin slices cut from these specimens radiolaria are to be seen, though for the most part in a very poor state of preservation."* This perceptive comment underscores his ability to raise timely geological questions.

Along with G.T. Prior, Coomaraswamy discovered a new borosilicate of alumina and lime with magnesia and alkalies (including lithia) at Gangapitiya, near Ambakotte, which is about 18 kilometers east of Kandy, in 1902. Soon after making this discovery Prior and Coomaraswamy noted: *"Bands of this granulite, which is composed mainly of quartz and feltspar in finely granulitic and occasionally graphic structure, altrnate with bands of limestone upto 18 inches wide. Between limestone and granulite occur contact-zones consisting, next to the limestone, almost entirely of a colourless diopside but near the granulite of a mixture of diopside with other minerals, viz., blue spinel, a little apatite, occasionally scapolite on plagioclase and the blue mineral which is the subject of the present note. Hand-specimens may be collected showing this relation. The Serendibite occurs in sufficiently large amount to give deep blue colour to the contact-zones."* Incidentally, recall that Professor Pierre and Professor Marie Curie had isolated *Polonium*, the first ever element to be discovered by radiochemical analysis in 1898. And just as the Poland-born scientist Marie Curie christened the new element *Polonium* – the name that perpetuates the memory of Poland – so did Prior and the Serendib-born geologist Coomaraswamy term the new mineral *Serendibite* in order to confer enduring fame upon **Serendib** – an old Arab name for Ceylon.

Prior and Coomaraswamy stated that *Serendibite*,

$$Ca_2(Mg, Al)_6O_2[(Si, Al, B)_6O_{18}]$$

exhibits pleochroism, which is a variation in colour when rotated in plane polarized light. Thus it absorbs different wave-lenghts in different directions. In other words, as intially observed by Prior and Coomaraswamy, it can be defined in terms of the different vibration directions such as *"very pale yellow (almost colourless) to pale sky blue in one variety and pale brownish-yelow to deep indigo-blue in another variety."* And the recent pleochroic investigations made by the contemporary mineralogists confirm the original observations made by Prior and Coomaraswamy. For the recently observed pleochroic complexions of colour in respect of Serendibite are pale greenish blue, yellowish green, blue-green, greyish blue-green, pale yellow, pale yellow-green, pale brownish blue; pale blue, blue-green, pale yellow (almost colourless); and pale blue, blue, deep indigo blue, prussian blue, greenish yellow. Again

Prior and Coomaraswamy calculated that its cleavage is good; for when viewed along the atomic lattice and based on the relative strength of the bonds in this plane, Serendibite has a perfect cleavage parallel to 001. The two collaborators also observed that it is a triclinic system – that is according to the shape of the unit form, the three edges are of different lengths and they are not at right angles to each other.

In their authoritative work entitled [**Rock Forming Minerals, Single-Chain Silicates**, Professor W.A. Deer, F.R.S., (University of Cambridge), Professor R.A. Howie (University of London) and Professor J. Zussman (University of Manchester), Longman, pages 659 – 661, 1978] state that the paragenesis (which can be defined as the relationship of minerals expressed in terms of a time sequence) of Serendibite *"is restricted to skarns, in which boron metasomatism has been active, at carbonate with acid rock contacts. Thus at Gangapitiya, Sri Lanka, the locality from which Serendibite was first described (Prior and Coomaraswamy, 1903) it occurs in association with diopside, spinel, scapolite, apatite and plagioclase, in a contact zone between limestone and granulite, while at Johnsberg, Warren county New York (Larsen and Schaller, 1932), Serendibite with diopside and phlogopite, is present at the Grenvile limestone – Granite contact. At Riverside, California, Serendibite is developed in thin replacement bands of hydrothermal contact metamorphosed dolomitic marbles. Here the associated minerals include diopside, grossularite, vesuvianite and wollastonite (Richmond, 1939)".*

Coomaraswamy's greatest contribution to geology was his discovery of **Thorianite** $(ThU)O_2(He, Ce, La, Pb, Fe)$. Originally discovered in the Bambarabotuwa district of Ceylon in 1904, Thorianite can also be found in the black river and beach sands of Alaska, Madagaskar and Siberia. And paradoxically enough, Coomaraswamy made this great discovery in the process of scrutinizing the results of a colleague. Actually, Dr. D.W. Holland had wrongly indentified a quantity of a heavy black mineral occuring in cubic crystals as pitchblende or uraninite. In fact, Dr. Holland's error can be traced to the fact that Thorianite is isostructural with uraninite. Coomaraswamy's preliminary examination showed that it was a new mineral. Again he took the precaution of sending samples of this mineral to Professor Wyndham Dunstan, Sir William Crookes, Sir William Ramsay, and Professor Marie Curie, for further cheminical examination and analyses. These analyses confirmed Coomaraswamy's summary of his discovery in his own words: *"This new discovered mineral, peculiar to Ceylon, is of great commercial importance owing to the use of thoria in the manufacture of incandescent gas mantles. It occurs in very heavy black cubic crystals at Bambarabotuwa, where over a ton has been obtained. It is valued at 600 Sterling per ton. It is of great scientific interest too, on account of its chemical composition, one or more new elements being possibly present; it contains a large amount of occluded helium. Though radioactive, there is no more than a trace of radium present."* And the results revealed that these samples of Thorianite sent by Coomaraswamy to Professor Dunstan, Professor Ramsay and Professor Marie Curie consisted of Thorium Oxide, Cerium Oxide, Lanthanum and Didymium Oxide, Zirconium Oxide, Ferric Oxide, Lead Oxide, Silica and occluded helium.

Coomaraswamy discovered that Thorianite belonged to the isometric system (a synonym for the cubic system where the units are cubes with the three edges equal in length and at right angles to each other). He also worked out the hardness at 5.5-6 on the Mohs scale, named after the German mineralogist, Friedrich Mohs (1773 – 1839). The hardness of Thorianite, according to recent research is 7 on Mohs scale. Coomaraswamy had also accurately determined the specific gravity (which can be defined as the number expressing the ratio of the weight of a mineral to the weight of an equal volume of water) at 8.98 – 9.32, which is very nearly equal to the latest investigations done on recent samples (9.7 – 9.8).

Here I wish to refer to an interview I had with Professor D.N. Wadia, F.R.S., (Professor Darashaw Mosherwan Wadia), the Geological Adviser to the Government of India, at his office in New Delhi, on December 1, 1953. Professor Wadia was not only the most distinguished Geologist produced by modern India but had also been *loaned* to the Government of Ceylon (to use a *bureaucratic expression*) in 1938, for a few years. In fact, as the Mineral Adviser to the Government of Ceylon, he was entrusted with the task of bringing out the revised version of A.K.C.'s geological writings by the then Government of Ceylon. Dr. Wadia told me that he had brought out two publications; however, he could not complete the project as the Second World War had begun. He had brought several A.K.C. publications with him. And Dr. Wadia had allowed me to go through a file dealing with A.K.C.'s discovery of *Thorianite*. Unfortunately, those were the days when we did not have the facilities to photocopy documents! I had to content myself with taking down notes – especially from the letters, exchanged between Madame Marie Curie and A.K.C. This original file was then available in the Ceylon Mineralogical office. It also contains reprints of several papers, many Letters to the Editor (published in *Nature*), as well as A.K.C.'s correspondence with Dunstan, Curie and Ramsay, two Royal Society papers published in French and German journals. More importantly, this file enabled me to scrutinize seven interesting letters – A.K.C.'s letter to Marie Curie dated October 1, 1903; Marie Curie's reply to A.K.C. dated December 15, 1903; A.K.C.'s letter to Dunstan dated January 15, 1904; Professor Dunstan's reply to A.K.C. dated March 15, 1904; A.K.C.'s letter to Sir William Ramsay dated February 15, 1904, Sir William Ramsay's reply to A.K.C. dated April 15, 1904 and A.K.C.'s letter to Dunstan dated October 1, 1905. Indeed, in his letter of October 1, 1905, A.K.C. had rejected not only the Cameron thesis of *Thorianite* occurring in South India but also remarked that *"the geological formation of South India did not contribute to the existence of thorium - bearing mineral in South India."*

A good deal of the confusion in the wake of the *discovery* of *Thorianite* was due to the fact that several samples of this new mineral were sent by (a) Mr. Holland to Ramsay, and (b) by A.K.C. to Marie Curie, William Ramsay, Sir W. Crookes and Dunstan, at different points of time. The confusion was made worse by the fact that the correspondence was conducted by sea-mail.

The intitial confusion persisted even when A.K.C. did the preliminary research on the new mineral. Holland had wrongly identified the new mineral, since *Thorianite* is isostructural with *uraninite.* Although A.K.C. had accepted Holland's identification in the first instance (for A.K.C. felt that Holland was familiar with the geology of Bambarabotuwa district), he became increasingly sceptical of Holland's explanation. This led A.K.C. on the trail, at the laboratory as well as in the field. In fact, the Coomaraswamy discovery of *Thorianite* has to be understood at four levels: For A.K.C. did his field work as well as his Lab. experiments at the same time. While Holland had already sent the samples for chemical analysis to the scientists of England, A.K.C. obtained not only his own samples of the new mineral in the bed of Kuda Pandi-Oya, but also conducted chemical experiments at the Lab. He sent the samples to the various labs, in England and France for the verification of his own results. He also knew that the European scientists – who enjoyed greater facilities ranging from more sophisiticated equipment to inter-disciplinary expertise – would enlarge the scope of the subject. A.K.C. made four points in his letter dated October 1, 1903, to Madam Marie Curie: (a) A.K.C. commended Holland's initiative in getting a sample of the new mineral which inspired his interest in the subject; (b) the sample sent by him was based on his own field investigation in the bed of Kuda-Pandi-Oya; (c) he had worked out a provisional chemical formula – $(ThU)O_2(He, Ce, La, Pb, Fe)$ and (d) he made a plea for doing an exhaustive analysis (by physicists, chemists and mineralogists) on this subject. For A.K.C. had correctly realized that science is infinitely perfectible.

In her truly historic reply (personally, I believe that it is one of the most famous letters in the history of modern science) dated December 15, 1903, Madame Marie Curie complimented A.K.C. on two counts – the accuracy as well as the beauty of the Coomaraswamy fomula:

$$(ThU)O_2(He, Ce, La, Pb, Fe).$$

She was also appreciative of A.K.C. for **seeing a clear chemical picture of the new mineral** in his mind. I cannot imagine a greater tribute to a young scientist working in a sister discipline. **Furthermore she suggested that the new mineral ought to be named after him as 'Coomarite'**. For some of the minerals discovered during that period – Allanite, Fergusonite, Gekielite and Baddeleyite were named after their discoverers. And finally she underscored the need for a more sophisticated analysis of the new mineral by the well-known German scientist Hillebrand. Here it is worth remarking that even the latest editions of the two celebrated Reference Publications of our time – the *Encyclopaedia Brittanica* and the *Encyclopaedia Americana* – do not contain the correct formula of *Thorianite*, which is a pity!

At this point, it is necesary to refer to A.K.C.'s retraction of the original Uraninite paper in *Spolia Zeylancia* (August 1904, p.59) in the wake of Dunstan's announcement in *Nature* (March 31, 1904). Again, a careful study of the literature of the period – A.K.C.'s letter to Madame Marie Curie, October 1, 1903, Marie

Curie's reply to A.K.C. dated December 15, 1903; A.K.C.'s letter to Dunstan dated January 15, 1904, Dunstan's reply to A.K.C. dated March 15, 1904; A.K.C.'s letter to Sir William Ramsay dated January 2, 1904; and Ramsay's reply to A.K.C. dated April 15, 1904, the paper **Thorianite, a New Mineral from Ceylon** by Dunstan and Blake published in the *Proceedings of the Royal Society, London* (May 1905) and Ramsay's Letters published in *Nature*, London (April 7, 1904 and April 14, 1904), give us a different perspective of the entire question. Furthermore, it is not necessary to draw any conclusion from A.K.C.'s *retraction.* For A.K.C. was aware of the chemical composition of *Thorianite* even in 1903. Incidentally, Dunstan and Ramsay suggested the name *Thorianite* for the new mineral since A.K.C. was reluctant to accept Marie Curie's suggestion that it could be named **Coomarite.** Dunstan certainly suggested that the new mineral ought to be named *Thorianite.* Also, it is not correct to maintain that Dunstan pronounced *Thorianite* as a new mineral. Indeed one is reminded of Ramsay's initial confusion which resulted in a subsequent retraction. In his first letter (*Nature*, April 7, 1904) Ramsay was unable to confirm that the new mineral contained Thoria. In his second communication, Ramsay modified his earlier statement. An interesting analysis of Ramsay's Letters to *Nature*, A.K.C.'s account of the discovery of *Thorianite* based on the official publication of the Government of Ceylon and Hillebrand's method (to which Marie Curie had referred to, in her reply of December 15, 1903 to A.K.C.) form the subject of a definitive Royal Society paper on *Thorianite* by Dunstan and Blake (*Proceedings of the Royal Society*, London, 1905). A.K.C. had sent a note on his preliminary investigations (which helped Dunstan to verify the formula) as well as a copy of Marie Curie's reply to Dunstan and Ramsay. Incidentally, Ramsay's initial mistake is a trifle surprising; but one cannot, however, magnify it.

The foregoing comments highlight not merely the question of priorities in the history of science but also lead us to an assessment of two memeorable chapters in the hisotry of modern science – the *Raman Effect* and the *Racah Polynomials.* To begin with, the *Raman Effect* was discovered in 1923 by Professor C.V. Raman, who found new frequencies in the scattered radiation which were characterisitic of the scattering medium. In other words, the *Raman Effect* may be described as the scattering of light from a solid, liquid or gas with a shift in wave-length from that of the usually monochromatic radiation. This great discovery announced in February 1928, modestly described as a new type of radiation by Professor Raman and happily christened as the *Raman Effect* by the well-known German Physicist Pringsheim, won international recognition for Raman, including the Hughes Medal of the Royal Society and the Nobel Prize for Physics in 1930.

It is interesting to note that Smekal in 1923, H.A. Kramers and W. Heisenberg in 1925 and P.A.M. Dirac in 1927, explained mathematically that scattered light should contain other wave-lengths besides the wave-length of the incident light and that these new wave-lengths ought to be characterisitic of the particles scattering the incident beam. However, it is Raman who is regarded as the sole discoverer of this phenomenon for a variety of reasons which are given below:

In fact, Raman's association with this discovery began while voyaging through the Mediterranean and Red Sea in 1921. And Raman questioned the validity of Lord Rayleigh's thesis that *the much-admired blue of the Mediterranean Sea is simply the blue of the sky seen by reflection.* It was clear to Raman that the blue of the Mediterranean Sea – observed through a polarising Nicol Prism at the Brewsterian angle in order to quench the surface-reflection – was due to the scattering of the sunlight by the molecules of the water. It struck Professor Raman even on board the ship that the Einstein-Smoluchowski concept of thermodynamic fluctuation – which was developed to explain special optical phenomena near the critical point – could be extended to explain molecular diffraction in liquids. It was the subject of a famous paper published by Raman in the *Proceddings of the Royal Society, London,* in 1922. Again in a monograph on *The Molecular Diffraction of Light,* published in 1922, Raman argued that the interaction of the photon with a molecule must reveal itself by a change in colour. Furthermore, Raman realized as early as in 1922 that *Maxwell's field equations must be altered to introduce the quantum of action* – a phase of Quantum Electrodynamics which was developed subsequently by Pauli, Dirac and Heisenberg.

Not surprisingly, while examining the scattered track with a direct vision spectroscope on February 28, 1928, Raman discovered that the classical and modified scattering appear in the spectrum as separate regions – with a distinct dark region between them – a strikingly clear demonstration of a change of wave-length in the scattering of Light. And Albert Einstein rightly commnted that *"C.V. Raman was the first to recognize and demonstrate that the energy of a photon can undergo a partial transformation within matter."*

One can also refer to Giulio Racah at the other end of the spectrum. Giulio Racah, an eminent Israeli physicist, is considered a pioneer in the study of Complex Atomic Spectroscopy – a rigourous mathematical formalism which is currently known as the Racah-Wigner Algebra in the mathematical ambience of the Quantum Theory of Angular Momentum. There is an Institute named after him in Israel.

Commenting on their newly discovered polynomials which they chose to christen as the *Racah Polynomials*, Richard Askey and James Wilson observed: *We will call these polynomials as Racah Polynomials. Racah was unaware of the exrstence of these polynomials, but he was the first to find an orthogonality relation ⋯. Since almost all the classical orthogonal polynomials are named after a rediscoverer rather than the original discoverer, we felt that it was appropriate to err in the opposite way and name the polynomials after the first person to treat them, even if he was unaware of the orthogonal polynomials buried in his results.* This quotation is from *A Set of orthogonal polynomials that generate the Racah coefficients or 6-j symbols,* Richard Askey and James Wilson, *SIAM Journal of Mathematical Analysis,* **10** (1979) p.1008 – 1016.

It was characteristic of Coomaraswamy's self-effacement that he called the new mineral as *Thorianite* instead of linking it with his own name. In fact, several mineralogists had associated themselves with their discoveries such as stated before. Indeed in 1925, Professor Millikan had experimentally verified the hypothesis of the Austrian physicist Victor Hess that a radiation far more energetically significant than the most powerful emanations of radium was bombarding the earth from different directions in space. Naturally impressed, *The New York Times* editorialized on November 12, 1925, that this radiation could be termed Millikan rays in an attempt to assosciate them *with a man of such fine and modest personality.* However, Millikan had modestly characterized them as *Cosmic Rays.*[1] And pleasantly enough, a new mineral discovered by D.B. Pattiartchi has been named after Coomaraswamy as **Anandite.**

The history of the gains from serendipity reinforces the validity of Louis Pasteur's contention that *chance favours the prepared mind.* Yet, in the ultimate analysis, it is possible to argue that serendipity is a creative combination of the different areas of scientific inter-relatedness rather than an individual discovery. It is true that Max Planck worked out the Quantum Theory while investigating the radiation from black bodies. It is equally true that Rutherford arrived at the structure of the atom in the wake of his study of the properties of alpha rays. However, one of the historic gains from serendipity was derived from the first phase of the inter-relatedness of twentieth century science. For it inspired Niels Bohr to realize that some of the major scientific developments of the twentieth century such as Lord Rutherford's memorable work on alpha rays leading to the conclusion that a chemically inert element like helium is an important factor in the constitution of the atomic system of Uranium, Thorium and Radium, Max Planck's theory of radiation as well as the investigations of great chemists – Willaim Ramsay's discovery of the chemically inert gases like Helium, Neon, Argon, Krypton and Xenon; Otto Hahn's investigation of the radioactive decomposition of Thorium resulting in the discoveries of Radiothorium and Mesothorium and Professor Marie Curie's celebrated work in isolating Radium as well as deriving the properties of nuclear chemistry from pitchbelnde – could all be synthesized into a theory of the atom.

The four Ceylon Administration Reports compiled by Dr. A.K.C. (1903, 1904, 1905 and 1906), in whose preperation he was ably assisted by Mr. James Parsons, are a mine of information, ranging from the contour and the nature of the soil of Ceylon to the mineral wealth of the island. Furthermore, these Reports (which contain the usual sketch maps, modes of occurrence of graphite, a glossary of terms, a scheme of the new Mineral Gallery and the feedback derived from field work in various districts) are notable for their stress on two fundamental discoveries: the two different varieities of Coomaraswamy terminology – *Charnockite Series* in order to classify the rocks of Ceylon, as well as the *gneissic structure* to identify the nature of the rocks of Ceylon.

In order to understand the Coomaraswamy contribution to the geological structure of Ceylon, it is necessary to refer to an analysis of the geological map of South India, south of latitude 16 degrees. Actually, eighty percent of India is occupied by Archean rocks. The other portions of the area cover unfosiliferous sedimentary Cuddapah and Kurnool formations, forming the shape of a crescent and resting unconformably on the eroded upturned edges of the ancient crystalline rocks; the Lower Gondawana rocks (of Barakar age) in the Godavari Basin; formation of Upper Gondawana patches generally intercalated with marine Cretaceous rocks, along the east coast; and an isolated outlier of basaltic lave-flow of the Deccan trap age. Bordering most of these formations, is a coastal belt of younger Tertiary and recent formations.

Just as we owe the advancement of knowledge relating to the Geology of India to Mr. R.B. Foote, Sir Thomas Holland, Mr. C.S. Middlemiss, Dr. Walker and Dr. L.L. Fermor, so did Ananda Coomaraswamy extend not only the frontiers of the state of geology in Ceylon but also determined the inter-relationship of the various members of the geological series which encompass the rocks of India and Ceylon. Here we note that Mr. R.B. Foote had worked out the inter-relationship of the different members of the Archean group and also mapped the Dharwar formations; Sir Thomas H. Holland made three major contributions – the discovery of the Charnockite family of genetically related pre-cambrian plutonic intrusive rocks, which are widely distributed in the state of Tamil Nadu, the Elaeolite syenite series occurring at Sivamalai in Coimbatore as well as an exhaustive account of the economic importance of the pegamatites in India; Mr. C.S. Middlemiss who catalogued the ultra basic intrusions of Salem and Corundum occurrences of Coimbatore and Salem; Dr. Walker who successfully mapped the Kondalite series (graphite-garnet-quartz-sillinanite-schists) of the Kalahandi area comprising Vijayawada and Ganjam; and Dr. L.M. Fermor, who discovered the Manganese-bearing Kodwite between the Chota-Nagpur and Eastern Ghat facies of Archeans.

The gneisses which could be found in most of the areas of India are biotite gneissose granites giving rise to low rounded hillocks and shaping themselves into hillocks and table-lands in Coimbatore, Wynaad, Bellary, Arcot and Vizagapatnam districts. Again the Charnockites form the hill masses of South India including the Nilgiris, Shevaroys and Palnis and also occur as bands in Coimbatore, Salem, Vynaad, South Arcot, Vizag and Ganjam districts. It is also worth noting that these districts are noted for accessory minerals such as graphite, ilmenite and monazite as well as building-stones and road-metal. Furthermore, switching to the Charnockites, the exposures can be seen in the neighbourhood of the city of Madras: (a) St. Thomas Mount, (b) the adjoining Pallavaram Hills and (c) the rocks exposed at Mahabalipuram which carry an aplitic variety of Charnockite; these rocks also carry quartz crowded with acicular inclusions and micro-perthitic felspar.

Commenting on the Ceylon Administration Reports, Mr. Frank Dawson Adams wrote in a 1929 issue of the *Canadian Journal of Research*, (pp.430 – 431): *"In these Administration Reports, Coomaraswamy has given the fullest and most accu-*

rate accounts which have appeared of the geology of the ancient crystalline rocks of Ceylon, and most of this information is also to be found distributed through certain papers by him in all other publications ···. *Coomaraswamy having visited all parts of Ceylon had a much wider acquaintance with the rocks of the Island than any of his predecessors. He states that the term gneiss used in the sense of a rock having a* gneissic structure *is applicable to most of the rocks of the Island. If the term* granulite *is used, as it is by many writers to designate a gneiss containing garnet, a large proportion of the crystalline rocks of the Island ought to be termed granulites. He also mentions that the Crystalline rocks of Ceylon belong in large part at least, to the* Charnockite Series *which Holland described in Southern India* ···". Actually Coomaraswamy's assessment of the *gneissic structure* of the ancient Crystalline rocks of Ceylon is accepted today. At this point recall that *"Gneiss is a venerable rock name, derived long ago from a Slavonic word meaning* nest *and applied by the mediaeval miners of Central Europe to the foliated rocks in which the metalliferous veins occur; the rock, so to speak, being the* nest *of the ores"* (Holmes, *Physical Geography*, 1978). Interestingly, the term was coined by Werner to describe crystalline rocks, which constitute the product of high grade regional metamorphism. Here it is also worth noting that though a major part of the Wernerian terminology is dated, *gneiss* is still a contemporary geological term. Thus the coarse foliation or banding characteristic of gneisses shapes itself into the gneissic structure.

To the historian of science, the Ananda Coomaraswamy – Otto Hahn correspondence relating to the discovery of Radiothorium is of interest. In addition to the analyses done by Professor Wyndham Dunstan, Mr. G.S. Blake, Professor Marie Curie and Sir William Crookes, Sir William Ramsay was engaged in investigating the same subject. Sir William Ramsay had also asked his student Otto Hahn to seperate Radium from a sample of Thorianite which was sent by Coomaraswamy. In so doing, Hahn discovered a new material, a highly active form of Thorium which he called *Radiothorium.* This discovery impressed not only Sir William Ramsay but also enabled Otto Hahn to acquire an international reputation.

Coomaraswamy was honoured both in Engand and elsewhere. He was elected a Fellow of the Geological and Linnean Societies of London and also received the D.Sc., of the London University in 1906. And his outstanding work gave us not only two gains from Serendipity – *Serendibite* and *Thorianite* – but also helped in raising the state of geology to the status of a scientific discipline in **Serendib** (Sri Lanka).

Locksley Hall
852, Poonamallee High Road
Chennai - 600 010
June 1997.

A. Ranganathan

List of References[1]

I. Contributions to the Geology of Ceylon

1.a *Note on the Occurrence of Corundum as a Contact-Mineral of Pont Paul near Morlain (Finistere),*
Quarterly Journal of the Geological Society, London, **57** (1901) 185–188.

1.b *Silification of Crystalline Limestones,*
Quarterly Journal of the Geological Society, London, **58** (1902) 16–19.

1.c *The Balangoda Group,*
Quarterly Journal of the Geological Society, London, **58** (1902) 418–422.

1.d *Intrusive Pyroxenites, Mica-Pyroxenites and Mica-Rocks & Charnockite Series or Granulites in Ceylon,*
Geological magazine, London (1905) 363–369.

2. *Note on the Scenery of Ceylon,*
British Association for the Advancement of Science, London (1902) 613–614.

3. *Origin of the Crystalline Limestones of Ceylon,*
Quarterly Journal of the Geological Society, London (1902) 375–378.

4. *The Crystalline Limestones of Ceylon,*
Quarterly Journal of the Geological Society, London, **58** (1902) 399–424.

5. *The Point-De-Galle Group (Ceylon): Wollastonite-Scapolite Gneisses,*
Geological magazine, London, **58** (1902) 680–689.

6. *Report on the Occurrence of Cassiterite (Oxide of Tin),*
Government Printing Office, Colombo (1905)

7. *Serendibite, A New Borosilicate of Ceylon,*
G.T. Prior and A.K. Coomaraswamy, *Mineralogical Magazine*, London, (1903) 224–227.

8. *The Crystalline Rocks of Ceylon,*
Spolia Zeylancia, Colombo, (1904) 105–111.

9. *Mineralogical Notes,*
Spolia Zeylancia, Colombo, (1904) 57–64.

[1]This is a Select List of References only

10. *The Rocks and Minerals of Ceylon,*
Colombo Museum, Guide to the Collections, Edited by A. Wiley, third Edition, Colombo (1905); *Spolia Zeylancia*, Colombo (1905)

11. *On Ceylon Rocks and Graphite,*
Quarterly Journal of the Geological Society, London, **56** (1900) 590–615.

12. *Mineral Resources,*
Official Handbook of the Ceylon Court, Colombo (1904)

13. *Recent Marine Clays at Kuchaveli, Ceylon,*
Spolia Zeylancia, Colombo (1904) 199–200.

14. *Report on Thorianite and Thorite,*
Government Printing Office, Colombo (1904) 1–4.

15. *Note on the Occurrence of Corundum as a Contact-Mineral at Pont Paul (Finistere),*
Journal of the Geological Society, London (1901)

16. *Occurrence of Radiolaria in Gondwana Beds near Madras,*
Geological magazine, London, (1902)

17. *Observations on the Tiree Marble, with Notes on others from Iona,*
Quarterly Journal of the Geological Society, London, **59** (1903) 91–104.

II. Mineralogical Survey, Ceylon: Administration Reports

18. 1903, Report published in 1904, L1–L12.

19. 1904, Report published in 1905, E1–E21.

20. 1905, Report published in 1906, E1–E18.

21. 1906, Report published in 1907, E1–E11.

III. Comments

22. Letter to the editor, *The Occurrence of Thorium in Ceylon,*
W.R. Dunstan, *Nature*, London, **69** March 31, 1904, 510–511.

23. Letter to the editor, *A New Mineral from Ceylon,*
Sir William Ramsay, *Nature*, London, **69** April 7, 1904, 533–534.

24. Letter to the editor, *A New Mineral from Ceylon,*
Sir William Ramsay, *Nature*, London, **69** April 14, 1904, 559.

25. Letter to the editor, *A New Mineral from Ceylon,*
T.A. Henry, *Nature*, London, **69** April 14, 1904, 559–560.

26. *Thorianite, A New Mineral from Ceylon*,
Wyndham R. Dunstan and G.S. Blake, *Proceedings of the Royal Society*, London, May 1, 1905, 253–265.

27. *A Variety of Thorianite from Galle, Ceylon*,
Wyndham R. Dunstan and B. Mount Jones, *Proceedings of the Royal Society*, London, April 11, 1906, 546–549.

28. *Serendibite* $Ca_2(Mg, Al)_6O_2[(Si, Al, B)_6O_{18}]$,
W.A. Deer, R.A. Howie and J. Zussman, Longman Group, London, 1978, 659–661.

29. *Anandite, A New barium Iron Silicate from Wilagedera*,
North Western province, Ceylon,
D.B. Pattiaratchi et. al., *Mineralogical magazine*, London, **30**, March 1967, 1–4.

30. *The Traditional theory of Literature*,
Ray Livingston, University of Minnesota Press, 1962.

Part I

Contributions to the Geology and Mineralogy of Ceylon

11. NOTE on the OCCURRENCE of CORUNDUM as a CONTACT-MINERAL at PONT PAUL, near MORLAIX (FINISTÈRE)

By A.K. COOMÁRASWÁMY, Esq., F.L.S., F.G.S. (Read January 9th, 1901.)

Q. J. of the Geo. Soc., London, **57** (1901) 185 – 188.

In the neighbourhood of Morlaix the schists and quartzites of the Lower Devonian are extensively developed. The are much altered, and often injected by intrusive granite. They altered rocks are described by Prof. Barrois[1] as quartzites micaceés and leptynolithes. The minerals found in the latter include biotite, muscovite, andalusite, sillimanite, quartz, iron-oxide, chlorite, garnet, staurolite, zircon, graphite, and rarley felspar. The quartzites micacés contain large grains of irregular quartz (not elastic), with flakes of dark and white mica, zircon, iron-oxide, graphite, and garnet.

Among the intrusive masses, that of Pont Paul is of especial interest. The rock is a granite rather poor in quartz; the minerals include orthoclase, microcline, oligoclase, quartz, hornblende, biotite, apatite, zircon, sphene and allanite[2] (the last two not noticed by myself). The granite is much decayed for a depth of many feet. Included fragments of dark micaceous rocks are to be found in it; they are about the size of a man's fist, and not very abundant. Some may be found *in situ*, and dug out of the crumbling granite, others by searching the floor of the shallow quarry. These included fragments resemble the Devonian leptynolithes of the district. They contain abundantly small tabular hexagonal crystals of the blue corundum, first noticed by Prof. Barrois in 1887.

Fig.1:- *Isolated crystals of corundum (washed in hydrofluoric acid). see p.186.*

[1]Bull. Soc. Géol. France, ser.3, vol.xiv (1887) p.890

[2]Michel Lévy & Lacroix, 'Note sur un Gisement fracais d'Allanite' Bull. Soc. Min. France, vol.xi (1888) p.65.

[Striations on the basal plane are faintly indicated. From a photograph: × about 9.]

Exmination of thin sections of the included fragments shows that the minerals certainly present are biotite, muscovite, corumdum, plagioclase, andalusite, pyrite, magnetite, sillimanite, green spinel, and zircon. Orthoclase and quartz may possibly occur.

The corundum is very abundant, forming conspicuous crystals which exhibit hexagonal or rectangular outlines, and contain comparatively few inclusins of iron-oxide, very much more rarely sillimanite and zircon, also gas-inclusions. The crystals in a thin section are colourless, or have a faint blue tinge, unequally distributed. In the latter case there is slight pleochroism from pale blue for the ordinary ray, to very pale greenish-yellow for the extraordinary. Optical anomalies were not observed. Cleavage parallel

186 *Mr. A.K. Coomára-swámy on the Occurrence* *[May 1901*

to *r* (1011) is not infrequent; the cleavage-cracks become more conspicuous in polarized light, as is often the case with calcite. Traces of a parting parallel to *c*(0001) are also sometimes seen. The rock appears a little bleached in the neighbourhood of the corundum, that is, colourless minerals rather than mica are usually in contact with it; but this is not invariably the case. Less pure corundum occurs also in a finely granular form, mixed with other minerals from which it has not, as it were, been able to free itself.

Isolated crystals, detached from the matrix, and washed in hot hydrofluoric acid, are tabular in form, being combinations of the hexagonal prism and basal plane, the latter characteristically striated

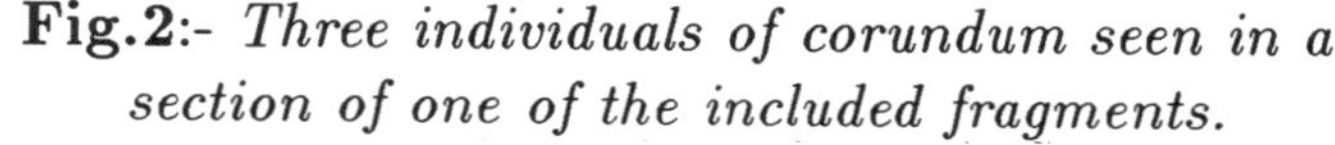

Fig.2:- *Three individuals of corundum seen in a section of one of the included fragments.*

[On the left hand is one with hexagonal outline, cut approximately parallel with the basal plane. A zone of lighter coloured minerals separates the three crystals from the darker part of the rock. From a photograph; ×27.5.]

(see fig.1, p.185) and sometimes slightly stepped. One of the largest measured about 1.7 mm. in greatest diameter by .5 mm. in thickness, and weighed about .005 gramme. The crystals have a beautiful sapphire-blue colour.

The dark mica is strongly pleochroic, from warm dark-brown to pale straw-colour.

Andalusite is fairly common, occasionally showing faint characteristic pleochroism, or nearly rectangular cleavage with diagonal extinction. The greater part is, however, ill-defined, and not easy to distinguish.

A minute mosaic of colourless grains, with low refractive index and weak interference-colours, fills up the background; and in many cases these show the multiple twinning of plagioclase. Twinning on the albite and pericline plans is found: in one case also apparently Carlsbad twinning occurred. Among others, the following nearly symmetrical extinction-angles were measured from (presumably) albite twin-lamellae:— 12.11, $12\frac{1}{2}$.13.

Some few individuals which appear untwinned, and extinguish approximately parallel to their greatest length, may be orthoclase.

Fig.3:- *Individual of corundum seen in a thin section of one of the included fragments*

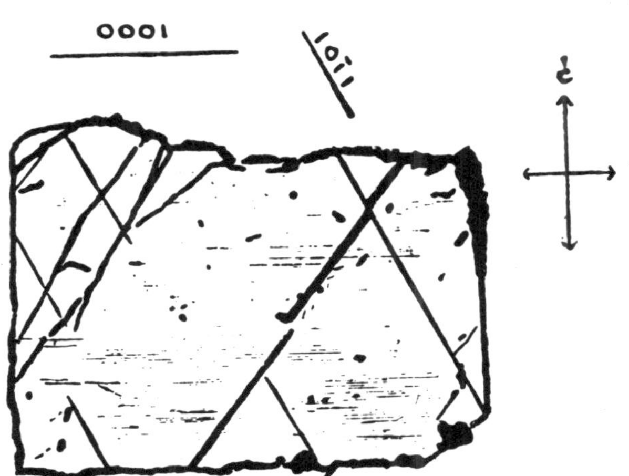

[The crystal is cut at right angles to the basal plane, which is slightly stepped owing to alternation with a rhombohedral face. The faint horizontal shading indicates approximately the distribution of the pale blue colour. The sharp rhombohedral cleavage-cracks are drawn, as if seen between crossed nicols; in ordinary light they are only with difficulty distinguishable. The few inclusions consist of iron-ore, a little sillimanite, a grain or two of zircon (?), and gas-inclusions. ×38.6.]

It is difficult to be certain about the occurrence of quartz. I feel sure, however, that very little or none is present, for close examination of many tiny grains in the colourless mosaic which suggested possible quartz, revealed faint multiple twinning, or a portion of a biaxial interference-figure was obtained. The numerous inclusions of the colourless minerals render their examination somewhat troublesome.

Iron-ore is abundant in scattered grains, and also an inclusions in the other minerals, including corundum.

Sillimanite is locally abundant, elsewhere scarce or perhpas absent. Where abundant, it forms sheaves and aggregates of much elongated colourless crystals. The characteristic transverse parting is seen, the segments being often separated by a portion of the enclosing mineral.

Green spinel occurs in irregular, rather small grains, or granular aggregates.

A few fairly weel-shaped crystals of zircon were notice.

In a slice passing through the actual contact, there appears to be some injectio of felspathic material. Prof. Barrois noticed that some of the inclusions were penetrated by very fine veins of the enclosing granite (especially rich in hornblende). In view of the considerable

188 *Occurrence of Corundum as a Contact-Mineral.* *[May 1901*

amount of felspar found in the inclusions we may well suppose that some has been drived from the surrounding granite, for the metamorphism has evidently been very intense. Doubtless the inclusions were never in a state of igneous fusion; but there has been a sufficient amount of molecular freedom to admit of the formation of comparatively large, sharp-edged crystals of corundum, without many inclusions. The original sedimentary rock was probably poor in silica.

The presence of corundum in the enclosing granite has not been noticed; it would be interesting to examine the decayed rock, especially in the neighbourhood of the inclusions, in order to see whether any corrosion thereof has led to the presence of corundum in the granite.[3]

It is interesting to compare this occurrence of corundum as an undoubted contact-mineral in included fragments, with those described in Southern India by Mr. C.S. Middlemiss,[4] which are supposed to result from the metamorphism of foreign inclusions.

[3]Compare K. Buxz, 'On the occurrence of Corundum produced by Contact-Metamorphism on Dartmoor' Geol. Mag. 1896, p.492

[4]'Prediminary Notes on some Corundum Localities in the Salem & Coimbatore Districts' Rec.

In conclusion, I have great pleasure in expressing my indebtedness to Prof. Barrois, who not only very kindly directed me to the Pont Paul locality, but also gave me full permission to describe the rocks. I have also to thank Prof. Bonney for allowing me to examine his slides of other altered rocks from Morlaix.

DISCUSSION.

Mr. R.D. OLDHAM remarked that, in Southern India, more than one of the known occurrences of corundum had been recognized as being in the marginal portion of an intrusive mass. In one instance a direct causal connection appeared to have been established by Mr. T.H. Holland. The corundum of Sivamalai occurred in a coarse-grained felspar-rock intrusive in an elaeolite-syenite, by whose fusion an excess of alumina seems to have been introduced into the intrusive rock. On the subsequent solidification of the last named the excess of alumina crystaxllized out as corundum.

Mr. PARKINSON called attention to a specimen on the table from one of the ellipsoidal masses found in the Charnockite Series of the Salem District (Madras Presidency). The mode of occurrence of these lenticles and the large amount of corundum that they contain make it extremely probable, according to the published accounts of the district, that they are altered inclusions of some foreign rock. He thought the specimen of interest in connection with the Author's paper.

Geol. Surv. India, vol.xxix (1896) pp. 40-46; see also T.H. Holland, 'Corundum' pt. i of 'Economic Geology' in the reissue of the 'Manual of the Geology of India' Calcutta, 1898, pp.18 & 39-43, and Mem. Geol. Surv. India, vol.xxviii (1900) 'The Charnockite' Series pp.183 & 234.

V.— CONTRIBUTIONS TO THE GEOLOGY OF CEYLON: II. SILICIFICATION OF CRYSTALLINE LIMESTONE.

By A.K. COOMÁRASWÁMY, B.Sc., F.L.S. F.G.S., Director of the Mineral Survey of Ceylon.

(Q. J. of the Geo. Soc., London, **58** (1902) 16 — 19.

The occurrence of small quantities of chert and opal, usually in or near exposures of crystalline limestone, but very often in fragments or boulders not quite *in situ*, is not unusual in Ceylon. For some time the rogin of these siliceous rocks remained obscure; observations made within the present year (1903), however, enable me to give a more detailed account of their mode of occurrence. I have had the advantage of my colleague Mr. James Parsons' company in examimining many of the exposures, and have been able to discuss with him the problems raised.

A band of chert occurs *in situ* on the path descending from the am̃balam just $\frac{1}{4}$ mile W.N.W. of Uduwela trigonometrical station (about 3 miles south-east of Kandy), and about 4 to 5 yards below the fourth of the six bands of limestone which are crossed in descending the hill[1]. The width of the band of chert is about 5 feet; it includes a number of varieties, all with good conchoidal fracture. These are : homogenous green opal; homogeneous brown chert

(these two with very smooth fracture); brown mottled chert, with phlogopite, spinel, and graphite ; green chert, with abundant mica and graphite, and less frequent spinel ; and whitish decomposed chert, with the same accessory minerals in addition to blue apatite.

Specimens of limestone with identical accessory minerals occur quite near. On its north side, the chert band appears to pass into decomposed limestone which shows green spots suggestive of partial silicification. It was not quite certain that these apparent transition types occurred *in situ.* There were, however, many specimens which could only with difficulty be definitely named as chert or limestone. A thinner band of brown chert occurs on the path a little below the main band.

[1]The locality can be identitied on the map, Q.J.G.S., vol. lviii (1902), pl. xiii, but the position of the bands of limestone is not correctly indicated there.

Partially silicified crystalline limestone. ×22.
C, carbonate (dolomite); M, mica ; S, spinel; remainder, opaline silica.

Five thin sections of these cherts were prepared. One of the green opal (1056) shows merely a green, structureless, homogeneous, isotropic rock. The brown chert (1052) consists of chalcedonic silica, in characteristic spherulitic aggregates ; there is a colourless transparent base in which are scattered very numerous tiny ferruginous aggregates which give the brown colour to the whole rock. Certain cracks are filled with characteristic chalcedonic infiltrations. Rounded spots containing fewer of the ferruginous specks, and appearing rather dark between crossed nicols, call to mind the appearance presented by structureless radiolarian casts ; the presence of radiolaria is, however, quite out of the question. The greenish micaceous chert (1055) consists of opaline and chalcedonic silica in roughly equal proportions, enclosing numerous individuals of well-preserved phlogopite and a flake of graphite. In another, very similar specimen (1054) the mica is much hydrated, and silica has been deposited between the laminae, which are swollen and displaced. In the mottled chert (1053) chalcedonic chert is much more abundant than chalcedony, and the accessory minerals include graphite and abundant and characteristic spinel, colourless in the

thin section but pink in the hand specimen. In none of these slides are any remains of carbonates to be found.

Some specimens collected from blocks resting on crystalline limestone, but not quite *in situ*, on Upper Rajawela estate (about 11 miles from Kandy on the Teldeniya road), about a third of a mile E.S.E. of Rajawela trigonometrical station, were also sliced. Of these one (1074) evidently consisted of partially silicified limestone; the section showed disintegrated and corroded crystals of dolomite embedded in an isotropicsiliceous matrix, in which an abundance of hydrated phlogopite and a few grains of spinel are also found.

The silica has penetrated along the cleavage cracks of the carbonates, with every appearnce of corrosion. The dolomite individuals are thus broken up into irregular fragments, often more or less rhombohedral, and these graduate into the smallest specks which remain scattered in the siliceous base, sometimes indicating by their disposition the rough outline of the original carbonate.

Another specimen (1073) from this locality consisted entirely of brown chert and resembled No.1053.

The amount of chert present in any locality is always small, and quite insignificant in comparison with the total amount of crystalline limestone present; nor can the occurrences of chert be followed for any distance. They are also met with in other parts of Ceylon, e.g. in the Uva Province, although their connection with crystalline limestone is not always traceable; but there is no direct evidence of their occurrence as a replacement of any other rock.

From the foregoing observations I conclude that these opaline cherts result from the alteration of crystalline limestone, the carbonates being dissolved and replaced by opaline or chalcedonic silica, or a combination of the two. Very possibly the pure siliceous rocks, free from accessory minerals, do not so directly replace the limestone, but are siliceous deposits similar to the chalcedony deposited in cracks in the other cherts, which must already have had time to harden and develop cracks, previous to the introduction of a further supply of silica. The silicification is probably the result of the presence of heated waters containing silica in solution introduced after the consolidation of the crystalline limestone in its present form. In other words we have here a metasomatic transformation. A number of hot springs are known to occur in Ceylon, e.g., at Badulla, Alupota, Babule, and Bibile, in the Uva Province[2] ; and near Koggala, Magam Pattu, in the Southern province [3]. The occurrence of these springs lends support to the probability of such alterations having taken place.

It is of interest to notice the bearing of these observations on the origin of cherts in general [4]; we are here dealing with cherts which are certainly of inorganic origin. The mica, spinel, and graphite met with in the chert are proof that the original rock was a crystalline

Kennard & Warren — Tufa Deposit in Totland Bay. *19*

limestone quite similar to those still met with in large quantity; no source of abundant

[2]Uva Manual, by H. White, Colombo, 1903, p.82

[3]Ceylon Administration Reports, 1902, Survey Department, p.B.30.

[4]For a discussion of this question, see C.A. Raisin, Proc. Geol. Assoc., xviii (1903), pp.71—82.

silica can be found in these rocks, so that we are driven to conclude that it has been introduced from without. It is simplest to suppose that the silica was introduced in solution in the waters of hot springs. It may have been deposited at first in the colloid form and subsequently have become chalcedonic in parts; or the two forms of silica may have been deposited more or less simultaneously. It seems likely, however, that at least to some extent, there has been a transformation from opal to chalcedony.

In conclusion, the cherts described represent a secondary condition of a rock originally different, viz. crystalline limestone; the silica has been introduced from without, and is of inorganic origin; the silica has been chemically deposited, chiefly in the colloid form, and replaces the carbonates which have been removed in solution.

VII.CONTRIBUTIONS TO THE GEOLOGY OF CEYLON:

III. THE BALANGODA GROUP.

By A.K. COOMÁRASWÁMY, B.Sc., F.L.S., F.G.S., Director of the Mineralogical Survey of Ceylon.

(Q. J. of the Geo. Soc., London, **58** (1902) 418 — 422.)

The name 'Balangoda group' is proposed for a series of granitio and pegmatite-like rocks, intrusive in, but distinct from the Charnockite series; first met with in the Balangoda district, but evidently widely distributed over a large area between Balangoda and Hatton. The rocks are best described as granites, but occur most often in rather narrow dykes, after the manner of pegmatites. Yet there is no reason for separating the smaller from the larger masses, and the term granite is applied to both. The group (of which a more detailed account will ultimately be needed) includes in particular zircon granite, allanite granite, magnetite granite, and granite without conspicuous accessory minerals; as well as the probably similar rocks in which the hitherto unlocated minerals geikielite, baddeleyite, rutile, fergusonite, thorite,[5] thorianite, etc., may be looked for; and the vein of pegmatite at Gampola, which consisted of quartz, felspars, and biotite, with apatite, ilmenite, tourmaline, and the new mineral described as thorianite [6] as accessory minerals.

These granites are intrusive in the Charnockite series, and though frequently occurring in lenticular masses (Denagama) with a disposition parallel to that of the foliation planes of the charnockites, have often been observed to transgress these foliation planes and to behave as intrusive rocks. Contact phenomena have not, however, been observed, except perhaps in a slight tendency to a peripheral fineness of grain in the intrusive rocks. At the junctions granite and charnockite are usually welded together, there being no absolutely hard line of separation, although the junction may be called sharp; in the case of the larger masses no good junctions have been seen.

A description of the rocks is given below, with special reference to the localities where they can be seen:–

Zircon granite:— This rock was seen *in situ* at several points, and is the best known member of the group. The finest and longest exposure is on Massena estate, six miles from Balangoda; here a considerable mass of granite, fully two miles in length and

averaging perhaps a hundred yards in width, runs, parallel to the Charnockite foliation-strike, along the trough-like strike valley of the Massena Oya, in which the main part of

[5]W.Dunstan, Nature, 1904, p.510.

[6]"Spolia Zeylanica", vol.i, pt.4(1904), p.112; and *Nature*, loc.cit.

the estate is found. The rock occurs in enormous masses, both *in situ* and in great boulders, amongst which the Massena Oya finds its way. Two rows of curiously weathered masses stand out of the swamp below the bungalow; good specimens very rich in zircon can be collected here. The rock usually shows no trace of foliation in small specimens ; but foliation (vertical, with strike about 15° N. of W.) is evident in an exposure *in situ* near the middle of the estate, near the 'lines'.

The rock consists essentially of quartz, felspar, and biotite, with zircon and ilmenite as characteristic accessory minerals, and apatite as a microscopic constituent. The content of zircon vaies greatly from specimens in which a crystal can hardly be found to others

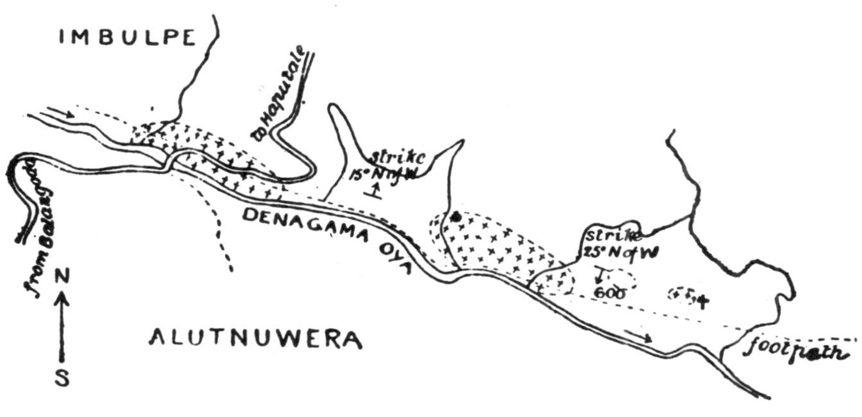

Fig.1:- *Rough sketch-map of zircon granites near Denagama Oya bridge, near Balangoda, Sah, Ceylon,* × × × *zircon granites;* † *zircon granites well foliated. Scale, 5 inches =* $1\frac{1}{2}$ *miles.*

in which it forms a noteworthy proportion of the rock. The zircon occurs in moderately good, idiomorphic, stumpy, prismatic crystals, generally terminated, the forms *m* (110), *a* (100), *p* (111), *e* (101) being certainly present; the colour is hair-brown, often quite pale; in thin section no colour is seen ; the longest individuals exceed $\frac{3}{8}$ inch in length, but most are somewhat smaller. The rock itself is realtively fine-grained, the quartz and felspars rarely exceeding in diameter the length of the longest zircons. The felspars include orthoclase (sometimes microperthitic) and an acid plagioclase with e.g. near to that of orthoclase. The biotite occurs in ragged crystals, macroscopically nearly balck, but brown in thin section, and with pleochroism from warm brown to pale straw colour. Ilmenite is common, occasionally moulding the zircons, and generally partly altered to 'lencoxene'. Apatite is fairly abundant in six-sided terminated prisms about $\frac{1}{40}$ inch in length. The general structure is hypidiomorphic, only the zircon and apatite occurring in well-developed crystals.

Another exposure (Map, Fig.1) is seen near the bridge over the Denagama Oya, about 6 miles from Balangoda on the Haputale road. Small idomorphic zircons ($\frac{1}{16}$ inch) are scattered sparingly in the rock, which resembles that of Massena estate, but is of somewhat coarser grain. Ilmenite is common, the zircon quite scarce. Other exposures are found along the line of strike on the left bank of the stream. The granite seems to from a series of lenticular masses. At one point the rock becomes well foliated (as if by pressure), the small crystals of biotite being closely packed and sweeping round the augen orthoclase. Well-formed zircons $\frac{3}{8}$ inch in length are to be found.

A similar foliated zircon granite is exposed by the roadside $1\frac{1}{2}$ miles further on, just beyond the 91st milepost; the zircons vary from $\frac{1}{16}$ to $\frac{1}{8}$ inch in length, and are of the usual light-brown colour.

No junctions were observed in the instances above referred to.

Near Haldummulla and about 18 miles from Balangoda a small granitie dyke was seen in the Weli Oya valley about half a mile above the road, crossing foliation in the Charnockite series and containing a few minute zircons just visible to the naked eye.

In the Bamberabotuwa district a large number of small intrusive pergmatites or granite dykes were examined on Hopewell estate (15-16 miles from Balangoda), where they are well exposed, and cross the foliation of the Charnockite series in all directions; only two were found to contain minute crystals of zircon sparingly distributed. Zircon may occur in others, but so rarely as to be overlooked. This exhausts the list of localities where macroscopic zircons have been seen *in situ.*

Zircons of all sizes (up to $1\frac{1}{2}$ inches or more) occur abundantly in every stream and river gravel, and are found in quantities when gemming operations are carried on, joining a large proportion of the heavy residue (*nambu*[7]) at the bottom of the gemming-basket: the clear-coloured varieties are of value as gems; the remainder is rejected. The irregular zircons described by Mr. Spencer [8] occur in this way in various parts of the Balangoda and Bamberabotuwa districts. Near Kondrugala zircon is very abundant in large individuals. Well-developed twins on $e(101)$ are found. With the zircon are associated 'thorianite', thorite, ilmenite. Large zircons have also been found in gemmings from the Hatton district. Zircon is also abundant over a wide area in the Southern Province in the Galle, Matara, and Morawaka districts, and no doubt also at Rakwana; the same rocks may be expected to be met with in these districts.

Allanite granites.— Allanite granites are well exposed in two places in the Balangoda districts. It is some years since Mr.W.D. Holland discovered a granite or pegmatite dyke,

[7] A Sinhalese term which may with advantage be adopted.

[8] *Nature*, April 14th, 1904, p.575.

crossing foliation in the Charnockite series, in the bed of the Wewel Dola near the lower end of his estate of Dik Mukulana, and containing allanite in some abundance. The determination of allnite was confirmed

by a partial analysis made for Mr. Holland, and by Mr. G.T. Prior, to whom a sample was submitted. The granitie dyke is composed of quartz, felspar, hornblende, allanite, biotite, pyrite.

The felspars include porphyritic orthoclase and also a series of smaller individuals of orthoclase and plagioclase (some of the latter are porphyritic like the orthoclase), forming with quartz the finer-grained portions of the dyke. Allanite and bornblende occur in varying amount, both being locally very abundant. Biotite is scarce. Pyrite occurs chiefly in secondarily deposited films.

A better exposure of allanite granite is found in the lower part of Denagama estate, about seven or eight miles from Balangoda. A conspicuous dyke, three to four feet thick, crosses the left branch of the stream which runs through the tea below the path, and forms a conspicuous ledge inclined at a low angle to the foliation. The granulites are inclined at a very similar angle, but it can be seen clearly that the dyke does not keep strictly to the foliation planes; moreover, a few short processes, six inches to a foot in width, project into the rock underlying the dyke, clearly showing the intrusive character of the latter. The dyke is coarse-grained, and consists mainly of orthoclase (porphyritic idiomorphic individuals often about 3×1 inches), quartz, and biotite, the latter in long thin crystals (measuring e.g. $9 \times 1 \times \frac{1}{4}$ inches) scattered in all directions through the rock.

In these two dykes the allanite is very unevenly distributed, being in places very abundant, and elsewhere almost or quite absent. The allnite forms thin tabular idiomorphic as well as more irregular individuals; the largest attain a length of three inches, those of medium size measure about $1 \times \frac{1}{2} \times \frac{1}{4}$ inch. A curious point as that the allnite seems to form a centre for radiating cracks in the rock, giving it a rather conspicuous appearnce, of which a diagram is given in Fig.2 (Denagama).

The allanite is macroscropically black (in thin section brownish olive-green), and has a resinous lustre and conchoidal fracture ; hardness about 6; sp. gr. 3.2 to 3.5 ; before the blowpipe it intumesces strongly.

Magnetite granite.— A small dyke 2 inches wide and of the usual character, but containing irregular individuals of magnetite about $\frac{3}{4}$ inch in diameter, was seen in the bed of the Wewel Dola at Dik Mukulana. Another dyke 6 inches wide, containing similar magnetite, was observed on Hopewell estate.

Tha Gampola pegamtite[5]— This rock consisted mainly of orthoclase, quartz, and biotite, and contained apatite, tourmaline, ilmenite, and uraninite ('thorianite') as accessory minerals.

Granites without conspicuous accessory minerals.— These are of fairly general distribution in the Balangoda district, sometimes occurring in the form of dykes (usually less than three feet in width) in very considerable abundance. It is possible that a much more extended search might reveal the presence of macroscopic zircons in some of these rocks; for the most part, however, they

are similar to the granitic rocks described above, but without the characteristic accessory minerals. There is a considerable exposure of reddish granite on the Ratnapura road about a mile below Balangoda, and this is known as a locality for 'graphic granite'. A tendency to graphic structure was noticed in many of the rocks already described. Smaller masses (dykes) are common at Dik Mukulana (11 miles from Balangoda) and on Hopewell estate,

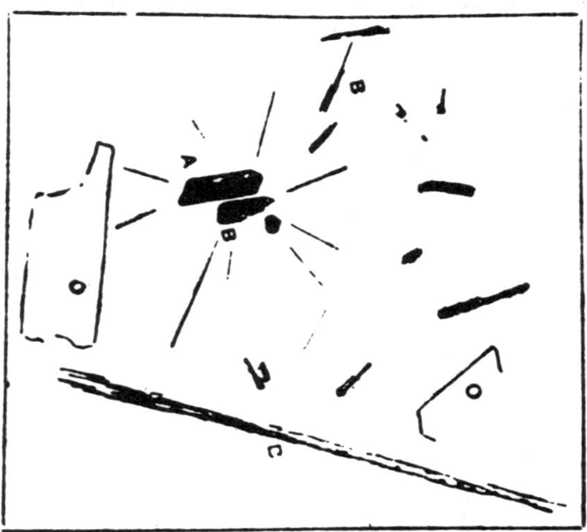

Fig.2- Structure of allnite granite; Denagama estate, Balangoda. *A*, Allanite; *B*, biotite; *O*, orthoclase; remainder quartz and felspar. Present scale, $\times 5\frac{1}{2}$ times.

15 miles from Balangoda , and at many other points, these granites (as well as others containing zircon) are also well exposed on Herimitegala estate, about 8 miles from Balangoda.

The list of types of rock belonging to the Balangoda group actually met with is now exhausted. There can be no doubt that many other varieties will be found, and it is evident that some are likely to be of great interest. Since, however, detrital zircon is exceedingly abundant, yet is known *in situ* in a few localities only, it is clear that the chance of finding any particular one of the other and rarer minerals *in situ* must be rather small.

[5] "Spolia Zeylanica", vol.i, pt.4 (1904), p.512.

V.— CONTRIBUTIONS TO THE GEOLOGY OF CEYLON:
4. INTRUSIVE PYROXENITES, MICA-PYROXENITES, AND MICRO-ROCKS IN THE CHARNOCKITE SERIES OR GRANULITES IN CEYLON.

By A.K. COOMÁRASWÁMY, B.Sc., F.L.S., F.G.S.,
Director of the Mineralogical Survey of Ceylon.

(PLATE XX, IN TEXT.)

Geo. Mag., London (1905) 363 – 369.

The object of the present paper is to provide a short account of a group of rocks to which the general name of *pyroxenite* may be given, inasmuch as they are essentially non-felspathic plutonic rocks, free from olivine, and of igneous origin, and with a nearly colourless monoclinic pyroxene as the chief or only constituent of many of the most conspicuous varieties [9] The exposures are widely distributed in the Central, Uva, and Sabaragamuwa provinces, and probably throughout Ceylon, but are individually of very small extent. The rocks are composed of the minerals diopside, amphibole, phlogopite (or biotite), scapolite, pyrite, sphene, and sometimes felspar, calcite, and spinel; the dominant mineral is the pale green or grey granular diopside, colourless in thin sections; the rocks are often entirely, or almost entirely, composed of this mineral; next in importance are green amphibole (dark in hand-specimens, but very pale in thin sections), and a pale golden-brown phlogopite, which sometimes forms a large proportion of the rock, and in extreme cases the whole bulk of sills or dykes several feet in thickness.

The rocks occur in sills and dykes of all sizes, from a fraction of an inch to six feet or more in thicknes, and their intrusive character is generally clear. Very similar or identical rocks occur also as nodular aggregates in crystalline limestones, and also near and along junctions of granulite and crystalline limestone[10]; but for the sake of brevity and clearness, only those occurrences which are less particularly associated with the crystalline limestones are described in the present paper. The pyroxenites are clearly intrusive in the charnockites or granulites ; although often occurring in beds or sills following the foliation planes of the granulites, they occur equally or more often as typical dykes crossing these foliation planes, and sometimes occupying joints; the figures illustrate examples of both modes of occurrence, which, moreover, are frequently associated.

[9]For a discussion of the nomenclature of similar rocks, see C.H. Gordon "On the Pyroxenites of the Grenville Series in Ottawa County, Canada", Journal of Geology, vol.xii, No.4, 1904, pp.324, 325.

[10]For descriptions see A.K. Coomáraswámy, "The Crystalline Limestones of Ceylon," Quart. Jour. Geol. Soc., 1902, vol.lviii, pp.399-422.

This occupation of joints shows that the pyroxenic material must have been introduced after the granulites had sufficiently consolidated to allow of the formation of joints. This fact, together with another, viz., the occurrence of what seems to be a very ordinary mica pyroxene dyke cutting the zircon granite (itself belonging to the Balangoda group[11], and forming an intrusive sill

in the Charnockite Series or granulites) of Massena estate, Balangoda, appears to indicate that the date of the pyroxenite intrusions was long posterior to that of the intrusion and consolidation of the granulites. Nevertheless, careful examination of the junctions shows that there is never any sign of a chilled edge, but the pyroxenic rock passes, rapidly indeed, but never abruptly, into the granulitic rock next to it, and this even where the intrusive character is otherwise most distinct. There is, in fact, a rapid transition from one rock to the other. In as much as the pyroxenites are intrusive in rocks which had cooled sufficiently to allow of the formation of joints, these facts seem to show that re-melting must have taken place to some extent.

A zoned structure is very characteristic of the rocks, the zoning being marked by the distribution of mica in bands parallel to the edges of the dyke or sill, the individual crystals of mica being disposed in planes perpendicular to the edges of the dyke or sill, which is often symmetrical in strucutre about a median line, after the manner of mineral veins. The zoned structure is not always produced by mica; in Fig.III, Pl.XX, p.366, for example, the central darker zone is distinguished by its content of hornblende, and mica is not present.

The pyroxenites seem to be in some way connected both in mineral composition and in actual distribution with the occurrence of crystalline limestones, though the exact relation is not perfectly clear. One curious feature is the presence of carbonates in parts of the dykes, especially at Ella (Uva) and Kadawat Oya (Sab.); when it becomes abundant, the central zone is sometimes indistinguishable from an ordinary crystalline limestone with abundant accessory minerals.

A more detailed account of various exposures (especially those illustrated) will now be given; the localities described have been visited at various times during the last two years.

1. At Ella (Uva) are interesting exposures (Fig.I, Pl.XX, p.366) in the bed of the Kitul Ela, just below the point at which it is joined by the small stream from near the Rest House. The main part of the intrusive rock consists almost entirely of granular greyish-green diopside (1152);[12] in parts scapolite and a little granular sphene are also present, and more locally hornblende or mica. A section (1153) through the junction

[11] A.K. Coomáraswámy, Geol. Mag., No.482 (1904), pp.418-422.

[12] The numbers in brackets refer to thin sections in the author's collection

with granulite at the point X in the figure shows granulite (a rock composed of quartz and orthoclase with a little pale-green (diopside) passing very rapidly but with no loss of continuity into a granular rock consisting of colourless diopside with scapolite and a few rounded grains of sphene. At one point in the exposure a large crystal of diopside projects from the pyroxenite into the granulite, merely an exaggeration of the usual interlocking of minerals at the junction. The inner portion of the pyroxenite (at *h*) becomes rich in carbonates, and the relation to the rather impure granulites at *g*

is not quite clear. The bands of intrusive pyroxenite following the granulite foliation planes tend to pass into lenticles, as if by a process of pinching, the lenticular froms being exactly followed by the granulite foliation planes; this fact, and the junction phenomena, makes it seem likely that the granulites were not in a completely consolidated state when the pyroxenic material was introduced. Similar 'pinching' is shown in Fig.II, Pl.XX, p.366, at X.

In the bed of the Talatu Oya at Uduwela, near Kandy, there are several exposures of pyroxenic rock. Fig.II shows sills of pyroxenite (nearly pinched out at X) combined with dykes along joint planes. A section (1067) of the junction at X shows the intrusive rock to consist internally of an aggregate of colourless diopside with a peripheral hornblendic zone passing into the granulite, in which plagioclase and diopside are the dominant minerals, and pyrite is also common. A section (1069) of the pyroxene further from the edge shows only colourless diopside with a trace of plagioclase and green spinel. A junction section (1068) at ϕ (Fig.II) showed orthoclase diopside rock (granulite) passing into a diopside plagioclase rock (pyroxenite) with accessory pyrite, sphene, and apatite. The (subordinate) plagioclase has an ophitic habit, enclosing more than one individual of diopside, which is also intergrown with it in radiating root-like fashion. In the central part of the dyke mica becomes abundant, and is arranged in planes perpendicular to the edges of the dyke, as indicated in Fig.IIAPl.XX, p.366. On other slide (1066) from the place marked XX shows an aggregate of diopside and amphibole with a little accessory pyrite.

Some smaller dykes at Uduwela are very micaceous; a section (1149) of one about an inch wide shows granulite (quartz-orthoclase-plagioclase rock with a little diopside) passing successively into diopside rock with a little plagioclase, diopside rock, and diopside-mica rock. The mica is in hand-specimens the most conspicuous mineral, and is arranged in planes perpendicular to the edges of the dykes.

3. Fig.III shows a zoned pyroxenite dyke cutting the foliation of the Charnockite Series, exposed in the bed of a stream entering the Badulla Oya from its right bank, about 8 $\frac{1}{2}$ miles from Badulla and near Nanla. The outermost zone (about 3 inches wide) shows under the microscope (1147) a compact intergrowth of colourless diopside (partly also in granular forms) with subordinate plagioclase. The diopside intergorwn with plagioclase exhibits root-like forms, and the intergrowth is on a very fine scale,

so that the plagioclase is not easily recognizbale. The central one or two inches of the dyke are clearly marked by the presence of dark hornblende, which in the hand-specimen can be seen moulded on idiomorphic grey-green diopside in ophitic wire. In the thin section (1148) the hornblende appears pale bluish green, with pleochroism a pale staw, b bluish-green with a brown tinge, and c bluish-green. This is the usual amphibole characteristics of the rocks now described.

Geol. Mag. 1905. *Dec. V, Vol. II, Pl. XX.*

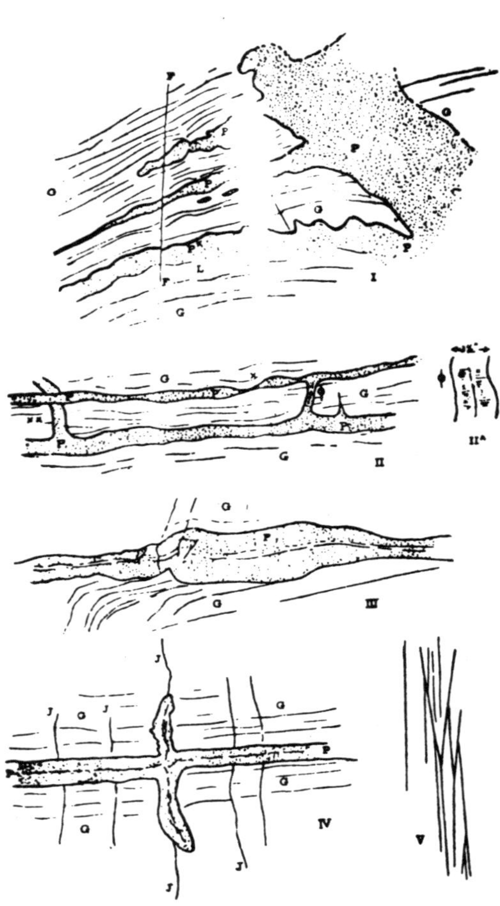

Intrusive Pyroxenites in the Ceylon Granulites

In illustration of Mr. Coomáraswámy's paper.

4. There are good exposures of pyroxenites at Kadawat Oya bridge, between Belihul Oya and Halpe (near the 99 $\frac{1}{2}$ mile-post, Sabaragamuwa. A number of pyroxenic sills and dykes occur, and very heavy hornblendic and pyritic types are characteristic. The dykes vary from mere threads up to 1 $\frac{1}{2}$ yards in width. Some show very clearly penetration of the pyroxenic material along joints (Fig.IV). Several dykes have quite narrow central zones, consisting mainly of pyrite; in one case large individuals of pyrite occured at intervals along this line. Some of the larger dykes have central portions composed largely of carbonates with abundant accessory minerals (amphibole, phlogopite, spinel).

A section (1122) from the point marked X in Fig.IV shows a rock consisting almost entirely of pale sea-green diopside with a very little pyrite, apatite, and phlogopite; the diopside has a very carious structure, the cavities being occupied by semi-ophitic plagioclase. The whole dyke is about 6" wide. A section (1123) of another dyke, 6" - 8" wide, proves it to consist mainly of very pale sea-green diopside, with some strongly pleocharoic phlogopite, yellowish-green amphibole, green spinel, pyrite, apatite, and calcite. The diopside tends to granular forms, moulded by the mica and spinel, the latter being apparently almost the last formed mineral, occupying interstices amongst the grains of pyroxene. As the junction with granulite is approached the mica dies out, and some felspar and pink garnet come in. One other dyke, very dark in colour, consisted mainly of hornblende (especially in the central zone), pale bluish green in thin section (1150), and diopside with a trace of plagioclase occupying interstices ; the hornblende has a strong tendency to crystalline form, even when in contact with diopside (a reverse relation is found in the central zone of the dyke shown in Fig.III, Pl.XX,p.366 see description above).

5. 'Thread dykes' and others slightly wider. These very narrow dykes appear as dark lines crossing foliation on the surface of rock exposures (Fig.V). Under the microscope there appears nothing more than a rapid concentration of hornblende along a zone perhaps $\frac{1}{8}$ inch wide (1103). These very narrow dykes are often in more or less parallel groups of two or three (Fig.V); sometimes they intersect at acute angles. Occasionally they widen out into dykes an inch or more wide, clearly visible in the field, but with such indefinite margins as to be hardly recognizable in hand-specimens.

A few additional examples may be briefly referred to.

6. A sill of grey-green pyroxenic rock 6-8 feet thick is exposed in the river bed below the road, about 10 $\frac{1}{8}$ miles from Badulla on the Bandarawela road. The rock (1109) is a coarse aggregate of diopside and scapolite with scattered idiomorphic sphenes; other varieties consist practically entirely of diopside, individuals of which reach $\frac{3}{4}$ inch in width. The rock is very tough. Somewhat similar rocks are found, not quite *in situ*, on the slope below the 69th mile-post on the Badulla Nuwara Eliya road, and some of these are very handsome coarse aggregates of diopside, amphibole, and scapolite, with

calcite and idiomorphic sphene (often over $\frac{1}{8}$ inch

368 *A.K. Coomáraswámy — Intrusive Pyroxenites, Ceylon.*

long). The magnesian silicates and scapolite are idiomorphic towards the calcite and present a handsome appearance in weathered specimens.

In the Uma Oya below there are numerous exposures, both of 'thread dykes' and of intrusive pyroxenic and micaceous rocks, as well as of coarse felspathic pegmatites, with which their relations are somewhat confused.

7. Of pure mica rocks or veins there may be instanced those described in the (first) "Administration Report of the Mineralogical Survey", Colombo, 1904. Analyses of mica will be found in Professor W.R. Dunstan's "Reports on the Results of the Mineral Survey [of Ceylon] in 1903-4", London, 1905.

One example of a mica rock or vein may be quoted from the (second) "Administration Report of the Mineralogical Survey", Colombo, 1905, p.20. A large vein of brown phlogopite is exposed in the Mineran Oya, just within the boundary of Sherwood estate (about nine miles north of Balangoda), in the south-west corner. The vein is about 5 feet wide, and strikes 20° N. of E. and hades 70° S.S.E., and crosses the foliation of the granulites, which are inclined at low angles. The Oya has for some distance cut its way along the mica vein, which is exposed in the bed of the stream. The largest crystals are 18 inches in diameter, but somewhat flawed.

8. A fairly typical zoned mica pyroxene dyke, from 2 inches to 1 foot wide, is found cutting the zircon granite (Balangoda group) of Massena estate, Balangoda. As the zircon granites are themselves intrusive in the granulites,[13] the pyroxenic intrusions would thus appear to be removed from the date of intrusion of the material composing the Charnockite Series, by the period required for the intrusion of the Balangoda group.

Canadian analogies. The rocks described closely resemble the pyroxenic rocks associated with the apatite deposits of Canada, described by A. Osann ("Notes on certain Archean Rocks of the Ottawa Valley", Geol. Surv. Canada, part O, Ann. Rep., vol.xii, 1902), and by C.H. Gordon ("Pyroxenites of the Grenville Series in Ottawa County, Canada", Journal of Geology, vol.xii, No.4, 1903, pp.316-325), who states that the evidence there too "clearly warrants the conclusion that the pyroxenic rock is intrusive". Detailed descriptions and further references will be found in the papers mentioned.

Summary. A group of pyroxenites, in which the minerals diopside, phlogopite, hornblende, and scapolite predominate, while sphene, plagioclase, pyrite, apatite, and spinel are often accessory, occurs in small sills and dykes intrusive in the Charnockite Series or granulites, and in one case apparently also in a zircon granite of the Balangoda

[13] A.K. Coomaraswamy, "The Balangoda Group": Geol. Mag., 1904, pp.418-422

group. The pyroxenic intrusions have usually a zoned or laterally symmetrical structure analogous to that of mineral veins. The intrusive material never shows a chilled edge, but

there is a rapid transition from the pyroxenite to the granulite. The course of the small sills and dykes of intrusive pyroxenic material is usually determined by the foliation planes and joints of the granulites or Charnockite Series.

EXPLANATION OF PAGE-ILLUSTRATION, PLATE XX (p.366).

INTRUSIVE PYROXENITES IN THE CEYLON GRANULITES.

Fig.1. - Intrusive pyroxenite at Ella (Uva); area shown about 6 × 4 yards. P, pyroxenite; G, granulite; F, fault.

Fig.2. - Pyroxenite sills and dykes, Uduwela (Central Prov.); length shown about 8 yards. P, pyroxenite; G, granulite.

Fig.3. - Pyroxenite dyke with well-marked symmetrical structure, exposed in feeder of Badulla Oya, near Naula; average thickness, 6 inches, P, pyrcxenite; G, granulite.

Fig.4 - Intrusive pyroxenite following foliation planes and joints in granulites (also in Figs.1 and 2), Kadawat Oya bridge (Sab.); thickness, 6 inches. P, pyroxenite; G, granulite; J J, joints.

Fig.5. - 'Thread dykes', exposed in feeder of Badulla Oya, near Naula; average thickness, $\frac{1}{8}$ inch.

KANDY, June 19, 1905.

5. Note on the Scenery of Ceylon.

By A.K. Coomaraswamy, B.Sc., F.L.S., F.G.S.

(British Association for the Advancement of Science, Transactions of Section C, Report –1902, London, 1903, 613 — 614.)

It is probable that Ceylon has been exposed to continuous denudation since very early Palaeozoic times. The foliation of the crystalline rocks has had a marked influence in determining the directions of the river valleys and the general configuration of the country. The foliation strike is usually from N. to N.W. Many rivers have a similar direction, as examination of a map will show. A north-and-south strike-valley runs from Wattegama to Dambulla, followed by the railway as far as Matale. N.N.W. valleys are conspicuous N. of Hunasgiriya. The Mahaweli Ganga valley, south of Peradeniya, is a strike-valley; another good example is the valley running N.W. from Hatton. Each of these valleys is followed by the railway. Other conspicuous N.W. valleys (probably strike-valleys) are those between Kurunegala and Matale; and N.E. of Adam's Peak and S. of Ratnapura. A small area east of Kandy has been examined in detail, and shows a diagrammatic system of strike-valleys, with others at right angles thereto, the strike being here, however, more nearly east and west than is usually the case. The bands of limestone may have had some effect in determining the actual positions of these valleys. Of course *all* valleys in Ceylon are not strike-valleys; thus the Mahaweli Ganga valley crosses the general strike below Gettembe (E. of Peradeniya), forming a series of rapids.

A characteristic feature of the scenery of Ceylon in many parts is its precipitous character; the seemingly 'bedded' granulites form mural escarpments and dip slopes, as if they were a series of sedimentary rocks.

The above remarks apply only to the mountainous districts which occupy the south central part of Ceylon. A low coastal plain fringes the island, partly of alluvial and partly of raised beach origin ; sea cliffs are absent or very unusual, and even any coast exposures of rock are not common. In the north a greater area is flat and low, and the scenery resembles that of Southern India. Isolated hills of gneiss (Dambulla, Sigiri, &c.) rise conspicuously from the plain.

VII.—ORIGIN OF THE CRYSTALLINE LIMESTONES OF CEYLON.

By A.K. Coomáraswámy, B.Sc., F.L.S., F.G.S.

It is generally supposed that the crystalline limestones which occur amongst the schists, associated with orthogneisses, in various parts of the world, are altered sedimentary limestones whose accessory minerals and crystalline strucutre have been developed by simple contact and dynamo-metamorphism; and no doubt many crystalline limestones are of sedimentary origin, and owe their peculiarities to these agencies. In other cases the peculiar mode of occurrence of such rocks or their manner of association with igneous rocks (orthogneisses or nepheline-syenites) has led to the suggestion of other theories.

Professor Högbom[14] has described the crystalline limestone which is associated with the elaeolite-syenite of Alnö, where all transitions occur from pure limestone to calciferous and normal elaeolite-syenite. Intergrowths of calcite with nepheline, aegirine, and felspar were noted. Professor Högbom (p.109) thinks that, whatever the origin of the limestone (it seems not likely to be an altered sedimentary one), it is quite certain that it has been melted and taken up by the magma on a large scale without decomposition, and that on consolidation, calcite has crystallized out of the magma in just the same way as the other minerals.

Professor Judd[15] has suggested that the crystalline limestones of Burmah (which in many respects very closely resemble those of Ceylon) result from the alteration of the basic lime-silicates of the pyroxene-gneisses. My observations show that this is a theory not applicable to the crystalline limestones of Ceylon. This is shown by the field relations and by the fact that the Ceylon granulites are

almost always exceedingly fresh. The calcite and lime-silicates found near the junctions are not the result of alteration of minerals composing the granulites. Contact-metamorphic relations between the limestones and granulites are everywhere indicated.

Dr. Callaway has suggested that the crystalline limestones of Bodwrog and Porth Trecastell in Anglesey[16] and of Ceylon[17], may have originated by "segregation from plu-

[14]Geol. Fören. Stockholm. Förh., 1895, vol. xvii, pp.100 and 214.
[15]Phil. Trans.Roy. Soc., 1896, vol. clxxxvii A, pp.151—228.
[16]Rep. Brit. Assoc., 1887, p.706.
[17]Geol., Mag., 1902, p. 285.

tonic rocks during deformation". That the Ceylon limestones have not been formed thus is shown by the contact-metamorphic relations between them and the granulites, and the freedom from deformtion of the latter, with which they are inseparably associated.

Mr. Holland[18] has suggested that the crystalline limestones of Burmah may have existed in a state akin to fusion. Calcite has been claimed as an original mineral in nepheline-syenite by more than one observer[19]. The present writer has described what is probably an original intergrowth of calcite with quartz[20].

I have recently given some account of the crystalline limestones of Ceylon,[21] showing that a vast series of granulites, various in chemical and mineralogical composition, is there most intimately associated with a smaller but widely distributed series of crystalline limestones occurring in bands of various widths, whose foliation (dependent on variations in coarseness of grain, structure, and mineral composition) is parallel to that of the granulites and to the boundaries between the two rocks. The contact phenomena and the transitions between the two types of rock are also described.

As to the real *origin* of these limestones, not very much can be said. We might regard them as altered sedimentary or even tufaceous rocks, which the intrusive granulite has melted and partially absorbed. The wollastonite-scapolite gneisses of Galle might, on this supposition, result from the total absorption of a large mass of limestone by the granulites. It may be objected, however, that there is little or no evidence of the existence of other altered sedimentary rocks such as would probably have accompanied the limestone,[22] and that the limestones occur in different areas. Sometimes abundantly and in wide bands, sometimes in narrow bands far removed from other rocks of the same sort.

A.K. Coomáraswámy — Crystalline Limestones of Ceylon. *377*

Whatever their original character may have been, the limestones must have existed before the intrusion of the granulites. I think it is clear, however, that *in their present condition* the two rocks are 'contemporaneous', and this is shown by their intimate associations and similar behaviour, and by the transitions from one rock to the other which sometimes occur. I suggest that the limestones themselves have existed in a state akin to

[18]Mem. Geol. Surv. India, 1901, vol.xxx,pt.3, p.175.

[19]Holland : Mem. Geol. Surv. India. loc. cit., p.197. Adams : Amer. Journ. Sci., 1894, vol.xlviii. p.14.

[20]Quart. Journ. Geol. Soc., 1900, vol. lvi, p.606, also pl.xxxiii, fig.1.

[21]Quart. Journ. Geol. Soc., 1902, vol.lviii, p.399.

[22]I have met with rocks almost composed of dark mica in one or two localities (Yatirawana, near Wattegama, and Talbot Town, Galle) associated with the granulites. Almost any mineral occurring in the granulites may, however, be sometimes met with in this way as a mian constituent of a local variety. It is however, quite true that some varieties rich in garnet and biotite are not altogether unlike much altered sedimentary rocks, though I do not myself regard them as such. Lacroix has described some rocks in which andalusite, sillimanite, and corundum occur, but unfortunately the localities in Ceylon are unknown. Possibly these minerals indicate the existence of altered sedimentary rocks, but this is by no means necessarily the case.

fusion[23] and have behaved much as igneous rocks, and that the carbonates entering into the constitution of the granulites near the contacts are, as they appear to be, original minerals[24]. That the limestones have existed in a state akin to fusion is shown, I think, by the appearances of flow-structure (foliation)[25], the occasional porphyritic strucutre, the aggregates of accessory minerals (which correspond perhaps to the basic secretions of some igenous rocks), the intergrowths of calcite and dolomite, and by the field relations of the limestones and charnockite series. The accessory minerals have crystallized out as if from a magma. The limestones may have been softened before the intrusion of the granulites (which greatly exceed them in amount), or have become so as a result thereof. Each rock has been affected near the contact by the absorption of material from the other, the granulites showing the more conspicuous modifications. This transference of material is shown in the granulites near the junctions by the appearance of minerals rich in lime, such as scapolite, sphene, diopside, phlogopite, and calcite, and in the limestones by the appearance of minerals such as diopside, amphibole, phlogopite, and spinel, and very rarely scapolite and felspar. The common occurrence of spinel as a contact mineral seems to result from the appropriation of all available silica in the formation of lime-silicate minerals, leaving the oxides to form spinel.

The whole process must have taken place under conditions of great pressure. Differential movements during consolidation have produced the conspicuous foliation. To some extent the limestones and granulites have been rolled out together during consolidation, and in this way sills of granulitic rock in the limestone have been broken and portions separated, sometimes as lenticles containing among the strike the unbroken portions of the sill, sometimes to form irregular patches or snaky twists of pyroxenic rock in the limestone, sometimes to be found as quite isolated masses, surrounded by the enclosing limestone.

To summarize, I wish to suggest —

(1) That contact-metamorphic relations between the limestones and the charnockite series are everywhere clearly indicated.

378 *Notices of Memoirs.*

(2) That it is not altogether easy to regard the limestones as altered sediments.

(3) That whatever their origin they have actually existed in a state "akin to fusion", and as far as their present characters are concerned, are contemporaneous with the charnockite series.

(4) That limestones and the charnockite series have suffered but little from deforming earth-movements since their final solidification.

---

[23]The direct use of the term 'molten' is avoided, inasmuch as we can know but little of the conditions of matter under very great pressure, and it is inadvisable to use the same terms in the description of conditions which, though analogous, must yet be very different.

[24]Cf. J.J. H. Teall, Geol., Mag., 1886, p.349.

[25]It is evident that the limestones have not (except very locally) suffered from deforming earth-movements since the development of the accessory minerals, whose varying abundance is largely the cause of the apparent foliation.

The CRYSTALLINE LIMESTONES of CEYLON.

By ANANDA K. COOMÁRASWÁMY, Esq., B.Sc., F.L.S., F.G.S.

(Read March 12th, 1902.)

[PLATES XIII & XIV.]

Geo. Mag., London, **58** (1902) 399 – 424.

CONTENTS.

I. INTRODUCTION.

In 1860 Baron Ferdinand von Richthofen[26] recorded the occurrence of limestones in Ceylon. He observed the intimate associations of 'gneiss' and granular limestone, and the transitions between the two.

Mr. A.C.Dixon[27] has given a short account of the beds of limestone as follows:— The
'beds run through the gneiss in a somewhat parallel direction, striking generally north-west by north to north, and having various angles of dip, from 10° to 40°'.

He refers to five bands.

*'The first is one which outcrops a few miles this side of Balangoda, and runs north-north-west, occurring again at Hanuwala. The second runs through Dolosbage and Maskeliya; probably the bed occurring at Bilhul Oya is continuous with this. The third outcrops under the Great Western [Mountain] on the Great Western Estate, and is continuous to the north-north-west with the Wattegoda and Medakumbura dolomites, and probably also with the beds at Gampola and Kurunegala. A subsidiary bed—or it may be an outlier of this— occurs near the Pussellawa rest-house. The fourth bed outcrops largely at Wilson's Bungalow, Glen Devon, Dumbara, and Matale. The fifth occurs in the Badulla District'. (*Op.cit.pp.42-43*).*

[26]'Bemerkungen über Ceylon,' Zeitschr. Deutsch. Geol. Gesellsch. vol.xii (1860) p.523.
[27]Journ. Roy. Asiat. Soc. Ceylon Branch, vol.vi (1880) p.39

I may say that the identity of these bands over such considerable distances requires confirmation.

Under the name of cipolin, Prof. A. Lacroix[28] has described some specimens of crystalline limestone from Ceylon, which form part of the De Bournon collection belonging to the Collége de France and that of Leschenault de la Tour in the Muséum d'Histoire Naturelle, Paris. The specimens were chiefly collected 7 leagues east of Kandy among acid genisses. The minerals calcite, dolomite, blue apatite, phlogopite, spinel, pyrrhotine, and chondrodite were

400 Mr. A.K. Coomáraswámy on the [Aug.1902,

noted. Other cipolins from Cornegal (Kurunegala) were examined, and consisted largely of yellow calcite. They contained elliptical masses, reaching a foot in length, made up of a mixture of various minerals. One such was formed of calcite, oligoclase, green pyroxene, pyrite, quartz, and sphene; another of green amphibole, calcite, scapolite, sphene, oligoclase, pyrite, and zoisite. These nodules thus present a composition not altogether unlike that of the pyroxenic gneisses of the same districts.

In 1898 Dr. Max Diersche[29] examined a specimen of limestone from 'Queen's Palace' (Anuradhapura?) which Prof. Zirkel had collected; he found in it calcite, dolomite, olivine, phlogopite, and rutile.

In 1900 Dr. E. Ch. Schiffer[30] described in some detail a granular limestone from Wattegama, from the railway-cutting near the level-crossing, giving analyses of dolomite, apatite, phlogopite, hydrophlogopite, green serpentine (probablyl derived from forsterite), and white serpentine, recording also spinel, pyrite, and magnetite.

In the same year the present writer [31] noted the occurrence of limestones at Hakgala, Kandy, etc., referring especially to large pyroxenic masses with much blue apatite, occurring in the Hakgala limestone[32].

Mr. John Parkinson[33] has described a section exposed in a railway-cutting near Matale, showing crystalline limestone in which occurs a band of rock resembling pyroxene-granulite which he (I think, rightly) supposes to be an intrusion in the limestone (see p.408). He suggests also that limestone-material has been absorbed by the igneous rock.

[28]'Contributions á l' Etude des Gniess á Pyroxéne et des Roches á Wernerite', Bull. Soc. fran c, Minéral.vol.xii(1889) pp.336-44.

[29]'Beitrag zur Kenntniss der Gesteine u. Graphitvorkommnisse Ceylons' Jahrb. d.k.-k. Geol. Reichsanst. vol. xlviii (1898) p.271.

[30]'Chemische Untersuchungen eines körnigen Dolomits aus dem Gneiss von Wattegama in Ceylon,' Inaug-Diss. Munich, 1900.

[31]'Ceylon Rocks & Graphite,' Quart. Journ. Geol.Soc. vol.lvi (1900) p.600.

[32]This is the locality 'near Newara Eliya' referred to by Prof. A.H. Church in 'Nature' vol.lxiii (1901) p.464.

[33]'Notes on the Geology of South-Central Ceylon,' Quart. Journ. Geol. Soc. vol.lvii (1901) pp.204-206.

Prof. E. Weinschenk[34] refers to the occurrence of these coarse-grained dolomites and cipolins, and regards the forsterite, chondrodite, phlogopite, and spinel which they contain as contact-minerals. He associates these rocks with the peculiar andalustite-, sillimanite-, and corundum-bearing rocks described by Prof. Lacroix, and considers that they represent beds of unknown age into which the rocks forming the Granulitic or Charnockite Series were intruded, producing widespread, contact-metamorphism.

Dr. Fr. Grünling,[35] who collected the specimens of limestone described by Schiffer, in a general account of the results of his excpeiditon to Ceylon, suggests that the corundum and spinel found

in the river-gravels are derived from the dolomitic limestone; and, further, he saw in Colombo a specimen of calcite containing rather large crystals of zircon. I have never found corundum or zircon in the limestones, though spinel is very commonly seen in them.

The object of the present paper is to give a somewhat more complete account than has yet been attempted of the crystalline limestones and their relation to the (igneous) granulites (Charnockite Series) among which they are found. It is based on observations made in Ceylon in 1901, and refers especially to the Kandy and Hakgala districts, while limestones from Matale and farther north are referred to, and also some from Balangoda and Rakwane.

The crystalline rocks of Ceylon may be classified as follows:—

(1) Older Gneisses (orthogneisses).— These have suffered from earth-movements to a considerable extent, previous to the introduction of the extensive Granulite or Charnockite Series.

(2) Crystalline Limestones (cipolins of French authors)— These occur in bands or beds of very various width, the boundaries of which are parallel to the foliation (mineral banding) of the granulitic igneous rocks with which they are associated.

(3) Charnockite Series or Granulites:— These constitute the mountain-massif of South-Central Ceylon, and are probably intrusive in the older gneisses. (Dykes of a rock which may be connected with the Charnockite Series are clearly intrusive in the older gneiss at Ambalangoda). These rocks are specially characterized by mineral banding and variation in chemical and mineralogical compo-

[34] 'Zur Kenntniss der Graphitlagerstätten: III. Die Graphitlagerstätten der Insel Ceylon,' Abh. k.-bayerisch. Akad. Wissensch.vol.xxi(1901) pp.286-90. See also review in Geol. Mag. 1901, pp.175-76.

[35] 'Ueber die Mineralvorkommen von Ceylon,' Zeitschr. für Krystallogr. vol.xxxiii (1900) pp.233, 235.

sition. They are all transitions from pure pyroxene-rocks to pure quartz-rocks. The limestones are interbanded with these rocks, and almost inseparably connected with them. The rocks of the Point de Galle Group must be included in the Charnockite Series, and form a sub-division thereof. This group includes granulitic rocks, in part largely composed of pyroxene, wollastonite, and scapolite, and in part resembling types of rock more usual in Ceylon.

The rocks included in Divisions (2) and (3) have, since their final consolidation, remained almost unaffected by deforming earth-movements.

With regard to terminology, I have adopted the name Charnockite Series for the granulitic igneous rocks of Ceylon, on account of the convenience of employing a definite name for them, and because they resemble, more than they differ from, the Indian rocks to which this name has been applied by Mr. T.H. Holland. So far is this the case, that any considerable series of rock-specimens from the Nilghiri Hills, the Shevaroy Hills, or from near Madras and from near Madura, could only with difficulty be distinguished from such a collection from Ceylon. There are of course varieties forming part of this series in India that have not been matched in Ceylon, and *vice versa.* I feel, however, that there is some ambiguity in using the one term Charnockite for the whole series, and also for a special type of rock very characteristic of the series. Despite this objection, it seems more important to emphasize the resemblances between Ceylon and Indian rocks than to propose a new term.

I desire to thank Prof. T.G. Bonney, F.R.S., and Mr. G.T. Prior, M.A., for occasional assistance and kind advice, and the latter for his analysis of forsterite. I am indebted to Prof. Miers's kindness for crystal-measurements which were made by Mr. Graham at Oxford, and to Mr. Herbert Smith for measurements of clinohumite. In Ceylon I have to thank Mr. Shand (Rakwane) and Mr. Holland (Balangoda) for their hospitality ; and also Mr. and Mrs. Nock, who rendered my stay in the new bungalow at Hakgala very pleasant. Mr. W.C. Hancock, B.A., has taken an interest in some of the chemical analyses, which he has made with great care.

II. GENERAL DESCRIPTION OF THE LIMESTONES.

The crystalline limestones from beds, the boundaries and foliation of which are parallel to the foliation (mineral banding) of the neighbouring granulites (fig.2, p.406). The outcrops vary inwidth from a few feet to more than a quarter of a mile. The most general strike is from north to north-west, and it is sometimes quite constant for many miles.

The narrower bands of limestone are occasionally more intimately interbedded with the granulites than can be shown in the maps which illustrate this paper. The foliation of the limestone is marked by variation in coarseness of grain, in structure, and in mineral composition. Thus there are alternations of bands rich and poor in accessory minerals. The latter on the whole are more abundant near the contact with the granulite, but this difference is not always marked. Bands in which occur parallel intergrowths of dolomite and calcite may alternate with others in which ramifying intergrowths are found, or in which there is merely a granular mixture of carbonates, or dolomite or calcite alone. Calcite seems to predominate where accessory minerals most abound. The following accessory minerals are very characteristic: olivine, diopside, phlogopite, blue apatite, spinel, pyrite, and graphite; less common are amphibole, clinohumite, magnetite, scapolite, sphene, and orthoclase. I have never met with garnet or wollastonite associated with the limestones.

The limestones often contain nodular masses of silicates, recalling those that occur in the Glenelg Limestone.[36]. These knots or nodules weather out of the limestone, with which they are often connected by only a small neck, so that they are easily detached. They vary from less than an inch up to several yards in diameter, and their constituion is various.[37] Many are white, and consist entirely of diopside; a variety common in the Hakgala district consists of

diopside, blue apatite, and mica. The blue apatite 'from a locality near Newara Eliya'[38] (actually Lower Albion, near Hakgala, see map, Pl. XIV) occurs in a mass of this kind several yards in diameter, slightly quarried in the hope of obtaining mica of value. Carbonates are, as a rule, nearly or quite absent from these aggregates. The associations met with include:— Pure diopside (the commonest type); diopside, phlogopite, blue apatite (common); diopside, phlogopite, blue spinel (near Tataluoya); tremolite, calcite, blue apatite, colourless phlogopite (two-fifths of a mile south-south-west of Ulisna Muduna trigonometrical station, Pl.XIII); tremolite along (36 miles north of Kandy); clinohumite (near Gettembe); olivine (Harakgama, Pl.XIV); olivine, green spinel, calcite, pale phlogopite (Mount Olive, Pl.XIV); olivine, pale phlogopite, pink spinel, tremolite, calcite, raphite, blue apatite, pyrite, rutile (disused middle quarry at Gettembe); phlogopite alone (several localities).

The limestones vary much in coarseness of grain. The finer varieties are saccharoidal ; the coarsest-grained consist of individuals measuring over 4 inches along the rhombohedral edge. This exceptionally coarse variety forms a thin band of limestone, exposed by the roadside about a third of a mile east-south-east of Talatuoya Bridge. Speaking generally, the limestones are distinctly coarse-grained, and the individual accessory minerals are always clearly visible to the naked eye.

The limestones vary as much in purity; almost pure varieties have a specific gravity

[36]C.T.Clough & W. Pllard, 'On Spinel & Forsterite from the Glenelg Limestone,' Quart. Journ. Goel. Soc. vol.lv (1899) p.372.

[37]Here only those mineral-aggregates are referred to, that consist of accessory minerals characteristic of the limestones. It is not always possible, however, to draw a sharp line between these aggregates and inclusions of pyroxene-granulite in the limestone, which are peripherally rich in lime-silicates, diopside, phlogopite, and scapolite.

[38]A.H.Church, 'Nature' vol.lxiii (1901) p.464.

of 2.76 to 2.78; in very impure varieties (rich in olivine) the specific gravity may reach 2.92, and that of the silicate-aggregates is, of course, still higher.

Sometimes the carbonate-individuals are not uniform in size, but there are porphyritic crystals embedded in a finer matrix. In one case examined the larger crystals consisted of dolomite, the smaller of dolomite and calcite mixed, and intergrown. These porphyritic limestones occur at several points in the Hakgala district. Neither in them nor elsewhere are the carbonate-crystals idiomorphic.

The proportion of calcite and dolomite present varies considerably. It is always possible to separate the two carbonates by means of the Lemberg[39] stain, which may be applied either to a thin section or to a smoothed surface of rock. The only other distinctions between the carbonates of which I have made use in microscopic work, are the difference of refractive index and the twinning, which is usually but not quite invariably, characteristic of the calcite only.

Microscopic examination of the limestones shows that they have not suffered from deforming earth-movements since the development of the accessory minerals that characterize them.[40] In one case curved

cleavage-faces were noted on the carbonate-crystals (Herimitigala). For a case of slickensided limestone, see below.

In the field, the various bands of limestone crop out with considerable regularity. Variations of dip are more frequent than changes of strike, which are always gradual, with one exception referred to below. It must be emphasized also, from this point of view, that the limestones cannot be considered apart from the granulites (Charnockite Series in Ceylon), and that these do not appear to have suffered from earth-movements since their consolidation, so far as microscopic work and the greater part of my field-work go.

In one case, however, a difficulty was met with in the field-work. On the Hakgala-Bandarawela road, near the 56th mile-post, the limestone strikes in a west-south-westerly direction; a furlong farther on the limestone is much slickensided and sheared (the only case observed by me in Ceylon), and in the stream just beyond this point the granulites are clearly seen striking in a north-westerly direction; the limestone-boundary near by

[39] 'Zur mikroskopischen Untersuchung von Calcit, Dolomit u. Predazzit,' Zeitschr. Deutsch. Geol. Gesellsch. vol.xl (1888) p.357.

[40] It will be gathered that I am inclined to regard all these accessory minerals as original, that is, as contemporaneous with the carbonate rock-constituents.

Fig.1:- Geological Map of part of the

KANDY DISTRICT

by A.K. Coomáraswámy, B.Sc., 1902

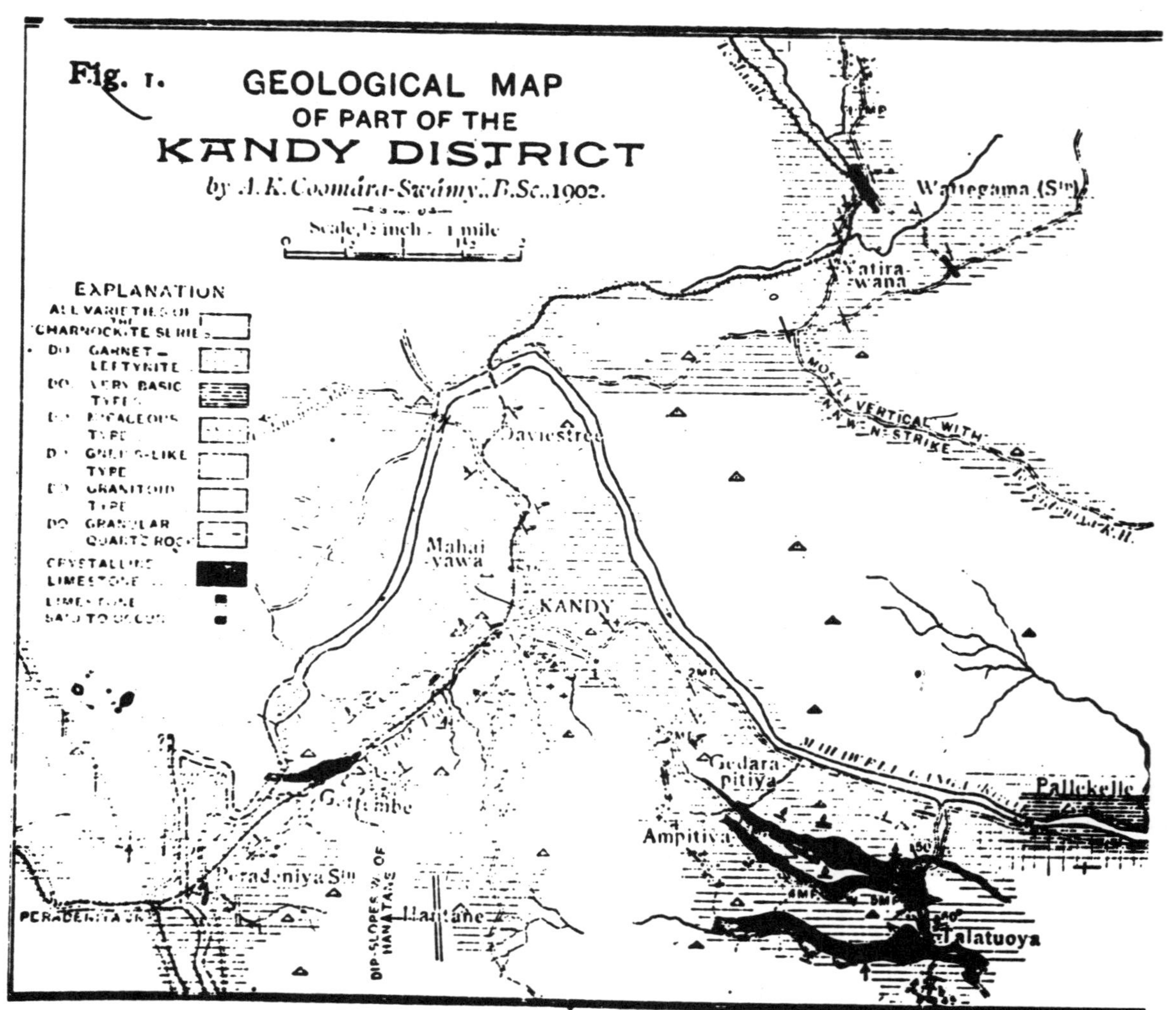

has the same direction. Appearances of shearing and crushing are found in the granulitic rocks south-west of this point also. The areas north-east of it are somewhat inaccessible, being overgrown with lantana. Some uncertainty was felt also in mapping the area north of the Padinawela wood (Pl. XIV). Farther west, in the Mount-Olive area, limestones and granulites behave quite normally, possessing a common strike, a little north of west. A fault has been inserted in the map, in order to call attention to the existence of these phenomena; but in all probability it does not represent accurately the state of affairs.

It will be seen, from the maps which illustrate this paper, that the limestones and granulites together have probably been thrown into gentle folds; work in the field, however, impresses one with the comparatively uniform structure of very considerable areas.

III. RELATIONS BETWEEN THE CRYSTALLINE LIMESTONES AND THE CHARNOCKITE SERIES.

With the exception of Mr. Parkinson's description of a section near Matale, to be referred to later, no account has been given of these relations. As stated above, the limestones and granulitic (igneous) rocks exhibit a common foliation, the direction of which is parallel to the boundary between the two rocks. The junction is often very easy to follow on the ground, though not often clearly enough exposed for detailed examination. The actual boundary seems never, or very rarely, to be abrupt; but there is either a zone of green rocks in which greenish diopside is the most prominent constituent, other minerals including greenish and brown micas and amphibole, green spinel, iron-ores, and sometimes calcite—separating the limestones from the ordinary granulitic rocks,—or there is a gradual transition between the latter and the limestones, marked by the gradual appearnce of calcite, accompanied by scapolite and sphene.

About 300 yards or so north of Talatuoya, on the gurudeniya minor road, the road-cutting on the east side of the road exposes a series of normal granulites, among which a band of limestone 8 feet wide is intercalated, there being a northerly dip of about 50°. A transition-zone 1 $\frac{1}{2}$ to 2 feet wide is seen on the north side, the southern boundary being less clearly exposed. North of this zone are normal granulites; south of it the rather rotten, coarsely crystalline limestone, dotted with dark pyroxene. Sections from the outer part of the transition-zone show a rock much like an ordinary, rather acid, pyroxene-granulite with quite fresh felspars (plagioclase and orthoclase very finely microperthitic), elongated quartz, pale-green augite, and iron-ores, but with also a varying proportion of calcite, sacpolite, sphene (individuals of the latter reaching 3 millimetres in diameter) and graphite; these minerals being evidently essential parts of the rock, and original constituents thereof. A small part of the calcite is probably secondary. A section much nearer to the limestone consists mainly of scapolite, calcite, and augite, but orthoclase-microperthite, plagioclase, quartz, sphene, and iron-ores are also present.

Fig.2:- *Section along AB in Pl.XIII: from the Mahaveli Ganga to south of Talatuoya*

[Horizontal scale; $2\frac{1}{2}$ inches = 1 mile.]

AA = Approximate level of the bed of the Mahaweli Ganga

M = Mahaweli Ganga	T = Talatuoya River	SS = Streams
R = Talatuoya-Gurudeniya road	R' = Kandy-Badulla road	R" = Kandy-Talatuoya road

The limestones are shown as black bands amid the (stripe 1) granular quartz-rock, leptynite, pyroxene-granulite, pyroxenite, etc.

A similar case was noted at Rakwane, where on the Allerton Estate, near a now abandoned mica-pit, some masses of limestone crop out. In the lower part of a large mass a junction with a dark pyroxene-granulite is seen, and patches and flecks of pyroxenic rock are scattered through the limestone and weather out from it. The limestone consists of calcite with diopside, orthoclase, scapolite, pyrite, phlogopite, and graphite and apatite as accessory minerals. In a section through a small pyroxenic patch, these accessory minerals were seen to be much more abundant. A section of a specimen next to the limestone at the junction referred

Vol.58.] *The Crystalline Limestones of Ceylon.* *407*

to showed much green augite and scapolite, with calcite (partly secondary), pyrite, little felspar, sphene. What is probably the same limestone is seen again in the Rakwane Ganga, Rangweltenne Estate, near by, and though very impure it has there been burnt for lime.

Fallen blocks below the Herimitigala Rock, on the estate of the same name near Balangoda, show a beautiful, coarsely crystalline limestone, with abundant crystals of diopside, scapolite, and some also of sphene. Some individuals of calcite measure more than 3×1.5 centimetres. There are gardual passages from this rock to a handsome green and white scapolite-augite rock, strongly recalling certain varieties of rock occurring at Galle. The country-rocks here are granitoid in aspect, and more coarsely crystalline than usual.

Cases in which a zone of green rocks is found are also common. In the stream-bed, north of the bridge at Talatuoya, it may be seen that a band of impure limestone is in contact with a wide zone of heavy green rocks in which the minerals diopside, biotite, greenish-brown hornblende and green spinel predominate (specific gravity=3.25, 3.31); sometimes calcite is present , associated with many large spinels. Similar rocks are seen on the Kandy-Talatuoya road, about 3 $\frac{3}{4}$ miles from Kandy at the top of the hill (see fig.1, p.404), also 3 $\frac{1}{2}$ miles from Kandy south of the limestone-band; and again south of the limestone-band at the little bridge over the Mahaoya, south of the Peradeniya level-crossing, where a specimen (of specific gravity 3.27) consisted of diopside, brown mica, and green spinel, the latter rendered nearly opaque by granules of iron-ore.

Near the boundaries of limestone-beds, thin bands of limestone and granulite are sometimes found to alternate, as at Gangapitiya, where in the moonstone-pit the impure limestone-bands are usually separated from the beds of acid granulite by successive zones of diopside alone, and of diopside, serendibite, and spinel (with sometimes scapolite or plagioclase), each from 2.5 to 5 centimetres broad.

Moreover, it is often found that the limestones themselves become very impure near the junction with the granulite rocks. Bands rich in mica, diopside, and spinel appear, as may be seen on the Kandy-Talatuoya road, about 5 $\frac{1}{4}$ miles from Kandy, and at Wattegama, east of the level-crossing, where some house-foundations afforded a section. Accessory minerals may, however, be quite as abundant in the main mass of the limestone, though here olivine (forsterite) is more characteristic.

It is not unusual to find in the limestones bands of pyroxenic rock, identical with the pyroxene-granulites, running parallel to the general foliation, and of various widths up to several yards. These bands are, as it were, sills of pyroxene-granulite in the lime-stone. When quite narrow, such rocks occasionally form irregular snaky twists and flecks in the limestone; a quite chaotic mixture is sometimes thus produced. The patches weather out from the limestone, and look like masses of rotten ferruginous sandsonte. A specimen of such a snaky twist had a specific gravity fo 2.95 and consisted of a granular mixture of augite, potash, and felspar, with small quantities of scapolite, sphene, iron-ore, biotite and calcite.

408 Mr. A.K. Coomáraswámy on the [Aug.1902,

The granular structure of these masses, especially if there is slight decomposition, causes them to shatter at a light touch of the hammer, so that it is sometimes difficult to collect specimens for slicing.

There are also occasionally to be found in the limestones, quite isolated, inclusions of pyroxene-granulite a few feet or yards in width; and it may happen that the foliation of the limestone bends around them, suggesting flow-structure. Such occur, for example on the Bandarawela road, beside the road, between the 56 $\frac{1}{2}$ and 57 milestones, and elsewhere in the Hakgala district (Pl.XIV).

Sections from the central portions of the bands and masses here mentioned possess the usual characters of the pyroxene-granulites, the minerals augite, plagioclase, orthoclase, quartz (one case only), amphibole, brown mica, magnetite, pyrite, and sometimes sphene and scapolite, being note.

Inclusions of quartz-rock, similar to that commonly occurring among the granulites are exceedingly rarely met with. In one or two cases, strips of quartz an inch or two wide were noted in association with, or forming part of, the pyroxenic inclusions. Bands of quartz-rock (with also a negligible quantity of pyroxene) an inch or two wide were noted in limestone in the bed of the Talatuoya, four-fifths of a mile north of the brdige at Talatuoya.

On the Hakgala map (Pl. XIV) it will be seen that some large ellipsoidal masses of rock belonging to the Charnockite Series are completely included in the limestone-bands, and there are others smaller. In several cases these consist of a coarse quartzofelspathic rock; in other cases ordinary acid granulites enter into the composition of these masses, which I regard as masses of intrusive rock in the limestone, less flattened laterally than is usually the case.

The exposure described by Mr. Parkinson[41] is seen in the railway-cutting south of the level-crossing, south of Matale railway-station. I have visited the section, and am in complete agreement with him (except that I cannot find here or elsewhere in Ceylon clear evidence of an augite possessing the pleochroism of hypersthene, my own sections of the 'band' exhibiting only the usual green augite). The cutting shows a horizontal band of pyroxene-granulite, with crystalline limestone above and below it. The band becomes laterally very rich in pyroxene, and is not sharply separated from the limestone. it is probable that there has been, as Mr. Parkinson supposes, some absorption of calcareous material at the junction.

It will be useful to refer in greater detail to two other localities in which bands of acid granulite or pyroxenic granulite are found in the limestone.

(1) Near the north-western corner of the Albion Estate, but a little north of the belt of acacias, and above and west of the Hakgala footpath, is a small section showing crystalline limestone with a

Vol.58.] *The Crystalline Limestones of Ceylon.* *409*

[41]Quart.Journ.Geol.Soc.vol.lvii(1901) p.204.

band of acid granulite therein; the limestones and included 'sill' being bearly horizontal. The band of granulite (specific grvity=2.63) is of a very usual type, composed mainly of elongated quartz, orthoclase, microperthite and plagioclase, with some green augite, magnetite, and abundant prisms of apatite. There are also brownish fibrous pseudomorphs, perhaps after rhombic pyroxene. This band of granulite, 2 feet wide, is separated from the limestone above and below by a zone of heavy, hard, dark rock, 3 to 6 inches thick (specific gravity=3.2 to 3.28), composed mainly of greenish-brown hornblende, with more or less colourless diopside, brown mica, green spinel, and scapolite. The limestone near the top junction contains much colourless diopside, and a little olivine, besides calcite and some pale phlogopite. A less pure band in the limestone was composed of phlogopite, diopside, calcite, and bluish-green spinel, the plates of phlogopite being slightly distorted as if by pressure. There are also elongated silicate-lenticles in the limestone, consisting of dark heavy rock (specific gravity=3.09, 3.31), similar in mineralogical composition to the contact-rock bordering the band of granulite. One measures about 2 feet by 6 inches, and consists of diopside, hornblende, and green spinel; in another phlogopite is abundant, in addition to these minerals.

(2) A small, quarry, on the right-hand side of the Kandy-Anuradhapura Road, just south of the 32 $\frac{3}{4}$-milestone, is of much interest. The quarry has been opened in a wide band of crystalline limestone along and over which the road runs. The limestone is of ordinary type, in part comparatively pure (specific gravity=2.83), with calcite, dolomite, and not very abundant olivine. Impure varieties, from which knots of rusty rock weather out, are also present. These contain diopside, pale phlogopite, calcite olivine and green spinel (specific gravity=2.85, 2.97). There are in the limestone two bands of granulite, striking north and south parallel to the general foliation and to each other.

The first appears as a dark band about a foot wide (fig.3), centrally hard and compact, and resembling a pyroxene-granulite, but

Fig.3:- *Interrupted sill of pyroxene-granulite in limestone, new Nalanda*

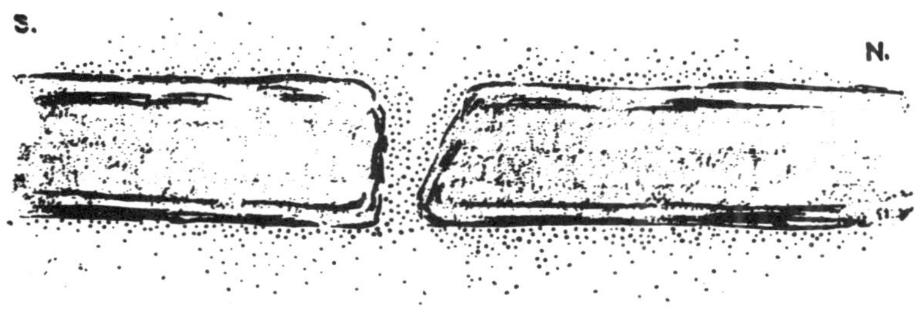

[A length of about 5 feet is shown]

laterally more micaceous. A thin slice from the central part (sp.gr.=2.96) shows much augite (very pale green to bright green), often with dark rod-like inclusions, and an equal amount of zoned plagioclase, most basic internally. Magnetite is abundant, and some pyrite is also present. Quartz is rare or absent, with the exception of a streak 2 millimeteres wide, parallel to the foliation, consisting of elongated quartz-individuals. Scapolite appears to occur. Other minerals present in quite small quantity are hypersthene, biotite, quantity and apatite. Without the scapolite, the rock would be an ordinary pyroxene-granulite. Another slice from a specimen nearer the edge

Fig.4:- *Band of acid granulite in limestone, separated from it by a 1-inch zone of diopside-rock, and passing into lenticles of corresponding character*

[A length of about 2 yards is shown]

of the band (sp.gr=2.99) contains much brown mica and hornblende, with green augite and plagioclase less abundant than before, and limpid scapolite tending to occur in plates moulding the other minerals. A third slice from the micaceous edge of this band (sp.gr.=3.16) is composed of diopside, phlogopite, pyrite, and scapolite. The diopside is partly in graphic forms standing out against the scapolite. The limestone at the junction (which is not abrupt) contains calcite, olivine, very little diopside, green spinel (black macroscropically), and much pyrite, the last confined to a zone 2.5 to 5 centimetres wide next to the pyroxene-granulite band. The last-named band, however, is not quite continuous, but is interrupted by a tongue of limestone a few inches wide, which crosses it, connecting the limestone on the east with that on the west of the band. This limestone-tongue forms thus a sort of dyke crossing the band of pyroxene-granulite.

In the second case a band of white acid granulite is seen in the limestone (fig.4), 2 yards west of the darker band just described. Here the acid granulite (possessing the usual characters) is separated from the limestone by a zone of hard, compact, greenish-grey rock, 1 to 1 $\frac{1}{2}$ inches wide, which borders it like a shell. This greenish zone (specific gravity=3.12, 3.25) is composed almost entirely of compact granular diopside, with some green spinel in parts, and peripherally a little interstitial calcite. Where the calcite is more abundant, the diopside and spinel are idiomorphic. After running for some yards as such, the band is continued as a series of lenticles, the structure of which is identical with that of the band. The band of acid granulite is thus, as it were, interrupted by transverse dykes of limestone. A specimen of acid granulite from the interior of one of the lenticles contained augen-orthoclase, measuring as much as 23×13 millimetres. A thin slice showed a fine-grained granular mosaic of orthoclase-microperthite, traversed by strings of individuals of elongated quartz, and containing irregular porphyritic individuals of microperthite which appear strained and perhaps fractured. The band and lenticles can be traced for quite 15 yards, appearing again beside the road, north of the exposure in the shallow quarry.

It is exceedingly interesting to compare these broken sills of granulite in limestone with the broken dykes of nepheline-syenite in the limestone of Alnö (described by Prof. A.G. Höbom),[42], the peculiar characters of which seem to correspond closely with those of the cases just described.

Before leaving this part of the subject, it may be useful to point out that the various bands of granulite met with in the limestones occur always as sills, and never as transverse dykes. If we suppose the granulite to be intrusive in the limestone, we must regard this parallelism as the result of injection accompanied by lateral pressure.

IV. INTERGROWTHS OF CALCITE AND DOLOMITE.

These are very characteristic of the crystalline limestones, especially in the Kandy and Matale districts, though rather unusual in the neighbourhood of Hakgala. Such intergrowths produce characteristic appearances on weathered surfaces.

Parallel intergrowths (fig.5, p.412).— It is very usual for the white carbonate-crystals composing the limestone to show a curious banding, well brought out on weathered surfaces. In such cases, there is a series of narrow parallel bands running across the crystal, diagonal to the cleavage, and alternately standing out and depressed. In the coarsest variety found, there were 15 upstanding bands to the half-centimetre. There are all transitions from this type to bandings of microscropic dimensions. Thin sections show that this appearance is caused by a parallel intergrowth of calcite and dolomite, in alternating planes parallel to the basal plane c(001).

[42]Geol. Fören. Stockholm Förhandl.vol.xvii (1895) pp.100,214.

Sections prepared parallel to the plane of intergrowth yield a negative uniaxial interference-figure in convergent polarized light. In most cases the calcite and dolomite have an identical optical orientation, in others the alternating bands show a slightly differing orientation.

Ramifying (graphic) intergrowths (fig.6, p.413).— On the weathered surfaces of other varieties of limestone, a vermicular tracery is often seen, frequently appearing to radiate from several centres. Fractured surfaces of such specimens present a 'lustre-mottled' appearnce. Thin sections show that these phenomena are due to ramifying intergrowths of calcite and dolomite. The radiating 'fingers' consist of dolomite, with a common optical orientation; the calcite-individuals are numerous and variously oriented, but each is

large enough to enclose poritons of more than one 'finger' of dolomite. Fig.6 is from a specimen in which the intergrowth is less finegrained than usual; it often sinks to quite microscropic dimensions, and such examples have been described by Prof. Lacroix[43] and Mr. Parkinson,[44] though without explanation.

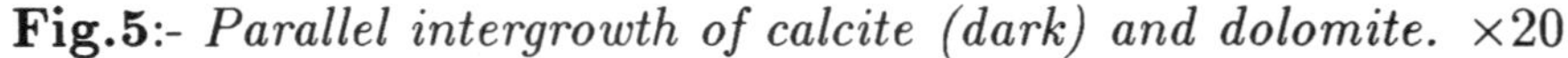

Fig.5:- *Parallel intergrowth of calcite (dark) and dolomite.* ×20.

[The shading represents the effect of staining with Lemberg's solution. This applies also to fig.6. TW=Twinning; CL=Cleavage.]

[43]Bull. Soc. franc. Mineral.vol.xii (1889) p.337 & fig.60.

[44]Quart.Journ.Geol.Soc., vol.lvii,1901) p.205.

In many of the parallel intergrowths the calcite and dolomite are of equal relative improtance; in others, and in most of the ramifying intergrowths, we susually find that calcite predominates, and should speak of the dolomite as forming parallel or ramifying inclusions in the calcite; it is exceedingly rare to meet with a reverse relation.

A conspicuous feature is the limitation of twinning to the calcite, the twinning-bands in it ceasing at the junction with dolomite, even when the two have a common optical orientation. Very seldom traces of twinning are found in the dolomite, and it seems perhaps to be not quite invariably present in the calcite, though usually so.

The distinctness of the intergrowths, even on a very small scale, shows probably that the carbonates consist of pure dolomite or pure calcite, and that intermediate types are absent.

---

414 *Mr. A.K. Coomáraswámy on the* *[Aug.1902,*

The finest examples of these intergrowths were obtained at a point three-quarters of a mile west of Ulisna Muduna trigonometrical station, along the footpath that descends the slope in a south-westerly direction (see map, Pl.XIII).

Fig.6:- *Ramifying or graphic intergrowth of calcite (C) and dolomite (D).* ×9.4

[MA=Magnetite; MI=Mica]

V. NOTES ON THE ACCESSORY AND CONTACT-MINERALS.

(1) Diopside.— Colourless monoclinic pyroxenes associated with the limestones and their contacts with the granulites are referred to as diopside. The monoclinic pyroxene of the pyroxene-granulites possesses always a green colour, though sometimes pale, and it is referred to as augite. The diopside of the green-rock contact-zones has sometimes also a pale-green colour.[45] Much diopside is present as the sole constituent of some white silicate-aggregates, or associated in them with other minerals, such as blue apatite and mica; it is also occasionally present as a constituent of the limestone-rock, with or without accompanying olivine.

(2) Olivine (forsterite).— Colourless olivine in crystals almost always clearly visible macroscopically (and even reaching 1 inch in length) is very commonly met with. Very often the limestone carrying olivine has a grey or dark-grey coloration, owing to the deposition of granules of iron-ore along the irregular cracks in the olivine. Alteration to serpentine is not very common, the crystals being very fresh.

Alternations of pure and olivine-bearing limestones give the effect of a white-and-grey banding. Such bandings are always parallel to the general strike. Olivine-bearing dolomites are well seen at Ampitiya, in quarries north-east of Ketawala trigonometrical station, and along the roadside west of the same point (Pl. XIII). An analysis of olivine from Ampitiya is quoted on p.415. Olivine is unusually abundant in and gives quite a dark colour to, limestones exposed on the Ambewela-Hakgala road, near the bridge over it, and near the 4th milestone.

Another case may be referred to in greater detail. Limestone occurs in a patch of jungle near the Hakgala Gardens, near the remains of an old kiln, where lime was formerly burnt (Pl.XIV). The limestone forms a sort of cave; it is doubtless a continuation of one of the bands which slope up from Harakgama. The minerals present are calcite, dolomite, olivine, phlogopite, pyrite, and pink spinel. The olivine in some specimens appears grey, in others yellowish; but these colours merely result from very slight quantities of iron-ore deposited along the cleavage-cracks. Grains treated for some hours with dilute hydrochloric acid become quite colourless. Such treatment serves to isolate the olivine, and seems to attack it very little or not at all. The grains obtained consist of separate flattened crystals, averaging 2 to 3 millimetres in length, and of more irregular aggregates. The flattened grains show in convergent

Vol.58.] *The Crystalline Limestones of Ceylon.* *415*

polarized light the emergence of an obtuse bisectrix. The optic axial plane is at right angles to the elongation of the crystals, which are thus elongated in the direction of the vertical axis, and flattened parallel to *b*(010).

[45] In one case a quite dark green pyroxene occurred in limestone (at Herimitigala) associated with scapolite. This pyroxene resembles the Tiree 'coccolite,' and probably belongs to the hedenbergite group.

CHEMICAL ANALYSES OF FORESTERITE AND SERPTENTINE FROM CEYLON

	I.	II.	III.	IV.	V.
	Per cent.	Per cent	Per cent.	Per cent	Per cent.
SiO_2 ...	42.55	41.16	42.8	39.25	39.88
$Al_2 O_3$ ...	0.23	...	...	1.78	3.18
$Fe_2 O_3$ ...	...	2.58	...	...	...
Fe O ...	2.36	...	2.6	...	...
Ca O ...	1.43	...	...	1.84	1.75
Mg O ...	51.97	52.60	55.4	39.00	38.64
			H_2O (hygroscopic) 0.6		
Loss on ignition	1.68		...	(H_2O) 18.13	16.55
			H_2O (combined) 3.2		
	100.22	100.14	100.8	100.00	100.00
Sp.gr. ...	3.14	3.13			

I. Analysis of forsterite from Hakgala, by Mr. G.T. Prior, M.A.
II. Analysis of forsterite from Ampitiya, by Mr. W.C. Hancock.
III. Analysis of forsterite from Kandy, by M.H. Arsandaux, Bull. soc. franc. Minéral, vol, xxiv (1901) p.475.
IV. } V. } Analyses of green and white serpentine from limestone, Wattegama level-crossing, by Dr. Schiffer, Inaug-Dissert. Munich, 1900

It is noteworthy that the olivine does not occur specially near junctions with the granulites, as does diopside, but is more common than the latter in the main masses of limestone.

(3) Mica.— Colourless to golden-brown micas, less often pale-greenish, are everywhere commonly found either scattered through the limestone (and then more or less idiomorphic), or accompanying malacolite, apatite, spinel, etc. in silicate-aggregates which weather out on the surface of the limestone. The most usual types have the colour and pleochroism of phlogopite, but are very pale, often colourless in thin section if not macroscopically. Greenish-brown micas, with the optical characters of phlogopite, are sometimes characteristic of impure bands in limestone near junctions with the granulites.

A dark-brown mica was found with hornblende in limestone at Wariapola.

Near Talatuoya several diggings for mica have been made: and it is probable that this mica occurs in connection with the junctions of limestone and granulite. It belongs chiefly to phlogopite, with various warm brown shades. The crystals are often idiomorphic, and not often more than 6 or 8 inches in diameter. Some have curious, flattened, black, lath-shaped inclusions, arranged parallel and perpendicular to the rays of the percussion-figure.

Quite dark greenish-brown or green micas (nearly uniaxial) occur with diopside, amphibole, and spinel in the heavy green rocks sometimes found between the limestones and granulites (Talatuoya).

(4) Spinel.— Pink and violet-pink spinels are very common

416 *Mr. A.K. Coomáraswámy on the* *[Aug.1902,*

scattered through the limestone, for example, in the neighbourhood of Talatuoya, and also at Hakgala. The crystals are often well-shaped octahedra; sometimes, as at Hakgala Limekiln, combinations of the forms (111) and (101) are found. The largest crystals obtained *in situ* reached about 5 millimetres in diameter; but I have received much larger specimens, with the locality of which I am not acquainted, although they probably occurred in limestone.

Blue spinel was noted as forming, with olivine and pale phlogopite, a small mineral-aggregate in limestone near Talatuoya. Pink spinel occurs in a micaceous mineral-aggregate in limestone in the middle disused quarry at Getembe. Green spinels (green or black macroscopically, often nearly colourless in thin sections) are common in the dark patches and mineral-aggregate in limestone and associated with diopside or amphibole in contact-zones, and with green diopside, amphibole, and mica in the heavy green rocks which are sometimes found between the limestones and the typical granulites (Talatuoya, etc.). Green spinels are rarely found scattered through the limestone, but seem to occur more definitely as a contact-mineral. Some varieties of limestone rich in green spinel are, however, to be met with, as at Talatuoya in the stream north of the bridge.

The presence of spinel as a contact-mineral seems to result from the appropriation of silica in the formation of lime-and magnesiasilicates.

(5) Apatite.— Blue apatite is everywhere common as an accessory mineral in the limestones, and associated with malacolite and phlogopite in mineral-aggregates in the limestone, as at Lower Albion, Hakgala. In the limestones it is present in the form of prismatic crystals, with rounded outlines, always big enough to attract attention. Isolated grains show (as was noted by Mr. Parkinson) a clear pleochroism, namely, for rays vibrating parallel to c sky-blue, or rays vibrating perpendicular to $\acute{c}$ pale claret (nearly colourless). In this sections the blue apatite appears colourless.

Dr. Schiffer has analysed a blue apatite from Wattegama, his results giving the calculated formula $(PO_4)_3(FICI)Ca_5$. Prof. Church has also investigated the blue apatite of Ceylon.[46]

[46]'Nature' vol.lxii (1901) p.464.

(6) Amphibole.— The amphiboles associated with the crystalline limestones are not especially abundant, but have been noted in several localities occurring as contact-minerals, silicate-aggregates, or scattered through the limestone.

Fibrous crystals of colourless tremolite, with an extinction-angle on (110) of about 17°, form a white silicate-aggregate in limestone not far from the 36th milestone on the Kandy-Anuradhapura road.

Colourless glassy amphibole occurs in nodular silicate-masses (with colourless mica and blue apatite) in limestone two-fifths of a mile south-south-west of Ulisna Muduna trigonometrical station (Pl.XIII). A variety has a very faint greenish tinge, and some glassy crystals scattered through the limestone itself are of a clear pale-green.

On a small pale-green individual, weathered out, Mr. Graham has determined the forms (011) (110) (010).

Mr. W.C. Hancock has analysed the colourless variety from this locality, with the following result:—

	Per cent.	
C (graphite) ...	0.30	
H_2 O (combined) ...	0.60	The presence of so much
Si O_2 ...	47.04	soda and alumina, presumably
Al_2 O_2(with a trace of iron-peroxide)	13.76	introduced from the
Ca O ...	13.39	granulites, is interesting.
Mg O ...	21.26	
Na_2 O ...	4.01	
	100.36	

Specific gravity=2.92

A dark-brown amphibole occurs in lumps of limestone with brown mica, in a field east of the railway near Wariapola, near Matale. The crystals are frequently idiomorphic. Mr. Graham has determined the form (100) (010) (110) (310) (011) (101). In thin sections this amphibole is quite colourless. Cleavage-flakes show an extinction-angle of about 21° on *m* (110).

Many varieties are quite colourless, even macroscopically; others are pale green, greenish-brown, or dark-brown, but colourless or pale in thin sections. Others, especially those occurring definitely as contact-minerals, show colour in the slides and have a normal pleochroism, not so strong, however, as that of the amphiboles which occur in the pyroxene-granulites themselves. Amphibole darker than any of those referred to above, and resembling the amphibole of the pyroxene-granulites, is found with green diopside, dark mica, and dark-green spinel, in the heavy green rocks which seem to separate the limestones from the granulites in some localities.

(7) Scapolite is quite unusual as an accessory mineral in the limestones themselves. It may also be met with in mineral-aggregates in limestones, as for example in an inclusion measuring 7 × 3 × 4 inches from Palugama (Pl.XIV), consisting of diopside, phlogopite, plagioclase, scapolite, and calcite. Another aggregate graphic fusion with scapolite and a few individuals of orthoclase. Probably such aggregates correspond to inclusions of pyroxene-granulite, which being small have been much modified by the formation of lime-silicates. Hence they are not included in the list of mineral-aggregates on p.403, which refers rather to aggregates of minerals especially characteristic of the limestones themselves.

Scapolite is, however, very characteristic of the peripheral parts of sills and inclusions of pyroxene-granulite in the limestones, where it is usually accompanied by sphene. Thus the peripheral portion of the pyroxene-granulite sill described on p.409 consists of diopside, phlogopite, pyrite, and scapolite; while the rock in contact with the Hermitigala limestone, as seen in fallen blocks, is a fine scapolite-augite rock. Many other localities could be named. In such

418 *Mr. A.K. Coomáraswámy on the* *[Aug.1902,*

cases the scapolite is sometimes granular, but more often it forms plates enclosing the other minerals which sometimes assume graphic forms. Sometimes there are numerous elongated inclusions arranged parallel to the vertical axis.[47]

Fig.7:- *Intergrowth of diopside (square and rectangular outlines) and scapolite, Lower Albion, Hakgala.* ×120.

[The section is perpendicular to the vertical axes.]

[47] See Lacroix, Bull. Soc. franc. Minéral. vol.xii (1889) p.99 & fig.6.

(8) Iron-Ores.— Pyrite in small quantity is a common accessory mineral in the limestones; and is sometimes abundant also in the mineral-aggregates, but it is not usually characteristic of these. It may occur as a definite contact-mineral, as in the quarry 2 miles north of Nalanda.

Magnetite is occasionally met with; it was noted in some quantity in a variety of limestone from the western quarry at Gettembe.

Dr. Schiffer's examination of pyrite from Wattegama[48] showed its composition to be pure Fe S_2, and he noted the forms (111) and (210).

(9) Clinohumite was found at Gettembe, 3 miles west of Kandy, and at Ampitiya, 3 miles east of Kandy. At Ampitiya yellow granules of clinohumite were found in coarsely crystalline limestone about a fifth of a mile north-west of Ketawala trigonometrical station. At Gettembe three pits have been opened for lime-burning, north of the road and below the hill ('Primrose Hill'). The middle pit is disused and small. Yellow clinohumite is very abundant in one

variety of limestone occurring in the western pit, forming a handsome dolomite-clinohumite rock; other minerals present in small quantity are olivine (?) and blue spinel. Other varieties of limestone in the same pit are quite without clinohumite, and have different accessory minerals, or none. The clinohumite forms large crystals with some approach to crystalline form, reaching 1 inch in longest diameter; smaller granular crystals are abundant. These are very diminutive, but have sometimes numerous well-developed crystalline faces. The grains are about twice as long as broad and slightly flattened. Examined between corssed nicols, they are found to extinguish always parallel to their length, whether lying on one side or on the other, whence it follows that they are elongated in the direction of the orthodiagonal axis *b*. In thin sections, it can be clearly seen that the axial plane is parallel to the elongation. The pleochroism is a fine orange-yellow, *c* pale chrome-yellow, and *b* like *c* or very slightly darker. Absorption $a > b \geq c$. In thin sections, basal cleavage is seen parallel to the elongation; isolated grains are also occasionally seamed by irregular cracks, more or less transverse to their length.

Mr. G.F. Herbet Smith has kindly measured one of the tiny crystals from Gettembe,[49] and identified the forms

$$c(001)c_2(107)c_3(103)c_4(101)n_3(113)i_2(014)i_3(012)r_4(129)r_6(127)r_6(125)r_6(121)$$

The measurements quoted in the following table.—

[48]'Chemische Untersuchungen eines körnigen Dolomits aus dem Gneiss von Wattegama in Ceylon,' Inaug.-Dissert. Munich, 1900.

[49]The determination of the mineral as clinohumite depends on the crystal-measurements referred to. Without these it would have been necessary to speak vaguely of 'chondrodite.' The mineral which occurs sparingly at Ampitiya is probably identical with the Gettambe clinohumite.

Form		Observed				Calculated (Dana, 6th ed.)			
		Azimuth.		Distance.		Azimuth.		Distance	
c_4	101	0°	0′	79°	15′	0°	0′	79°	11 $\frac{1}{2}$
		"	"	79	7	"	"	"	"
		"	"	79	9	"	"	"	"
		"	"	79	17	"	"	"	"
c_3	103	"	"	60	6	"	"	60	12
		"	"	60	7	"	"	"	"
c_2	107	"	"	46	11	"	"	46	20
h_3	113	47	40	68	30	47	13	68	44 $\frac{1}{2}$
r_4	129	65	17	54	9	65	10	54	11
		65	7	53	59	"	"	"	"
r_6	125	65	17	68	8	"	"	68	9
r_8	121	65	7	85	15	"	"	85	25
		65	17	85	22	"	"	"	"
r_5	127	65	6	60	45	"	"	60	42
		64	58	61	0	"	"	"	"
i_3	012	90	11	70	37	90	0	70	32
		90	2	70	30	"	"	"	"
i_2	014	90	11	54	40	"	"	54	45

420 *Mr. A.K. Coomáraswámy on the* *[Aug.1902,*

Mr. G.F. Herbert Smith says that

'besides these faces there are a large number (more than a hundred) of others, mostly very small. Some of these might, on calculation, be found to have simple indices. Most of them, however, have very complex indices and would be due to etching. It would seem as if, in this case, some of the original faces were still left unaltered.'

Mr. W.C. Hancock has analysed the clinohumite from Gettembe, with the following results.—

	Per cent.
$Si\ O_2$ ...	37.52
$Fe_2\ O_3$ (with a trace of alumina)	9.00
$Mg\ O$ ...	49.75
$Na_2\ O$ ...	1.44
F ...	1.02
$H_2\ O$ (hygroscopic) ...	0.50
$H_2\ O$ (combined) ...	1.00
	100.23

Yellow clinohumite was also found in the limestone in the bed of the Madde Ela near Gettembe, associated with what may be a colourless variety. In a thin section, one case of twinning was observed; in the material from the limestone-pit no twins were found. A colourless variety of chondrodite from limestone from Ceylon has been described and figured by Prof. Labroix[50]; the fiture is, however, very suggestive of olivine.

(10) Graphite.— Flakes of graphite occur commonly in some varieties of limestone, though quite absent in others. The flakes are often more or less hexagonal in outline, and have a seemingly basal cleavage corresponding to that of mica.

(11) Sphene was only noted as an accessory mineral in the Herimitigala limestone, but is common accompanying scapolite, calcite, etc., in the peripheral parts of pyroxene-granulites which show endomorphic modification in contact with limestone.

(12) Orthoclase was noted as an accessory mineral in limestone, on Allerton Estate, Rakwane; and once or twice inmineral aggregates in limestone, in the Hakgala district.

(13) Tourmaline occurs very sparingly in joints in the modular serendibite-diopside rock-bands, and still more rarely in the intrusive granulite, at Gangapitiya. Tourmaline from river-gravels is one of the commonest semiprecious gems of Ceylon, but localities where it occurs *in situ* are unknown.[51]

(14) Serendibite.— This name has been given by Mr. Prior and myself[52] to a mineral occurring at the Gangapitiya moonstone-pits,

about 12 miles east of Kandy. To reach the spot the Kandy-Badulla reiverside road is taken as far as the Pallekelle ferry, which is crossed, and proceeding along the road to the Teldeniya rest-house, the village of Ambakotte is reached. Immediately after passing this village, a road must be followed which turns to the right across the 'Ambakotte' and 'Ambakotte New' Estates. the pits are on the farther side of the latter, some 2 miles distant from the main road. Several bands of limestone and granulite are passed on the way. Limestones are well exposed near a pond about halfway between the road and the pits.

The pits show intimately interbanded acid granulite and impure limestone. The bands do not exceed a few feet in width, but are usually thinner. This is probably a

[50]Bull. Soc. franc. Mineral, vol.xii (1889) p.338 & fig.60

[51]For crystallographic details of Ceylon tourmaline, see V. von Worobiev Krystallographische Studien *über* Turmalin von Ceylon u. einigen anderen Vorkommen' Zeitschr. *für* Krystallogr. vol.xxxiii (1900) pp.263-454 & pls. viii.xiv.

[52]Paper read before the Mineralogical Society on February 4th, 1902 abstract in 'Nature' vol.lxv (1902) p.383.

case of 'lit-par-lit' injection in the outer part of a band of limestone which is crossed before the pits are reached. The acid granulite is of the usual type with much elongated quartz, and a tendency to graphic structure. The minerals are orthoclase-microperthite, orthoclase, quartz, and plagioclase. Large augen-orthocases provide the moonstone, to obtain which the pits were dug. Individuals may reach a size of 12.5 × 10 centimetres.

The granulite-bands are separated from the limestone by a contact-zone, first of compact granular diopside 2.5 to 5 centimetres-wide next to the limestone, and secondly of diopside, serendibite, green spinel, sometimes scapolite and plagioclase, 1.25 to 2.5 centimetres wide. Similar lime-silicate rock occurs in thin nodular bands which alternate with the granulite, corresponding to bands of limestone so thin as to have been completely altered to silicate-rock. This type of contact-action (formation of diopside, etc.) recalls the cases described on pp.409, 410. Alternate bands of acid granulite and limestone were noted also in another moonstone-pit near Talatuoya.

The serendibite has a dark bluish-green colour in the rock and with the spinel gives it a similar hue. It rarely occurs in easily recognizable crystals, but is usually much mixed and intergrown with the diopside. In thin sections, it is rendered very conspicuous by its remarkable pleochroism from very pale yellowish-green[53] to deep indigo-blue, and by the multiple twinning which is as characteristic as in a triclinic felspar. The crystals are somewhat flattened and elongated, and the twin-plane is parallel to the elongation. Crystal-faces parallel to the elongation are fairly well seen, terminal faces are rarer. The crystals have sometimes an ophitic habit, so that several portions not in contact extinguish simultaneously.

Other characters may be summarized as follows.— Optically biaxial, probably triclinic, perhaps monoclinic. Cleavage none; hardness=about 7; specific gravity=3.42. Refractive index nearly equal to that of diopside, double refraction weak.

422 *Mr. A.K. Coomáraswámy on the* *[Aug.1902,*

Mr.G.T.Prior's analysis gave the following results.—

	Per cent
SiO_2 ...	25.33
Al_2O_3 ...	34.96
FeO ...	4.17
CaO ...	14.56
MgO ...	14.91
K_2O ...	0.22
Na_2O ... / Li_2O ...	0.51
P_2O_5 ...	0.48
Loss by ignition ...	0.69

[53] In thin slices the spinel and serendibite are indistinguishable for some positions of the lower nicol.

F ...	trace
$B_2\ O_3$ ...	(4.17) by difference

	100.00

The corresponding formula is

$$10(Fe, Ca, Mg)O\ .\ 5Al_2O_3\ .\ 6SiO_2\ .\ B_2O_3$$

(15) Rutile.--- A brownish-yellow mineral with high refractive index, occurring with olivine, phlogopite, spinel, tremolite, etc. in a mineral-aggregate in limestones at Gettembe, is probably rutile.

(16) Zoisite is recoreded by Prof. Lacroix[54] as occurring in a mineral-aggregate in limestone from Kurumegala.

Garnet and wollastonite; the total absence of these minerals from the limestones and contact-modified granulites is noteworthy.

Corundum has to been found in the crystalline limestones as yet, although it occurs thus in Burma; and Dr. Grünling[55] has suggested that the Ceylon corundum may occur in the limestone.

EXPLANATION OF PLATES XIII & XIV.

PLATE XIII.

Geological map of the country between Kandy and Talatuoya,
on the scale of 3 inches to the mile.

PLATE XIV.

Geological map and sections of the district lying
south-east of Hakgala, on the scale of 3 inches to the mile.

DISCUSSION.

Mr. PARKINSON, after expressing his sense of the value of the paper and the care and elaboration with which the details had been worked out, described the contact of granulite and limestone which he had seen near Matale, and remarked that he believed that this section proved, firstly, that the granulite was intrusive, and secondly, that the intrusion had been attended by absorption of the limestone, which had locally modified greatly the composition of granulitic

[54] Bull.Soc.Franc.Minéral. vol.xii (1889) p.344.

[55] 'Ueber die Mineralvorkommen von Ceylon' Zeitschr. für Krystallogr. vol.xxxiii (1900) p.233.

GEOLOGICAL MAP
OF THE COUNTRY BETWEEN

KANDY & TALATUOYA (CEYLON)

by A.K. Coomáraswámy, B.Sc.
1902

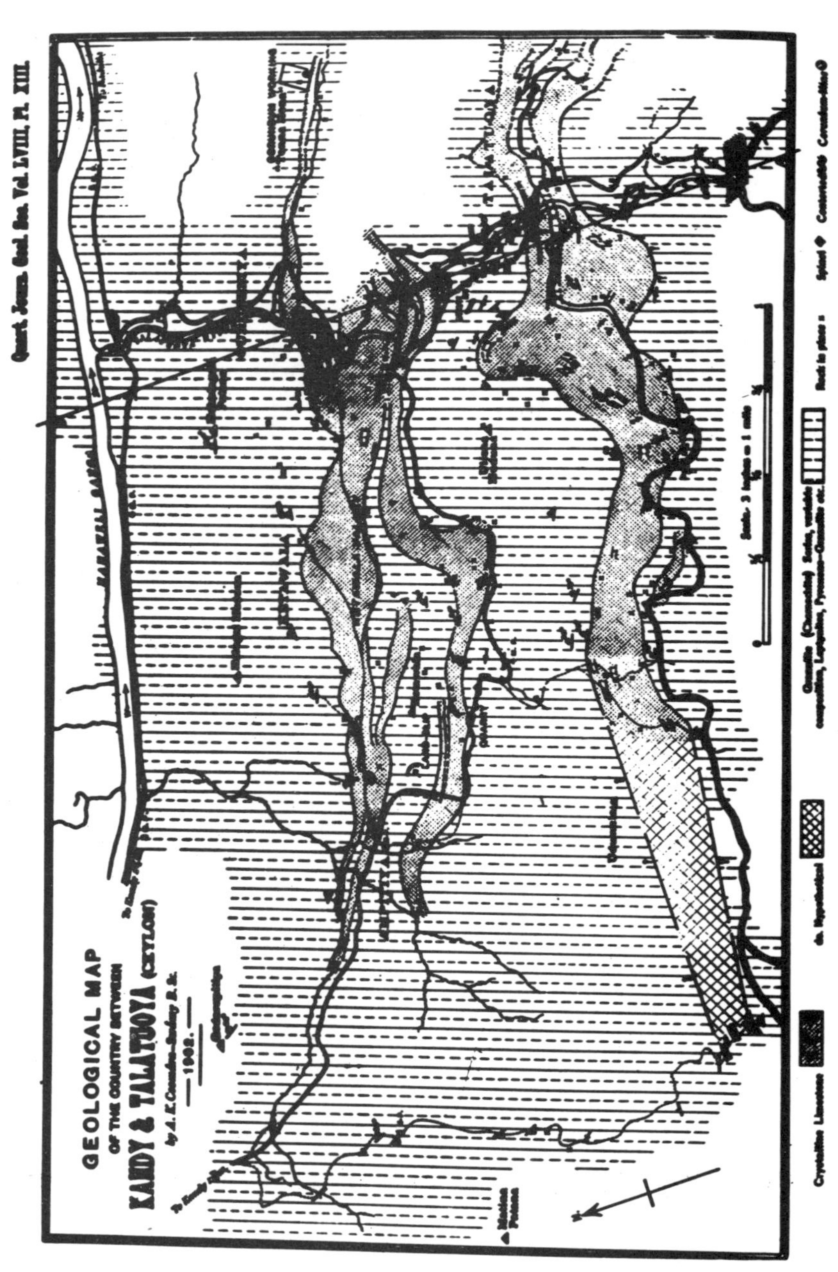

GEOLOGICAL MAP OF THE DISTRICT LYING SOUTH EAST OF

HAKGALA

Quart. Journ. Geol. Soc. Vol. LVIII, Pl. XIV.

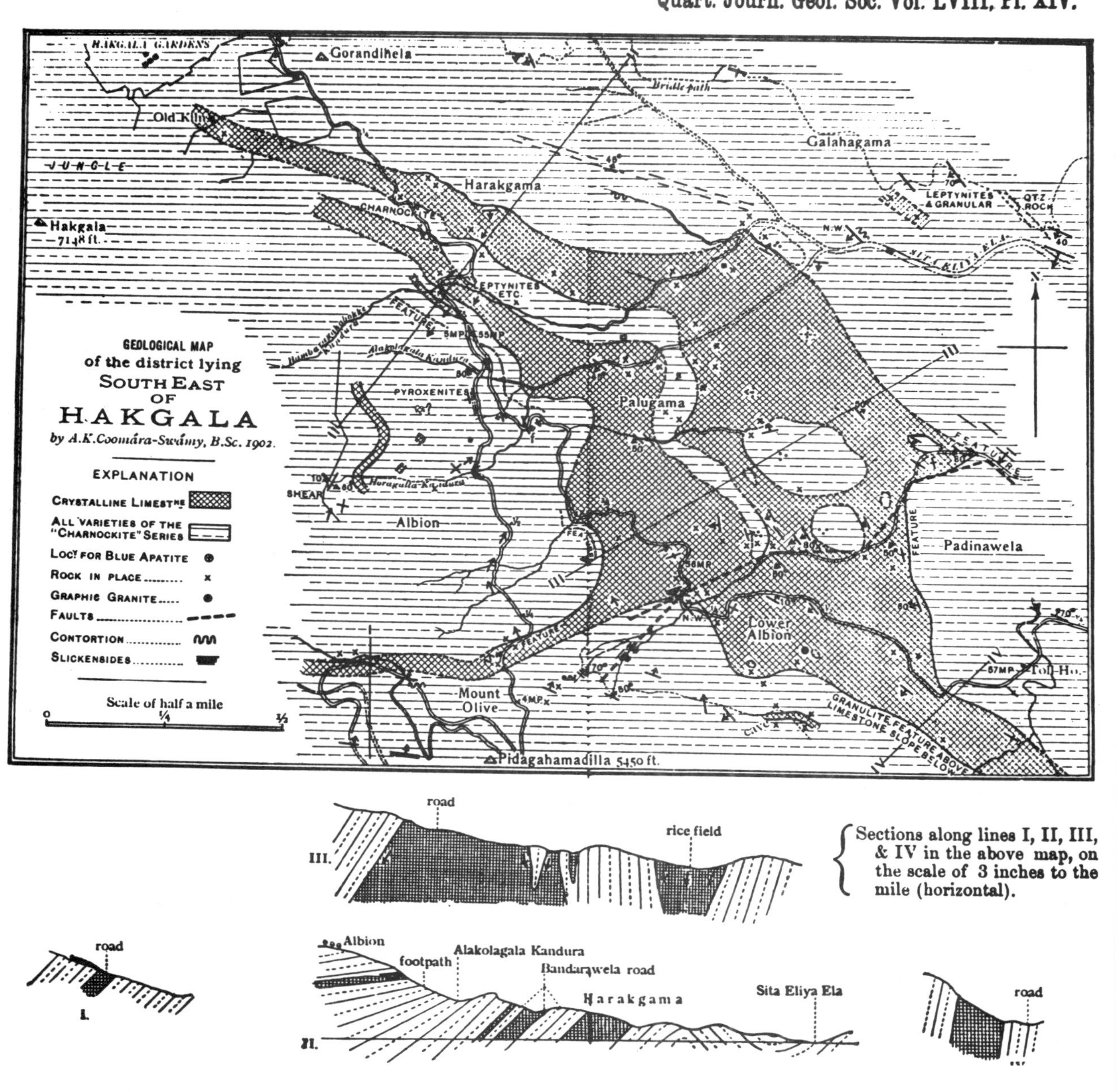

Prof. JUDD expressed his gratitude to the Author for bringing forward a description of a district so interesting to geologists. The rocks described were similar to those of Burma, except in the remarkable absence of certain minerals, such as corundum and its derivatives. He found great difficulty, as the Author did, in realizing that the charnockites could be intrusive in the limestones. In Burma and Ceylon alike, whatever might be the case in Southern India, the limestones were remarkably subordinate to the silicate-rocks, instead of the reverse being the case (as we should expect, if the latter were intrusive in the former). He agreed with the previous speaker as to the difficulty of imagining the limestones to have behaved as igneous rocks, and yet their relations with the igneous rocks were puzzling in the extreme. He referred to the occurrences of Glenelg and Tiree as affording fine illustrations of the part played in such a complex by calciphyres.

Mr. GREENLY remarked that in the Hebridian Gneisses of the North-west of Scotland there was also a great preponderance of igneous over what appeared to be sedimentary material. The Loch-Maree Group was now generally regarded as sedimentary, but it was a comparatively narrow zone, while from Loch Marce to Cape Wrath all appeared to be igneous. Limestones were a conspicuous feature of the Loch-Maree Group, but they were accompanied by graphite-schists, mica-schists, and other probably sedimentary rocks.

The AUTHOR, in reply to Mr. Holland, said that no doubt in most cases crystalline limestones were a result of the recrystallization of calcareous rocks under pressure; behaviour as a plastic medium did not involve a very high temperature, as had been proved by the work of Adams and Nicolson. But the limestones of Ceylon possessed a number of peculiar characters which, taken together, suggested to him that they had existed, in 'a state akin to fusion'. With regard to the interrupted sills, if this phenomenon was due to 'pinching' while both rocks were in a solid state (as Mr. Holland had suggested), why did the narrow lime-silicate contact-zones completely surround the lenticles, instead of occurring only on two sides of the granuite, as in the sill itself? Moreover, the accessory minerals in the limestones would show some trace of deformation if such powerful earth-movements had affected their matrix.

In reply to Mr. Greenly, the Author said that rocks composed mainly of biotite and garnet, which could be regarded as of sedimentary origin, were scarce; no rocks with kyanite, andalusite, or sillimanite were known *in situ*. Even if such exist, and are the remains of sedimentary rocks, the igneous rocks must greatly exceed them in amount. No graphitic schists had been found. Finally, the Author said that he had no wish to lay great stress on his 'igneous' theory, and he regarded the descriptive portion of his paper as of much more importance than the theoretical.

III. — March 12th, 1902. — Sir Archibald Geike, D.C.L., LL.D., F.R.S., Vice-President in the Chair. The following communications were read:—

1. "The Crystalline Limestones of Ceylon." By Ananda K. Coomára-Swámy, Esq., B.Sc., F.L.S., F.G.S.

The crystalline rocks of Ceylon may be divided into three series:—

(1) The Older Gneisses.
(2) The Crystalline Limestones.
(3) The Granulites (Charnockite Series) — pyroxene-granulite, leptynite, etc. A local subdivision of this series is the Point de Galle Group — wollastonite-scapolite-gneisses, etc.

The crystalline limestones of Ceylon are intimately asscoiated with the banded pyroxene and acid granulites (Charnockite Series).

They form bands with outcrops from a few feet to over a quarter of a mile in width, interbedded with the granulites. The limestones themselves have a banded structure (foliation) parallel to that of the granulites and to the boundaries. This foliation of the limestone depends on variations in structure, amount of accessory minerals, and relative proportion of calcite and dolomite. The grain is coarse, sometimes exceedingly so. Parallel and graphic intergrowths of calcite and dolomite are very characteristic. The most abundant accessory minerals are olivine, phlogopite, pink or violet spinel, diopside, pyrite, and blue apatite; less common are amphiboles, clinohumite, green spinel, etc. The most characteristic contact-minerals are diopside, amphibole, green spinel, and greenish micas; and, rather in the granulite than the limestone, scapolite, phlogopite, diopside, sphene. There occur also in the limestones, nodular mineral aggregates composed of characteristic minerals such as diopside, phlogopite, blue apatite, and spinel.

There are often transitions between the limestones and granulites. In some other cases a zone of green rocks (with diopside, dark mica, amphibole, and green spinel) intervenes. Bands (sills) of granulite of various width, down to less than a foot, may occur in the limestone, and are parallel to the foliation and general strike. These show peripheral transitions to the limestone by incoming of original calcite and the appearnce of lime-silicates, or are separated from it by a zone a few inches wide, in which the minerals diopside, amphibole, and green spinel are characteristic.

Some interrupted sills are described, and compared with the interrupted dykes of nepheline-syenite in the crystalline limestones of Alnö, described by Professor Högbom.

A sill may thus be continued along the strike as a series of lenticles. Elsewhere quite isolated masses of pyroxene-granulite occur as inclusions in the limestone.

Although the relation of the granulites to the limestones is on the whole intrusive, the two rocks in their present condition are essentially contemporaneous, and seem alike to have consolidated from a molten magma. The calcite occurring in the granulites near the contact has all the appearance of an original mineral. The foliation of the limestone is regarded as sort of flow-structure, and corresponds with that of the granulites to which it is always parallel. That the foliation does not result from the action of earth-movements on a solid rock is shown by this, that the very minerals whose variable distribution is one of its chief causes, have certainly not been affected by deforming earth-movements, nor are they such as to have been produced by these; moreover, in this respect a distinction cannot be made between the limestones and granulites, which would necessarily have suffered alike had they been subjected to deforming strains since the consolidation of the latter. The original nature of the limestones is less evident: they may have been sedimentary or tufaceous, and, if so, subsequently softened and metamorphosed; or possibly *ab initio* truly igneous rocks, and related to the charnockite-magma. Reasons for and against these views are given.

---

192 *Reports and Proceedings — Geological Society of London.*

The relations between the crystalline limestones and nepheline-syenites of Alnö have suggested to Professor Högbom that perhaps the limestones may have been a product of the nepheline-syenite magma there.

The author feels sure that the crystalline limestones of Ceylon have not arisen by the alteration of the basic lime-silicates of the pyroxene-granulites, although Professor Judd has advanced this theory in connection with the crystalline limestones of Burma, which seem to resemble those of Ceylon in many ways.

36. The POINT-DE-GALLE GROUP (CEYLON): WOLLASTONITE-SCAPOLITE GNEISSES. By Ananda K. Coomáraswámy, Esq., B.Sc., F.L.S., F.G.S. (Read June 18th, 1902.)

[Plate XXXIV — Map]

Geo. Mag., London, **58** (1902) 680 – 689.

I. INTRODUCTION.

Prof. A. Lacroix[56] has recorded the existence of wollastonite-and seapolite-bearing rocks in Ceylon. The specimens examined by him can hardly, however, have come from Galle, as they contain garnet, a mineral never found in the rocks of the Galle Group described below. It seems therefore probable that rocks allied to those of Galle, but differing from them in petrological details, will be found elsewhere in Ceylon.

I have on a former occasion [57] given a short description of the Galle rocks, and am now able to supply a more detailed account of the rocks and to describe their field-relations. There are excellent exposures at Galle, within the Fort on the western side, between the Signal flagstaff and the north-western bastion; outside the Fort below the north-western bastion; and south-west of the Victoria Park near the remains of an old powder-magazine. All these exposures can be easily visited in the course of one whole day spent at Galle.

II. GENERAL DESCRIPTION.

The Galle rocks differ from normal types belonging to the Charnockite Series in the following ways:—

(1) The presence of wollastonite, scapolite, and sphene.
(2) The existence of definite dykes and segregation-veins crossing the foliation.
(3) The absence of garnet, hypersthene, mica, and original hornblende, and bands of 'granular quartz-rock'.
(4) Somewhat coarser grain.

But resemble them closely in

(1) The variability of chemical and mineralogical composition.
(2) The conspienous mineral-banding (foliation).
(3) Common strike (about 10° west of north).
(4) The petrological characters of the acid types (especially as regards the felspars).
(5) The local tendency to graphic structures.

[56]'Contributions á PÉtude des Gneiss à Pyroxène et des Roches á Wernerite,' Bull. Soc. Minéral, Franc. vol. xii (1889) p.269 & pl.xii, fig.2.

[57]'Ceylon Rocks & Graphite,' Quart. Journ. Geol. Soc. Vol. lvi (1900) p.602.

The chief rock-types may be briefly tabulated as follows, though numerous transitions from one type to another occur:—

Basic......	1. Pyroxene-sphene-scapolite rock. Specific gravity=3.34
Intermediate	2. Rocks composed of pyroxene, scapolite, wollastonite, and graphite, iron-ores, and sphene, often with subordinate felspars or quartz[58] Specific gravity=2.99, 2.92
	3. Similar rocks, with abundant orthoclase-microperthite, or quartz, or both. Specific gravity=2.90, 2.83, etc.
Acid	4. Rocks formed of orthoclase-microperthite, orthoclase, oligoclase; with very little pyroxene, sphene, graphite; often rather coarser-grained than types 2 & 3. Specific gravity=2.64, 2.60.
	5. Coarse-grained quartzofelspathic rocks of granitoid aspect.

Types 1, 2, & 3 have usually a granular structure, types 4 & 5 a granulite structure. Types 2, 3, & 4 form the main part of the exposures.

There are also (6) acid rocks composed of felspars, quartz, and little augite, iron-ores, apatite, zircon, etc., occurring as dykes crossing the foliation; and (7) coarse wollastonite-orthoclase-quartz-pegmatites.

The acid types (4 & 5) are of somewhat later origin than the basic, and though generally interbanded, behave on the whole in an intrusive manner towards them. The latest-formed dykes and segregation-veins (6 & 7) crossing the foliation are also of an acid type.

Foliation.— The alternation of grey and white rocks of types 4 & 5 with green and green-and-white rocks of types 2 & 3 produces a conspicuous banding. (Fig.1, p.682; see also Quart. Journ. Geol. Soc. vol. lvi, 1900, p.602, fig.2.) At the junctions of the different bands there is a gradual but rapid change from one type to the other. Quartz, when present, is usually in individuals elongated parallel to the foliation; flakes of graphite are similarly oriented, and when locally abundant may even produce a plane of parting. The general strike varies little. Occasionally sharp contortions are to be found, the axes of the sharp folds being parallel to the strike. These local sharp folds are the result of pressures acting during consolidation.

Dykes and segregation-veins.— A few dykes of normal charnockite-like type are found, crossing the foliation at right angles or obliquely. Such dykes are form 1 to 4 feet wide, and consist of grey and greenish-grey rock, not especially coarse-grained, in which very fine-grained orthoclase-microperthite is a predominant mineral, with quartz, a little augite, and sometimes apatite, sphene, and iron-ores as accessory minerals. They have a specific gravity varying from 2.62 to 2.67. One such dyke showed traces of a banded structure parallel to the edges. At the junctions with the matrix,

the minerals interlock, and chilled edges are not found, showing the contemporaneous character of these 'dykes,' which are similar both to normal charnockites and to the acid varieties of the Galle Group.

Fig.1: - *Foliation in the Point-de-Galle Group, Galle Fort: alternation of lime-silicate-pyroxene and quartzofelspathic types of rock.*

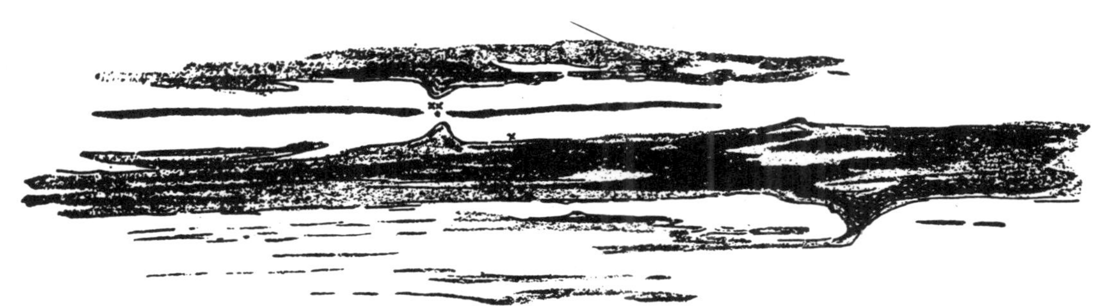

[At ×× the specimen, figured in Quart. Journ. Geol. Soc. vol. lvi (1900) p.603, was obtained; and at × the rock figured in the present paper, fig.7, p.688.]

More characteristic are the coarse pegmatite-patches and intrusive veins crossing the foliation (type 7). In these the component elements attain a large size and are frequently idiomorphic, with the exception of the wollastonite, which occurs in large masses moulding the other minerals (except pyrite, which however is rare in these veins). In such segregation-veins the minerals wollastonite, orthoclase, and quartz predominate, pyroxene being present in less quantity. Calcite and apophyllite occur occasionally, but may be secondary. The largest wollastonite - individual measured 36 × 15 inches; quartz 11 × $5\frac{1}{2}$ inches; orthoclase 24 inches in greatest diameter; a hexagonal prism of quartz, embedded in wollastonite, had a 3-inch prism side. The pyroxenes do not exceed a few inches in diameter, and are not always present. Scapolite was not found in the veins. Some veins consist almost entirely of wollastonite; one such, 2 to 4 inches wide, was noted on the shore between the Eolus and north-western bastions. Occasionally in these pegmatite-veins druses are found, partly filled with calcite. When weathered out, these sometimes yield idiomorphic orthoclase or apophyllite.

Vol.58.] Wollastonite-Scapolite Gneisses of Ceylon. 683

Weathering.— The felspars are generally very slightly cloudy, and not quite so fresh as in most of the inland charnockites. The scapolite exhibits often a fibrous structure due to incipient decompoisition. The rocks generally speaking are, however, very fresh. In some specimens the wollastonite has been replaced by a minute quartz-calcite mosaic.

Relation to the Charnockite Series.— It will be seen from the map that the rocks of the Galle Group occupy a belt or zone with normal members of the Charnockite Series on each side. No actual junctions are exposed, and probably no sharp contact could be recognized, even if complete exposures were to be seen. Probably there is rather a gradual transition; bands of acid rocks in the Galle Group are quite like acid members of the Charnockite Series, showing the close connection between the two grpups of rocks.

III. NOTES ON THE MINERALS.

Wollastonite,— never idiomorphic, but moulding other minerals; by no means always present; occasionally replaced by a quartz-calcite mosaic. Very large individuals in segregation-veins. It is interesting to note that Prof. Lacroix emphasizes the original character of the wollastonite in his Ceylon rocks, which, though probably not from Galle, are of very similar type.

Scapolite (wernerite),— constantly with small rod-and-lath-like inclusions parallel to the vertical axis. When decomposed has a cloudy look and fibrous structure. More frequently present than wollastonite. Never in large individuals, and not found in segregation-veins. Like the wollastonite, evidently an original mineral ; often accompanying quite fresh felspars.

A graphic intergrowth of scapolite and orthoclase-microperthite (fig.2, p.684) occurred at the junction of a small quartz-felspar-pyroxene vein with a green-and-white wollatonite-scapolite-pyroxene rock. It is interesting to compare this intergrowth with the intergrowths of scapolite and diopside, and scapolite and orthoclase, which were found in a variety of rock included in the crystalline limestone of the Hakgala district.[59]

Felspars,— usually orthoclase, with very minute microperthitic structure; often showing a peripheral zone of intergrowth with very fine vermicular quartz. Perhaps microcline-microperthite is sometimes present. The microperthitic structure is usually very fine; in other cases, the plagioclase-inclusions are larger and distinctly twinned. Large and small idiomorphic orthoclase-individuals, without inclusions, occur in the segregation-veins. Some from a drusy cavity were kindly examined by Mr. Richard Graham, who

Vol.58.] *Wollastonite-Scapolite Gneisses of Ceylon.* *685*

says that 'the forms present are too rough to measure except with the hand-goniometer; they are probably

$$m(110), b(010), z(130), c(001), o(\bar{1}11), n(021), y(\bar{2}01), ?d(241);$$

[59]A.K. Coomáraswámy, 'Crystalline Limestones of Ceylon' Quart. Journ. Geol. Soc. vol. lviii (1902) p.409.

Fig.2:-*Intergrowth of scapolite and orthoclase-microperthite, near the junction of a small vein with a more basic type of rock.* $(\times 12\frac{1}{2}.)$

[1 = Quartz; 3 = Orthoclase-microperthite; 16 = Scapolite, with fibrous decomposition-products]

Fig.3:-*Microscope-section of a rock ilustrating type 1* $(\times 12\frac{1}{2})$.

[14 = Sphene; 16=Scapolite; 20=Pyroxene]

the combinations *mbzcony*, *mbconyd*, *mbcon* being noted. 'Sections $\|b$ give an extinction of about 5° and positive bisectrix central'.

Plagioclase is not very common in separate individuals, and when present is usually referable to oligoclase (in one case labradorite). The felspars, in appearance and behaviour, much resemble those of the Charnockite Series in general.

Augite,— pale sea-green in sections, dark bottle-green macro-scopically, probably always belonging to the hedenbergite group.[60] Cleavage well developed. Extinction-angles over 40°. Occasionally replaced by amphibole, chlorite, calcite, etc., but usually fresh. Rare in segregation-veins, where individuals do not exceed a few inches in diameter. On one individual (moulded by calcite) from a small pegmatite-patch, Mr. Richard Graham determined the forms $m(110), a(100), b(010), o(\bar{2}21)$, and $u(111)$.

Quartz,— irregular clongated individuals in the acid rocks, the elongation parallel to the foliation. Large idiomorphic individuals in the segregation-veins. Often opalescent in the coarse quartzo-felspathic rocks. No quartz-veins occur.

Sphene,— usually present, but never idiomorphic except where moulded by wollastonite in local pegmatitic patches.

Graphite,— generally present in small flakes, as an original mineral. Rare in more acid types, and not noted in segregation-veins. One very thin vein of graphite was observed following a line of very slight disturbance.

Iron-ores,— pyrite most usual; magnetite, more or less titaniferous, also occurs.

Zircon and apatite rarely observed.

Apophyllite,— small idiomorphic individuals in a drusy cavity in a segregation-vein. The forms $a(100), c(001), p(111)$ have been identified by Mr. Graham.

The order of crystallization of some of the minerals is not very clear. Sphene, graphite, zircon, apatite, and iron-ores are the earliest; wollastonite is usually the last mineral to crystallize. When augite, felspars, quartz, and scapolite are present together, the strucutre is generally granular.

IV. OTHER LOCALITIES in CEYLON.

Rocks perhaps allied to those of Galle were brought up from the shaft of a graphite-mine at Nilhene (Baddegama district). Specimens composed of augite, scapolite, orthoclase, quartz, wollastonite, graphite, sphene, iron-ores, and calcite in various proportions were

[60]See chemical analysis, Quart. Journ. Geol. Soc. vol.lvi (1900) p.604.

Fig.4:- *Microscope-section of rock illustrating type 2* $(\times 12\frac{1}{2})$

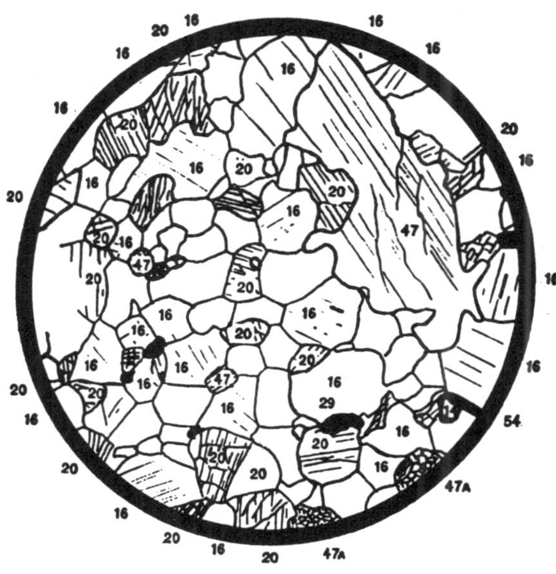

[14=Sphene; 16 = Scapolite; 20=Pyroxene; 29=Iron-ores; 47=Wollastonite; 47a=Quartz and calcite replacing wollastonite.]

Fig.5:-*Another rock illustrating the same type as above* $(\times 12\frac{1}{2})$

[6=Plaioclase; 14=Sphene; 21=Amphibole; 49=Calcite; 54=Graphite]

found, and have been briefly described by me in a former paper.[61] A quartz-calcite-micropegmatite occurred in one variety. No information concerning the field-relations of these rocks is available.

Rocks evidently similar to the Galle type occurred in a small exposure on the shore at Pittiwela, a few miles north of Galle. Specimens from this locality consisted of augite, scapolite, and sphene. The small exposure seemed to be quite isolated on the sandy shore.

Some charnockite-limestone contact-rocks in Ceylon resemble rock-types occurring at Galle, as, for example, the scapolite-augite rock with sphene, which occurs associated with crystalline limestone at Herimitigala [62] ; but the contact-rocks, on the whole, are of a rather different type, the absence of wollastonite being especially noticeable.

Prof. Lacroix's specimens indicate the existence of wollastonite- and scapolite-gneisses in some other parts of Ceylon.

V. CONCLUSIONS.

The Galle Group includes a series of rocks allied to, and part of, the Charnockite Series in Ceylon. Like those of that series, it is clear that the Galle rocks must be classed as orthogneisses, for they have evidently consolidated from a magmatic condition. The wollastonite and scapolite are original minerals, and certainly not decomposition-products from felspars or other minerals. They are often associated with feslpars which are in a perfectly fresh condition. This original character of wollastonite and scapolite in some gneisses rich in lime is well recognized. Wollastonite has even been found as an original mineral in eruptive rocks.

Possibly the richness in lime of some of the Galle rocks is due to the total absorption of a large mass of limestone (whether originally of sedimentary origin or otherwise) by a portion of the Charnockite Series. It may, however, have been an original character of the magma. Scapolite-pyroxene rocks are sometimes found at the junction of limestone and charnockite in Ceylon.[63] Wollastonite has not, however, been found in such situations. If the absorption-theory be accepted, the lime-silicate minerals must be regarded as endomorphic contact-minerals. In any case, the Galle rocks seem to show clear indications of progressive differentiation from basic to acid types ; the coarse segregation-veins seem to be the last product of such a process. There is no reason to regard the graphite-flakes as other than original and crystallized from the magma.

That the Galle rocks have not suffered from deforming earth-movements since their complete consolidation is evidenced by their microscopic characters, and by the segregation-veins and dykes which

[61]'Ceylon Rocks & Graphite,' Quart. Journ. Geol. Soc. vol. lvi (1900) p.602.

[62]'Crystalline Limestones of Ceylon,' *ibid*, vol.lviii (1902) p.407

[63]*Ibid.* pp.405 *et seqq.*

Fig.6:-*Microscope-section of rock illustrating type 3.* $(\times 12\frac{1}{2}.)$

[3=Orthoclase-microperthite; 14=Sphene; 16=Scapolite; 20=Pyroxene; 29=Iron-ores.]

Fig.7:- *Microscope-section of rock illustrating type 4.* $(\times 12\frac{1}{2}.)$

[1=Quartz; 3=Orthoclase-microperthite (and orthoclase?); 54=Graphite; × × = Direction of foliation.]

Quart. Journ. Geol. Soc. Vol.LVIII, Pl. XXXIV.

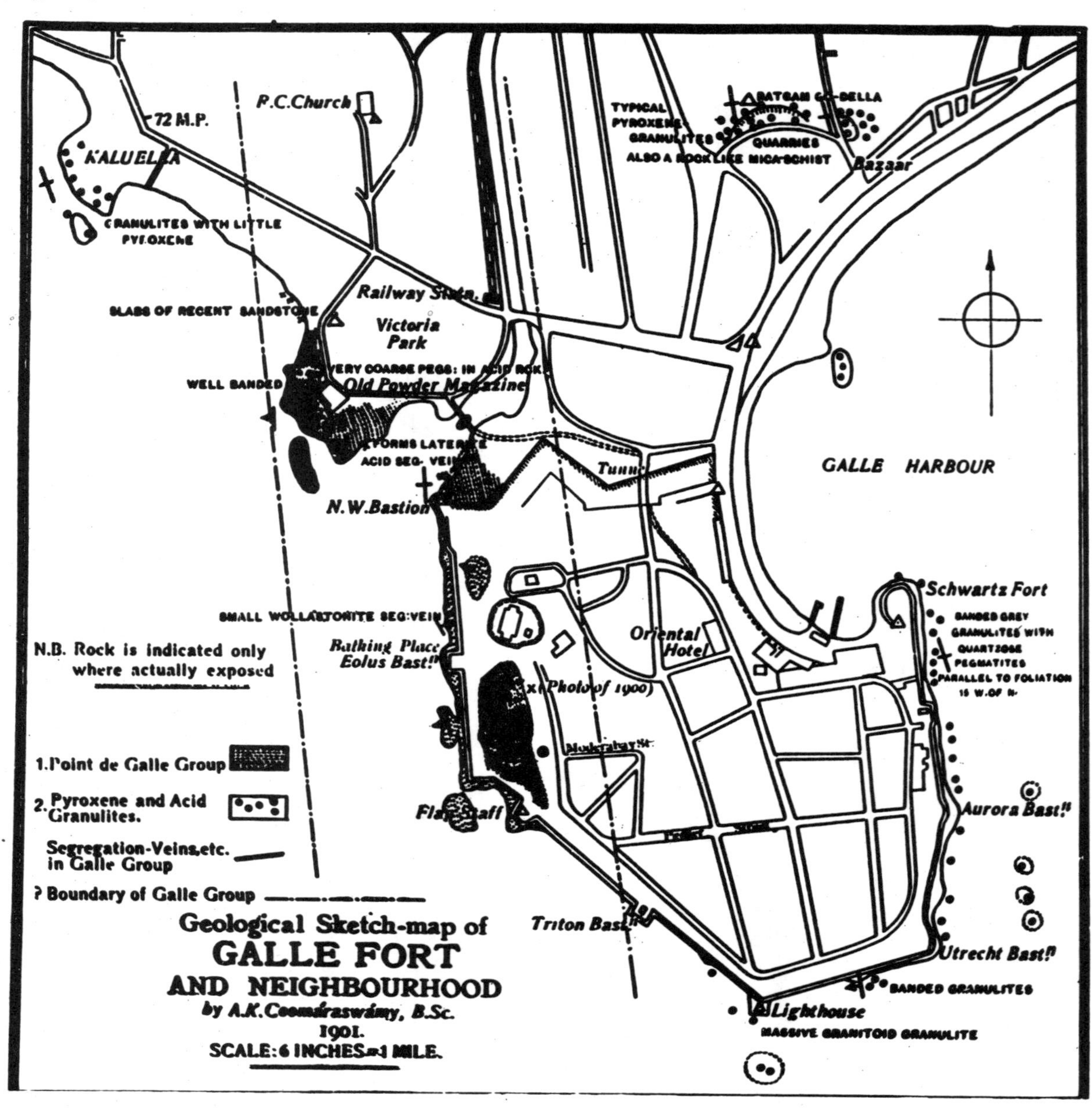

are never sheared or distorted, while the interlocking of the minerals at their junction with the matrix and their actual mineral composition, manifest their contemporaneous character and close connection with other rocks of the Galle Group.

EXPLANATION OF PLATE XXXIV.

Geological sketch-map of Galle Fort and neighbourhood, on the scale of 6 inches to the mile, showing the localities where wollastonite-scapolite gneisses are exposed.

DISCUSSION.

Prof. Groom asked whether it was not possible that some of the splendid gneisses described by the Author had been formed by the intrusion of the more acid materials, along planes of foliation of a schist produced by the shearing of a solid rock.

The Author thanked the Fellows for the kind way in which they had received his paper. In reply to Prof. Groom, he said that he thought the mineral-banding of the Galle rocks was a true fluxion-foliation. There was no evidence of foliation in the basic portions. Basic lenticles included in more acid rock were common in the Charnockite Series, and there were transitions from these through pinched bands to ordinary mineral-banding. Simple mineral-banding was chiefly conspicuous at Galle, and this seemed to point to the squeezing of a 'schlierig' rock in the course of consolidation, during which there was perhaps a successive introduction of basic and acid types, the final residuum forming segregation-veins.

REPORT ON THE OCCURRENCE OF CASSITERITE (OXIDE OF TIN) IN CEYLON.

By Ananda K. Coomáraswámy, B.Sc., Director, Mineralogical Survey.

Govt. Printing Office, Colombo (1905).

In the latter part of 1904 some *nambu* was brought to me while working at Gilimale, and it was stated that it had been obtained at Niriella (Pallepattuwa, Nawadun korale, Sabaraganuwa).

It was found to consist of cassiterite, and I determined to visit Niriella in the early part of 1905, in continuation of the mineral work in the Ratnapura District, but with a special view to this observation.

Two weeks were accordingly spent at Niriella in February, 1905. The cassiterite (alluvial tin) was found to occur in the bed of a very small stream called the Etunkahena-dola, at a point where gemming had been carried on ; no work is being done at present owing to a legal dispute. The deposits of *illam* are exceedingly limited; the proportion of *nambu* obtainable is also small; it consists of ilmenite, cassiterite, & c. The cassiterite is in the form of rather small well rounded or subangular fragments, quite black in colour and showing no trace of crystalline form. It is distinguished from ilmenite and rutile, to which it bears a considerable superficial resemblance, by its specific gravity of 6.91-7.0. Its pale brown streak, and by yielding globules of tin when fused on charcoal with sodium carbonate in the usual way. I was not able to discover traces of cassiterite elsewhere in Niriella; but I have received one or two specimens that must have come from other localities; one of these was stated to have been picked up on Gabbela estate. It was also reported in a newspaper last year that cassiterite had been detected at Hatton, but no specimens reached the Mineralogical Survey.

It is likely that in all cases the cassiterite is derived from isolated exposures of rocks of granitic character belonging to the Balangoda Group. Such rocks have been observed at several localities near Ratnapura, although the varieties exposed were not those yielding any valuable minerals.

It will be seen that all that can be regarded as established is that cassiterite (alluvial tin) occurs in a few localities, but so far as is known, only in quite small amounts. This is to be regretted, as its discovery in any quantity would be of great commercial importance. Both officers of the Mineralogical Survey are at present working in the Ratnapura District, and further discoveries may possibly be made.

ANANDA K. COOMÁRASWÁMY.
Ratnapura, March 18, 1905.

224

Serendibite, a new borosylicate from Ceylon.

By G.T. Prior, M.A., F.G.S., and A.K. Coomáraswámy, B.Sc., F.G.S., F.L.S.

[Read February 4, 1902.]

Mode of Occurrence.— This new mineral was discovered at Ganga-pitiya, near Ambakotte and about 12 miles east of Kandy, Ceylon. In this locality quarries have been dug for the well-known Ceylon moonstone, which occurs in large porphyritic crystals in an acid granulite.

Bands of this granulite, which is composed mainly of quartz and felspar in finely granulitic and occasionally graphic structure, alternate with bands of limestone up to 18 inches wide. Between limestone and granulite occur contact-zones consisting, next to the limestone, almost entirely of a colourless diopside, but near the granulite of a mixture of diopside with other minerals, viz. blue spinel, a little apatite, occasionally scapolite or plagioclase, and the blue mineral which is the subject of the present note. Hand-specimens may be collected showing this relation. The serendibite occurs in sufficiently large amount to give a deep blue colour to the contact-zones.[64]

Physical Characters.— The association of the serendibite with the diopside and spinel is of such an intimate nature that no definite crystals could be isolated from the material at our disposal.

Examination of fragments of the mineral and of thin slices of the rock in which it occurs, shows that it is biaxial and probably anorthic. Besides the marked pleochroism, from very pale yellow (nearly colourless) to deep indigo-blue, the most notable feature to be observed in thin slices is the remarkable polysynthetic twinning, which is quite as intimate as that of a plagioclastic felspar.

The mineral occurs for the most part in irregular grains. In parts of the rock, however; radiating detached grains of the serendibite are seen

Serendibite. A New Borosilicate from Ceylon. *225*

to belong to one individual crystal and to form a sort of ophitic structure with the diopside; while occasionally more definite crystals have been formed which show in thin slices traces of a prism or pinacoid, along which they are elongated and parallel to which occurs the twinning. Such crystals often show traces of two terminal faces. In the section

[64]Further details as to the mode of occurrence of the mineral will be found in Quart. Journ. Geol. Soc., 1902, vol.lviii.p.421.

of the crystal shown in the accompanying figure these terminal faces are inclined to each other at an angle of about 104°, and to the adjacent side planes at angles of about 138° and 122° respectively. In this crystal the extinction is nearly symmetrical and about 15° for the two sets of twin-lamellae, each of which shows the emergence of an optic axis nearly central: the pleochroism was from very pale yellow (nearly colourless) for vibrations parallel to the line of extinction to sky-blue at right-angles: compensation with the quartz-wedge took place at right-angles to the line of extinction.

Thin section of a crystal of serendibite between crossed nicols, one set of twin-lamellae being in the position of extinction. (Magnification about 12 diameters.)

As at present imperfectly determined (owing to the lack of crystals), the physical characters of serendibite are as follows:—

Biaxial, probably anorthic; polysynthetic twins with twin-plane probably a pinacoid; cleavage, none; fracture, subconchoidal; hardness, nearly that of quartz, $6\frac{3}{4}$; density, 3.42 (determined on 0.2710 gram); lustre, vitreous; colour, blue, varying in different varieties from pale sky-blue to deep indigo-blue; transparent; pleochroism, in one variety very pale yellow (almost colourless) to pale sky-blue, in another variety pale brownish-yellow to deep indigo-blue; refraction, near that of diopside, and between monobromonaphthalene and methylene iodide, about 1.7; double refraction, weak.

Chemical Composition.— The mineral is infusible: it is only slightly attacked by acids, even hydrofluoric; heated in the closed tube, it gives a little water with doubtful traces of etching from fluorine; fused with calcium fluoride and acid sulphate of potash it imparts to the flame the green colour characteristic of boron. The presence of boron was confirmed by fusing the powdered mineral with sodium carbonate, extracting the melt with water, acidifying with acetic acid and destilling with methyl alchohol; the distillate was treated with lime to fix the boracic acid and evaporated to dryness; the residue, treated with sulphuric acid, imparted a green colour to the flame of methyl alcohol, and dissolved in hydrochloric acid, gave the characteristic red colour to turmeric paper.

Material for analysis was carefully selected as follows. Fragments of the rock were crushed and the powder retained between sieves of 30 and 60 meshes to the inch; pure grains of serendibite were then picked out with a fine brush and examined, grain by grain, under the microscope and tested for pleochroism and double refraction. Of material picked out in this way only 0.3304 gram was available for the main analysis, and 0.2406 for the alkali determination. The amount of material at our disposal did not allow of a satisfactory direct determination of the boracic acid.

The result of the analysis (G.T.P.) is as follows:—

				Molecular ratios.
SiO_2	...	25.33	...	0.42
Al_2O_3	...	34.96	...	0.35
FeO	...	4.17	0.06	
CaO	...	14.56	0.26	0.69
MgO	...	14.91	0.37	
Na_2O, Li_2O	...	0.51*		
K_2O	...	0.22		
Loss on ignition	...	0.69		
P_2O_5	...	0.48		
F ?	...	not determined		
B_2O_3 (by diff.)	...	(4.17)	...	0.07
		100.00		

*Calculated as if wholly Na_2O : the presence of lithium, however, was clearly indicated by the spectroscope.

Serendibite. A New Borosilicate from Ceylon. 227

A provisional formula is

$$10(Fe, Ca, Mg)O \ . \ 5Al_2O_3 \ . \ 6SiO_2 \ . \ B_2O_4$$

This may be written

$$6(Fe, Ca, Mg)\ SiO_3 \ . \ 4MgAl_2O_4 \ . \ 2AlBO_3$$

suggestive of a combination of aluminium borate with molecules having the composition of diopside and spinel, the two minerals with which serendibite is so intimately associated. The mineral has some relationship with tourmaline in being a borosilicate of alumina, magnesia, and alkalies (including lithia). It has been formed as a contact mineral instead of tourmaline [65], probably owing to the basic conditions induced by the presence of the limestone. In many of its physical characters, colour, hardness, and density, the mineral is very similar to sapphirine.

It is hoped that more material (and especially crystals) may soon be available in order that the description of the mineral may be made more complete.

The name 'serendibite', given to the mineral, is derived from 'Serendib', an old Arab name for Ceylon.

[65]Tourmaline was also found very sparingly in the contact zones, but the few crystals observed occurred only in soft material apparently filling cracks in the harder rock.

THE CRYSTALLINE ROCKS OF CEYLON[66]

By A.K. Coomáraswámy, B.Sc.,

Director, Mineralogical Survey of Ceylon

The following paper is intended to give a brief account of the crystalline rocks of Ceylon, so far as at present known and studied.

The first question which arises is one of nomenclature. For a long time the rocks of Ceylon were somewhat summarily classified as *Gneisses*, or as belonging to "*metamorphic series*" is often used in its strict original sense and applied only to foliated crystalline rocks having the mineral composition of granites, and thus consisting essentially of orthoclase-felspar and quartz, with one or more of the minerals mica, hornblende, or augite. In this strict sense the term "gneiss" is only applicable to certain of the rocks of Ceylon. It has, however, been widely used in England and even abroad to desginate rocks characterized rather by the possession of a "gneissose" strucutre than by any particular mineral composition. In this extended sense the term is applicable to many or most of the rocks of Ceylon.

The name "granulite" has been extensively employed to designate foliated crystalline rocks of fine grain in which orthoclase, quartz, garnet, and more or less biotite, hornblende, or augite occur, but which are especially characterized by the presence of garnet.

This term is thus particularly applicable to the majority of Ceylonese rocks, and its use is perhaps to be preferred to that of the term "gneiss".

Mr. Holland has recently named the rock series of Southern India (which closely resembles that of Ceylon) the *Charnockite Series.* Ceylon and Southern India form together a well-marked petrographical province, and there is no doubt that the rocks of Ceylon must be considered as forming part of the charnockite series. It should be noted that the term "charnockite" is used to designate both a whole group of rocks and also one of the types of rock (having particular and definite characters) characteristic of the group. It is thus safest as a general rule to speak of

the foliated crystalline rocks of Ceylon as granulities, and as belonging to the charnockite series of Ceylon and Southern India.

[66]Spolia Zeylanica, COLOMBO, (1904) 105-111

With regard to the use of the term "metamorphic series", it should be recognized that the evidence tends to show that the mass of Ceylon rocks is rather conspicuous for the small evidences of metamorphism subsequent to formation which it affords. The term "metamorphic", embodying as it does a very definite and, generally speaking, erroneous view, is thus to be avoided.

If we come to inquire as to the real nature of the rocks composing the granulite or charnockite series, we are at first struck by the appearances in the field suggestive of a sedimentary origin (figs. 1 and 2). It is not surprising that the conspicuous bedded character varied mineral composition, and local abundance of alternating beds of limestone should have led to the supposition that such a gneisses series as we have here to deal with really represents the highly metamorphosed remains of some ancient sedimentary series. Closer examination of the rocks, however, reveals but little in support of such a view. Their mineral composition is found to indicate a chemical constitution differing in important respects from that of the great majority of sedimentary rocks, but similar to that of many well-known types of deep-seated igneous rock. Moreover, when we come to examine the minuter characters of the foliation or mineral banding so characteristic of the Ceylon rocks, we see that there is rather a lenticular than a definitely parallel structure (fig.2), and also that some definite relation is evident between the lighter-coloured (more acid) rock types and those of a darker colour—the more basic.

This definite relationship may show us ovoid lenticular masses of heavy dark rock included in a much lighter-coloured, more acid rock that seems to have flowed around them. In other cases the more acid types give other evidence of their intrusive relation to the more basic bands and lenticles. When the rock examined possesses a general dark colour and basic character we often find the more acid portions existing as irregular patches a few inches or feet in diameter, which send long tongues of acid material in parallel bands along the foliation planes, giving a banded character to a rock which would otherwise present a more uniform appearance. Such phenomena, together with others which a microscopic examination of the rocks reveals, are known to be characteristic of many rocks whose igneous origin is undoubted. It may then be assumed with safety that at least the greater part of Ceylon rocks, so far as at present known, are of igneous origin and that their well-banded appearnce (fig.2) results

Crystalline Rocks of Ceylon. *107*

from the drawing out by flowing movement of the various parts of a heterogeneous magma, probably during its gradual consolidation under conditions of great pressure.

The observed relations of the acid and basic varieties show that the latter types were amongst the earlier products of the magma, the more acid types as usual crystallizing later and having a more or less intrusive relation towards the more basic. The continuous flowing movement accompanying intrusion and consolidation has had the effect of drawing out together in bands the acid and basic types and preventing the former from transgressing the latter in the form of dykes.

Fig.1 - Foliation Planes in Granulites
(Charnockite series) Near Teldeniya, Central Province.

Fig.2 - Foliation in rocks of the galle group
(Charnockite series) Galle Fort. About 7 feet thickness shown.

No very definite nomenclature can be applied to the different varieties of granulite, which shade insensibly into one another, and are sometimes so intimately interbanded as to make it difficult to collect even a hand specimen of uniform composition. The following general types, however, include the more conspicuous varieties, beginning with the most acid:—

Amongst the rocks of the charnockite series are abundant beds of *granular quartz rock* of very various thickness possessing a mineral composition which varies in different cases from nearly pure quartz, through types containing small quantities of felspar and garnet and sometimes mica, to others which are at once recognizable as the more acid types of granulite. We are thus led to regard the quartz bands as genetically connected with the other members of the charnockite series, and in fact, as the last and most acid products of the magma.

A less acid group is that of the *leptynites*, rocks composed essentially of quartz and felspar (orthoclase microperthite predominating), with or without more or less garnet and biotite and accessory minerals such as graphite, zircon, and apatite. These light-coloured rocks bulk largely amongst the granulites. The *garnetiferous leptynites* are especially conspicuous, the bright red garnets in a nearly white rock giving a very handsome appearance. An especially characteristic feature of some leptynites is the great elongation of the quartz grains.

Many conspiciuous hills (*e.g.,* Bambaragala near Teldeniya, Central Province), with rocky summits weathered into "tors", consist almost entirely of a granular granite-like leptynite composed essentially of quartz and orthoclase microperthite, but often with a small amount of biotite as well. In such cases the rock seems to form a large ovoid mass amongst more basic types.

Typical charnockite consisting of quartz, felspars, and hypersthene has a very different aspect, being of a greenish colour and usually coarser grain, and having a greasy lustre. A considerable

108 *Spolia Zeylanica.*

range of types varying much in relative acidity have nevertheless a very similar appearnce in the field, and we pass by gradual transitions to rocks having the composition of *norites* (labradorite-hypersthene rock) and *diorites* (hornblende-labradorite rock). & c. Most of these less acid types might also be called *pyroxene granulites.* The most basic types occur chiefly as lenticles and "pinched" bands amongst medium and acid types : they include rocks of which the principal constituents are hypersthene, augite, garnet, basic plagioclase, & c. It is generally more or less evident that these most basic types are to be regarded as amongst the earliest products of the granulite magma.

Rocks of anomalous or unusual composition are sometimes met with, *e.g.,* a felspar corundum rock from near Kandy, and a garnetiferous leptynite in which spinel is very abundant; attention may also be called to the remarkable rocks containing corundum, sillimanite, & c., described by Professor Lacroix. There occur also in the Balangoda district certain zircon granites containing a large quantity of well-crystallized zircon; little is known as to the relations of these rocks; it is not unlikely that they may have some connection with the occurrence of a group of rare minerals (including uraninite, & c.) found in the Balangoda district; they are probably intrusive in the charnockite series.

Pegmatite Dykes, consisting essentially of orthoclase and quartz with mica or hornblende or both, are found occurring as dykes crossing the granulite foliation planes and as sills running along these planes. The pegmatites must be in many cases regarded as segregation veins; they appear to have crystallized before the cooling of the matrix, as the minerals are interlocked at the junction, which is rather a transition than an abrupt junction.

An important series of rocks not yet referred to is that of the *crystalline limestones.* These rocks are found intimately interbanded with the ordinary granulites, the beds of limestone varying in thickness from hundreds of yards down to a few inches. Many varieties are dolomitic, but there is great variation in this respect. Many interesting accessory minerals occur, such as phlogopite, forsterite, graphite, spinel, apatite, amphibole, clinohumite, & c. The nature and origin of these rocks is exceedingly obscure. Many would perhaps regard them as the highly metamorphosed remains of some ancient calcareous sedimentary series included in the vast mass of granulites, and owing their characters to the effects of contact metamorphism. There are, however, great difficulties in the acceptance of such a view. Amongst these may be mentioned the, on the whole, small quantity of the limestone compared with the vast bulk of the

granulites, and the still smaller quantity of other rocks possibly of a metamorphosed sedimentary character, the frequent occurrence in the limestones of fragments of peripherally modified granulite, and the strong general similarity between the characters of the crystallized limestones and those of igneous foliated crystalline rocks.

Putting aside, however, the doubtful question of origin, it is clear that as far as their present characters are concerned the crystalline limestones are contemporaneous with the granulites; and there can be little doubt that they existed in a state akin to fusion at the time of the consolidation of the latter, and that their accessory minerals crystallized under conditions resembling those accompanying the cooling of such plutonic magmas as have for example given rise to the granulites themselves. The foliation of the limestones must have been acquired at the same time and in the same way as that of the granulites.

A group in some respects intermediate between the limestones and granulites includes rocks composed essentially of one or more of the minerals diopside, mica (phlogopite, & c.), amphibole, scapolite, spinel, sphene, and calcite, with more rarely felspars, marcasite, serendibite, & c. In these rocks the silicates are usually idiomorphic when moulded by carbonates. Of the minerals mentioned diopside and mica are predominant. The rocks occur (a) as junction rocks between limestone and granulite, (b) as inclusions in the limestones, (c) as dykes in the granulites, these dykes being usually narrow and having a zoned structure, and (d) as bands of rock interbedded with the granulites. Mica of commercial value may occur in rocks of this type. The bands rarely exceed 6 ft. in thickness and are often narrow. Continuity between the types (c) and (d) can sometimes be observed. Some of the types of rock included here are very handsome, especially the coarse varieties in which scapolite is conspicuous.

Rocks of the *Galle group* are also in some respects intermediate between the limestones and granulites, inasmuch as they contain numerous lime silicate minerals and differ in other respects from the typical granulites, although in many ways resembling them; the presence of wollastonite (not yet found elsewhere in Ceylon) and of scapolite and sphene (in addition to a green pyroxene and the quartz and felspars characteristic of the granulites) and absence of garnet, amphibole, and mica are distinguishing characters of the Galle rocks. As in the granulites, there is a wide range of types, from basic to acid, and mineral banding is very conspicuous (fig.2).

110 *Spolia Zeylanica.*

A brief discription of the crystalline rocks is now concluded, but a few other points remain to be considered.

Not a great deal can be said as to the conditions under which the rocks were formed: they must have crystallized under great pressure, and of course the temperature must have been high; during crystallization the magma was probably rather viscous than fluid. There was evidently much flowing movement. The magma was probably far from homogeneous to begin with, and in addition to this the basic minerals crystallized in general before the acid; this heterogeneity combined with movement produced the conspicuous mineral banding. Probably the rocks crystallized at a great depth, the absence of even ancient sedimentary rocks overyling the granulites showing that denudation has long been at work in Ceylon. In India as we proceed southwards the palaeozoic rocks die out, and it is probable that the charnockite series is of prepalaeozoic or archean age.

Since their formation the rocks have suffered from earth movements, but seemingly not of a violent character, inasmuch as there is little evidence of deformation, nor has evidence of the existence of thrust planes or similar phenomena been noted.

An important series of pressures has thrown the granulites into a main series of anticlinal and synclinal folds, and to this system of folds we may give the name of *Taprobanian*; the axes of the synclines and anticlines, some of which are of great size while others are quite small, have a more or less constant trend about 30° W. of N. in the Kandyan districts, and perhaps more nearly north and south in the Badulla and some other districts; no evidence to date the movements is available.

This structural feature of the mountains of Ceylon is rendered conspicuous by its effect on the directions of river systems and even of roads and railways; every one interested in Ceylonese topography must be acquainted with the north-west and south-east valleys which are so conspicuous and which are due to the rivers cutting their way along the foliation planes of the granulites, forming *strike*[67] *valleys*, with corresponding strike ridges separating them. Valleys, usually of less importance, have also been made by streams that have cut their way along the joint planes and thus occupy *joint valleys* oriented approximately at

Crystalline Rocks of Ceylon. *111*

right angles to the strike valleys. Sometimes a stream occupies a strike valley for part of its course and a joint valley elsewhere, the change being indicated by a sharp turn. Perhaps the best illustration of the occurrence of parallel strike valleys is found in the Hatton district; in a map of this district the parallel valleys, along which run road and rail as well as stream, are most noticeable (fig.3).

There are no data vailable to estimate the actual thickness of the charnockite series in Ceylon (measured in a direction at right angles to the foliation planes), but it is clearly not unlimited, corresponding portions of the series being repeated by the foldings.

Fig.3-Strike Valleys near Hatton.

[67]The inclination of a rock to the horizontal is termed its *dip*; a line drawn at right angles to the dip on the foliation plane is called the *strike* of the rock; or the strike might be defined as idential with the outcrop on a horizontal surface. In fig.1 the rocks are dipping from left to right; the water line on the foliation planes serves to indicate the strike. Joints are divisional planes (other than foliation or bedding planes) occurring in a rock, and are usually approximately at right angles to the foliation planes.

MINERALOGICAL NOTES[68]

By A.K. Coomáraswámy, B.Sc.,

Director, Mineralogical Survey of Ceylon

I.-THE NEW MINERAL

In 1903 Mr. W.D. Holland (who has long taken an interest in the mineralogy of the Bambarabotuwa district) obtained a quantity of a heavy black mineral occurring in cubic crystals, which he naturally identified as uraninite (pitchblende). Samples were sent to Sir W. Crookes, Sir W. Ramsay, and others, and the whole amount available (some 5 cwt.) was purchased by the latter chemist. Samples were also sent through the Mineralogical Survey to the Imperial Institute for examination and analysis.

In a letter published in *Nature*, p.510, March 31, 1904, Professor Dunstan published the following analysis of two samples, made by Mr. G.S. Blake at the Imperial Institute:-

I			II		
Th O_2	...	76.22	TH O_2	...	72.24
Ce O^2	...	8.04	Ce O^2	...	6.39
La_2 O_3, Di_2, O_3			La O_2 Di_2 O_3	...	0.51
Zr O_2	...	trace	Zr O_2	...	3.68
U O_2	...	12.33	U O_2	...	11.19
Fe_2 O_3	...	0.35	Fe_2 O_3	...	1.92
Pb O	...	2.87	Pb O	...	2.25
Si O_2	...	0.12	Si O_2	...	1.34
			Insol.residue		0.41
		99.93			99.93
S.gr.	...	9.32	S.gr.	...	8.98

These analyses showed that the mineral was not pitchblende, and Professor Dunstand suggested the name of *horianite* for the new mineral.

In a letter in *Nature*, p.533, April 7, 1904, however, Professor Ramsay published the preliminery results of his examinations, stating that a much smaller percentage of

[68]SPOLIA ZEYLANCIA, COLOMBO (1904) 57-64

thorium occurred in the mineral, and that no appreciable amount of cerium, lanthanum, and didymium entered into its constitution; but the presence of one or more new elements was indicated. The mineral was strongly radioactive, but contained only a trace of radium, the radio-activity due to this source being certainly not 5 per cent. of the total.

The period of decay of the emanation appeared to point to the presence of a radioactive element closely resembling thorium X. Helium was yielded when the mineral was heated alone, 3.5 c.c. per gram being obtained; when fused with hydrogen potasium sulphate, it yielded 9.5 c.c. per gram.

In a subsequent letter (*Nature,* April 14, p.559) Professor Ramsay admitted that a considerable amount of thoria was present, in addition to the new elements, which may be identical with those recently discovered by Professor Baskerville. A partial analysis which was made in M. Curie's laboratory gave 79 per cent. oxides of rare earths (principally thoria) and 14 per cent, of uranium oxide. A rough preliminary analysis made by Professor Sir W. Ramsay is as follows: oxides like thorium oxide 76.4; uranium oxide 14.9; ferric oxide 6.1; lead, arsenic, & c., 2.0; insoluble (not S_1O_2) 0.7. At present the constituion of the new mineral and so also its commercial value cannot be regarded as definitely established; we must await the completion of further work in London.

With regard to the analyses made at the Imperial Institute I may say that I have not observed any trace of zircon occurring as an impurity in the mineral, and do not think the ZrO_2 found can be due to the presence of inclusions of zircon, but rather that the zirconia enters into its composition. On the other hand, many crystals have a slight amount of limonite attached to the surface, and even if these be carefully excluded there are others containing small cavities lined with limonite attached to the surface, and even if these be carefully excluded there are others containing small cavities lined with limonite which may in part account for the variations in specific gravity observed, so that a small percentage of iron oxide in the analyses may be regarded as an impurity.

The properties of the mineral not already quoted are referred to below:- Colour black, streak greenish gray. Hardness, 5.5-6. Crystallizes in the cubic system, crystals rarely exceed 7 mm. along a cube edge. Some crystals consisting of two interpenetrating individuals appear to be twins, but this has not been as yet confirmed by measurements. Fracture uneven, tending to conchoidal. Brittle, Lustre resinous to submetallic. Infusible, Dissolves readily in salt of phosphorus bead, vigorously giving off bubbles, doubtless of helium; the bead in O.F. is yellow when hot, green when cold, being the reaction for uranium. In borax bead yellow hot, pale greenish-yellow cold. Insoluble in acids.

Some account of the localities and mode of occurrence may be given next. The mineral occurs in greatest abundance in loose waterworn crystals in the bed of the Kuda Pandi-oya, a tiny stream near Kondrugala, Bambarabotuwa, Sabaragamuwa, where it often forms the bulk of the heavy residue ("Nambu") remaining in the gemming basket after washing. It collects also in hollows and potholes in the bed of the stream, which in its upper part is so small as not to contain running water in dry weather. It occurs also in fragments of a ferruginous conglomerate formed in the stream bed, and containing waterworn rock fragments and crystals of zircon, thorite, & c.

The mineral was traced to within a short distance of the head of the ravine, but no sign could be found of any rock containing it *in situ*, and a more extensive search for the matrix would probably be a long and expensive matter, owing to the thick soil, landslips, and thick jungle. I have little doubt, however, that it is derived from some granitic or pegmatite-like rock, such as those in which zircon and allanite have been found.

The mineral was detected in smaller quantities in washings from the Alupola-dola (some distance above the path) and in the Kuda-oya between Bataragala and the said Kondrugala path. Two or three crystals were also seen in gem washings from Mussena estate, but more could not be obtained. A single crystal was found in a washing taken from the Pita-ela between Walawe and Morahela estates near Balangoda.

Since uranite is *rarely* found in cubic crystals, it is most probable that the mineral recoreded by me as uraninite from Gampola *Spolia Zeylanica*, vol.I.part IV., 1904, is in reality the new mineral. It there occured in a pegmatite consisting mainly of orthoclase, quartz, and biotite, with apatite, tourmaline, & c., as accessory minerals.

I have seen other specimens from Ceylon of a mineral resembling this one, but regarded as uraninite; some of these are masive, and cannot be definitely calimed as uraninite or thorianite without chemical analysis; so that it is far from unlikely that new localities will be found where one or both may be met with in greater abundance.

In the Kuda Pandi-oya it is associated with quantities of ziercon (well-developed large crystals, with some excellent twins on e (101)), and with more waterworn crystals of thorite, a mineral which I at first identified as monazite. A crystal of the supposed monazite stated to be from Bambarabotuwa was sent to the Imperial Institute for examination, and Mr. G.S. Blake's analysis gave the following result, the mineral being identified as thorite:-

$Th\ O_2$	...	66.26	$Si\ O_2$	...	14.10
$Ce\ O_2$ & c.	....	7.18	$H_2\ O$	...	6.40
$Zr\ O_2$	...	2.23			———
$U\ O_2$	...	.46			99.89
$Fe_2\ O_3$	...	1.71			———
$Ca\ O$	...	0.35	S.gr.	...	4.98
$P_2\ O{-}5$	...	1.20			

The form of the crystals, however, shows that the mineral is not tetragonal; it appears to be orthorhombic or monoclinic. The oriented sections which I am having prepared will probably clear up this point. In any case I do not at present regard the mineral as thorite, although its chemeical composition is undoubtedly similar to that of thorite.

As has already been pointed out, the occurrence of thorium-bearing minerals (one said to contain a higher percentage of thoria than any mineral previously known) in Ceylon is of great scientific interest, and if they are present in sufficient amount will be of considerable commercial importance owing to the use of thoria for the manufacture of incandescent gas mantles. The 10 or 12 per cent, of uranium oxide alone gives a value of £20 or £30 a ton. The very small amount of radium present is of no practical importance.

II.-CORUNDUM, SILLIMANITE, & C.

Remarkable corundum-sillimanite rocks were found at Haldummulla (Uva). They occur all down the slope of Haldummulla estate, but not *in situ*, being derived from some point in the inaccessible hill above; the fallen blocks are found as far down as the western end of Kalupahane estate. The minerals sillimanite, corundum, orthoclase-microperthite, garnet, rutile, and ilmenite enter into the composition of these rocks. The principal types met with are; sillimanite-corumndum rock, sillimanite-garnet rock, and sillimanite rock; orthoclase-microperthite is present in subordinate and varying amount; ilmenite and rutile are accessory and do not occur together, but mutually replace each other; the corundum and garnet also do not occur together in the same rock. The corundum forms violet-coloured hexagonal crystals, usually less than $\frac{3}{4}$ inch in diameter; the crystals have often a tabular developement, the forms c(0001), a $(11\bar{2}0)$, and r $(10\bar{1}1)$ are characteristic; rhombohedral cleavage is well developed. The sillimanite occurs in parallel and radiating groups and single individuals, the latter generally diomorphic (prismatic, giving rectangular cross sections). The crystallization is much coarser than is usual for sillimanite; the largest crystals may reach a length of 2 inches or more and diameter one-tenth inch. the sillimanite has in the rock a pale gray colour, but is colourless in thin flakes. A more detailed account of these rocks will be elsewhere given.

Sillimanite has been met with in the garnetiferous leptynites in some abundance, over a large area between Bandarawela on the one hand and the Bambarabotuwa district (Sabaragamuwa) on the other; also in small amount in a garnetiferous rock forund at

Eraporuwa near Kolonna, Sab. The mineral is exceedingly rare in the Kandy District (where I have only quite recently discovered it; it is there found (1) in a curious rock from Dulmure, about 7 miles east of Kandy; this rock may be called a *garnet spinel leptynite*; and is remarkable for the minute graphic intergrowths of green spinel with felspar which characterize it; sillimanite also occurs, but very sparingly and is more conspicuous macroscopically than in a thin section; and (2) in a coarse garnet-leptynite, blocks of which occur by the roadside, but not *in situ*, towards the eastern end of Gregory's road (Upper Lake road, Kandy). In the Uva-Sabaragamuwa District referred to, however, sillimanite is a fairly common and characteristic mineral, though by no means invariably present. The sillimanite -bearing leptynites are characteristic and easily recognized rocks; the colourless, shining, perfectly fresh lath-shaped cleavagea surfaces of the sillimanite are very conspicuous in the slightly decomposed granulites. The sillimanite occurs in varying amount, but rarely if ever, however, to the total exclusion of felspar.

III.- PHLOGOPITE

Particulars of an almost colourless phlogopite mica from Ampitiya, near Kandy, are of sufficient interest to be recorded. The mica is found near a junction of crystalline limestone with granulite (a charactristic situation); the exposure is on the north side of the shallow valley between Kotawala hill and the main road about $3\frac{1}{2}$ miles from Kandy. The mica occurs in six-sided, but not very sharp-edged, tabular crystals, not exceeding 4 inches in diameter. A natural parting parallel to the rays of the percusion figure and to the edges of the crystal, is sparingly developed. The optic axial plane coincides with the leading ray of the percusion figure, being thus in the normal position for phlogopite. The rays of the percussion figure are inclined to each other at angles of not quite exactly 60*o*; the angle K between the two secondary rays (Holland, "Mica Deposits of India", 1902, p.18, fig.2) being from 60° to 62°, the other angles 60° to 59°. There are scattered, gray-coloured, hair-like and very thin lath-like inclusions, which are arranged in directions parallel and perpendicular to the rays of the precussion figure. The axial angle is small, so that the figure in thin flakes is apparently uniaxial and pieces of some thickness must be examined in order to determine the position of the axial plane. In thick pieces the mica has a greenish tinge, and is faintly pleochroic in shades of very pale brownish-green; thin flakes are quite colourless.

The following analysis was made at the Imperial Institute[69]

Al_2O_3	...	17.88	SiO_2	...	39.39
Fe_2O_3	...	0.21	H_2O	...	3.62
MgO	...	25.86	Moisture	...	0.84
Na_2O	...	1.09			
K_2O	...	9.90			98.79

[69]I am sorry that the names of the individual chemist or chemists by whom some of the analyses quoted were made have been withheld, and cannot therefore be given.

IV.-KYANITE

"On Sea-bottoms and Calcretes" J. Lomas, in Professor Herdman's "Report on the Pearl Oyster Fisheries", Roy. Soc., London, 1903.

Kyanite was found by mr. Lomas as a constituent of sands dredged by Professor Herdman off the coast of Ceylon in 1902; the mineral has not previously been recorded from Ceylon.

In Galle Bay were obtained the minerals, quartz, kyanite, corundum, rutile, tourmaline and mica; in Trincomalee Bay, quartz (magnetite), garnet, corundum, tourmaline, kyanite, mica; in Palk Bay, quartz, tourmaline, felspar, zircon, corundum, kyanite, mica, ilmenite; in various parts of the Gulf of mannar, quartz, ilmenite, magnetite, tourmaline, zircon, garnet, kyanite, rutile. These minerals have no doubt, as pointed out by Mr. Lomas, been brought down by rivers from the higher parts of Ceylon, and distributed by currents over the ocean bottom. The absence from these lists ofspinel and sillimanite is rather curious.

V -CHERT AND OPAL

I have recently shown (Geol. Mag., 1904, Dec.v., Vol.1,pp.16-19) that at any rate a part of the chert and opaline rocks which are locally, but in moderate abundance, distributed in the parts of Ceylon with which I am acquainted, are alteration products of crystalline limestones, the carbonates having been removed in solution and replaced by chalcedonic and opaline silica, so that we may find cherts containing the original accessory minerals (spinel, phlogopite, graphite, apatite) of the limestones, but showing no trace of the original carbonates. In other specimens relics of the partially removed carbonates are to be seen. This corresponds to what we know of many cherts that occur amongst sedimentary rocks in England, where it has been shown that the silica (whatever its source) has replaced the original carbonates, which appear to have been corroded and removed in solution. In Ceylon the process appears to have taken place long after the formation of the rock itself.

Mineralogical Notes. *63*

VI.-STEATITE (Talc.)

This mineral occurs in crystalline limestone, or rather dolomite, at Harakgama, Pata Hewaheta, Central Province, both in rounded and ovoid masses ledss than an inch in diameter, resembling amygdules, and also in small hexagonal prismatic crystals with good basal cleavage, pearly lustre, & c., but termianting irregularly (unlike the rather similar individuals of phlogopite mica in which the basal plane is always well developed). Appearances suggest a secondary origin for the steatite.

VII.-STILBITE

A rock specimen obtained from Nilhene graphite mine (near Baddegama) some 5 or 6 years ago was covered with small bright transparent crystals. These were examined by Mr. L.J. Spencer, M.A., and found to be stilbite, presenting the forms c(001), b(010), $f(\bar{1}01)$, and m(110), and twinned like Dana's figure 3. Stilbite has not previously been recoreded from Ceylon.

VIII.-SERENDIBITE[70]

This rare mineral, hitherto only known from the moonstone pits at Gangapitiya, Dumbara, Central Province, is found also in the pits at Attaragala near the 11th milepost on the Katugastota-Teldeniya road. The occurrence here is similar to that originally described; the pits are distant from those at Gangapitiya nearly four miles, along the same line of strike.

IX.-MOONSTONE

The following is an analysis of Ceylon moonstone (orthoclase felspar) from Gangapitiya, Dumbara, Central Province, made at the Imperial Institute. The material analysed was clear, colourless, and free from inclusions. The composition is that of an orthoclase rather rich in soda[71].

$Si\ O_2$	...	65.70	$H_2\ O$ (combined)	...	0.28
$Al_2\ O_2$	...	19.85	$H_2\ O$ (moisture)	...	0.10
$Fe_2\ O_3$	...	0.17			
$Na_2\ O$	...	5.20			99.34
$K_2\ O$	...	8.04			

X.-KAOLIN

A specimen from Alutwela, Teldeniya, Central Province, collected by Mr. James Parsons, was examined at the Imperial Institute with the following results. The material was of a yellowish pink colour; it contained small quantities of graphite and of ferruginous decomposition products. When mixed with water it furnished a paste which was only slightly plastic. It would only be suitable for the manufacture of common bricks.

[70] Mim. Mag., vol.XIII., No.61,1903

[71] It is noteworthy that the analyses of Ceylon orthoclase quoted in Hintze' mineralogy show no soda. The said analyses are, however, quite old.

Analysis

$Si\ O_2$	...	43.56	$H_2\ O$ (combined)	...	11.90
$Al_2\ O{-}2$	...	34.77	$H_2\ O$ (moisture)	...	5.63
$Fe_2\ O_3$	...	3.40			
$Na_2\ O$	...	0.36			99.90
$K_2\ O$	...	0.28			

XI.-SPHENE

This mineral is almost always present in rocks of the Galle group at Galle, and is then sometimes idiomorphic. It is common also in rounded grains in many limestone granulite contact rocks.

Some crystals were observed in a vein of pegmatite exposed in a small graphite pit near Talatu-oya (near Kandy, Central Province), those of Galle, but not containing wollastonite. The pegmatite consisted chiefly of quartz, orthoclase, and pyroxene very coarsely crystallized, with a considerable quantity of graphite in flakes and scales occupying cracks and spaces in the other minerals, and evidently deposited subsequent to their formation. One of the individuals of sphene was measured by mr. G.F. Herbet Smith and found to present the forms c(001), m(110), nd n(111).

XII.-MISPICKEL

Mispickel (arsenical pyrite) occurred in a quartz-felspar tourmaline rock sent in by Mr. W.A. Theobald from Little Valley, Deltota. The material is silvery white, of irregular from, and gives good reactions for arsenic. The greater part of the rock in which it occurs consists of an intergrowth of quartz with black toumaline-a type not infrequently met with; the presence of felspar (decomposed) is less usual. the arsenical pyrite has not been previously recorded for Ceylon.

On CEYLON ROCKS *and* GRAPHITE

A.K. Coomáraswámy, Esq., F.G.S. (Read June 6th, 1900.)

Q. J. of the Geo. Soc., London, **56** (1900) 590 – 615.

[PLATE XXXIII.]

CONTENTS.

I. THE PHYSICAL GEOGRAPHY OF THE AREA.

Ceylon is almost connected with India by the Manaar and Rameswaram Islands and the reef known as Adam's Bridge; there is no channel deep enough for large steamers to pass. Nearly fourfifths of its area is a flat 'low country' forming a maritime belt in the south, and covering a greater area in the north. The main mountain-massif, consisting of ancient crystalline rocks, is situated a little south of the centre of Ceylon in the South Central, Uva, and Sabaragamuwa provinces. In this area rise lofty and often precipitous mountains, of which the highest is Pidurutalagalla, 8296 feet above sea-level, while Adam's Peak reaches 7353 feet. In between these peaks stretch considerable tracts of table and or rolling grassy plains (Patanas), such as the Newera Eliya Plain, at an altitude of 6200 feet, or the Elk and Horton Plains, over 7000 feet. The rainfall in these districts may exceed 280 inches per annum, while it is only 50 inches in the northern flat, dry country. The rivers rise in the mountainous districts, running at first through rocky gorges, forming striking waterfalls, but meandering slowly through alluvial flats when they reach the low country.

Ceylon is essentially a continental island, and, as we should expect, shares to a great extent the flora and fauna of India. It is, however, remarkable that nearly 30 per cent of Ceylon phanerograms are endemic, and the same is true of 2 per cent of the genera

[13][5]. This fact indicates a not very recent separation from India, indeed it is evident that Ceylon has risen in quite late geological times. This is shown by the extensive areas of recent marine deposits which are found finging the coast, and in the north covering a larger area. Probably the same is true of Southern India. A depression of a few hundred feet affecting Ceylon and Southern India would submerge much of the low country (which

lies especially in the north), and would separate Ceylon from India more effectually than is now the case.

Thus endemic species which appeared while Ceylon was farther removed from India than at present, would not have been able to spread to the Peninsula as easily as would now be possible. The elevation may be comparatively recent, and is perhaps still going on.

Previous to 1887, remarks on the geology of Ceylon which have been made in books dealing with the island have been of a general character. The articles on the geology in the works by Sir Emerson Tennent [6] and 'An Officer of the late Ceylon Rifles' [9] are worthy of note. Since then, important contributions to our knowledge of Ceylon geology have been made by Sandberger [14], Lacrois [17], Walther [18], Melzi [29], and Diersche [30]. No geological survey is in progress in Ceylon, and it is much to be hoped that the Government will soon realize the importance of instituting one.

My thanks are due to Major-General F.T. Hobson, F.G.S., who kindly accompanied me to the Mahara Convict-quarry; to Mr. Edgar Ferdinand of Kurunegala, who showed me the Ragedara graphite-mines; also to Mr. Peter D'Abrew of Colombo, to Mr. Modder of Kurunegala, to Messrs. Casa Lebbe & B. Weerasiri of Kandy, to Mr. Kellow of Hakgala, to Mr. B. Seneviratne of Galle, to Mr. Hector van Cuylenberg of Colombo, to Messrs. A.A.C.W. Jayasekara & L.S. Amarasekara, and to Mr. Bell, Archaelogical Commissioner.

In England my thanks are especially due to Prof. Bonney, who has most kindly superintended the writing of this paper, which could not have been prepared without his assistance, and who allows me to say that the petrological determinations[6] made in it have his approval; to Mr. Teall, who has also very kindly given assistance and advice; to Mr. R. Graham, who kindly measured the crystals; and to Mr. S. Hastings, of Middlesex Hospital, who analysed a pyroxene.

[5]The numerals between brackets throughout this paper refer to the bibliographical list on p.612

[6][It may be mentioned here that details of pleochroism, extinction-angle, etc. are not given in the descriptions when they are quite normal, but that the determinations are based on the careful use of the usual methods, including examination in convergent polarized light. The figures given in parentheses after the word 'plagioclase' in many cases serve to indicate the highest measured extinction-angle from albite-lamellæ, in each instance.— July 2nd, 1900.]

II. THE RECENT DEPOSITS.

The greater part of the coast is fringed by coralline raised beach, and in the north a large area consists of bedded sea-sand and coralline limestone. In going by rail from Colombo to Galle, numerous diggings in the rubbly coral-limestone are seen; it is burnt for lime which is used in building or as a constituent of a chewing-paste. In some places considerable areas of pure sandy deposits are found, as in the Cinnamon Gardens, Colombo, where the soil is said to contain 98 per cent of silica.

Laterite (cabook), red or mottled red-and-yellow, occurs abundantly, often of considerable thickness, resting upon, and resulting from the decomposition of, the crystalline rocks. The less easily decomposable minerals, such as tourmaline and graphite, are sometimes found in it.

The rivers, as they approach the sea, flow through alluvial flats of their own deposition. Farther inland coarse gravels are found, or alternations of muddy and sandy or gravelly deposits. Thus the promontory, formed by the loop of the river flowing round the Peradeniya Gardens, consists of coarse crystalline gravel, which is found perhaps 50 feet above the present level of the Mahaweli Gorge. The Newera Eliya Plain consists of alternations of muddy and sandy or gravelly deposits.

With the exception of moonstone, all the gems for which Ceylon has long been so famous are found in drift-deposits, especially near Ratnapura and Rakwana in the Sabaragamuwa district. Here the gravels are very extensively dug, and gems of great value are discovered from time to time, while stones of less value are very abundant. The gem-stones include ruby, sapphire, white and yellow corundum and 'star-sapphires', chrysoberyl, cat's-eye, alexandrite, iolite, zircons of varied colour, tourmalines, blue and red spinels, garnet (precious and cinnamon-stone), and topaz.

Of the rough gems occurring in the gravels at Ratnapura and Rakwana, corundum is abundant and very often in good crystals which are quite large, reaching $4\frac{1}{2}$ cm. in diamter and 9 cm. in length. Hexagonal prisms and pyramids are the usual forms. Twinned crystals occur, but are rare. The smaller, or less-worn crystals have sometimes very brilliant unscratched faces and sharp edges. The following combinations were measured by Mr. Graham:-

$(11\bar{2}0)$ $(10\bar{1}1)$ $(22\bar{4}3)$ $(44\bar{8}3)$; (0001) $(01\bar{1}1)$ $(11\bar{2}1)$; (0001) $(22\bar{4}3)$ $(11\bar{2}0)$; (0001) $(10\bar{1}1)$ $(22\bar{4}1)$; (0001) $(10\bar{1}1)$ (4483).

Zircon is abundant, and usually of a characteristic brown colour, prisms and pyramids being the commonest form. Twins occur. A twin on (101) included the forms (110), (111), (101), (311). Other cyrstals showed various combinations of the forms (100), (110), (101), (111), (311), (331), (221) ; all these were measured by Mr. Graham.

Spinel occurs in rolled octahedra or irregular pieces.

Gold has been occasionally found in very small quantities in the sands of Ceylon rivers.

Apparently diamonds have never been found in Ceylon.

III. THE CRYSTALLINE ROCKS.

(1) Pyroxene-Granulites.)

This rock-type is widely distributed in Ceylon, occurring at Newera Eliya, Kandy, Ragedara, Dondra Head, Matara, Galle, and Colombo. The rock as seen in Ceylon is usually dark, and has a

characteristic greasy look. As a rule, foliation is not evident macroscopically, but it may appear in the thin slices. The minerals most frequently present include pyroxene (augite or hypersthene, or both), felspar (plagioclase, usually labradorite; more rarely orthoclase-microperthite), garnet, quartz, amphibole, magnetite, biotite, apatite, and zircon. Of these the pyroxene and felspar alone are essential constiuents; in the absence of quartz and garnet the rocks approach gabbro, while varieties consisting chiefly of pyroxene and garnet are near to eclogite. The structure is often granulitic, the minerals being much interlocked, and the quartz-grains elongated as in the normal granulites ; or it may be merely granular. Centric structures taking the form of pyroxene-felspar-intergrowths surrounding garents are very characteristic of some varieties. In the Ragedara granulite the felspar is orthoclase-microperthite, and the garnets are intimately associated with brown mica. The centric structures appear to have arisen in connexion with the corrosion of the garnets.

Among the specimens collected three types are recognizbale:—

(i) Plagioclase-bearing pyroxene-granulite with centric structures ;
(ii) Orthoclase-bearing pyroxene-granulite with centric structures ; and
(iii) Granular pyroxene-granulites without centric structures.

Type 1. Newera Eliya.— A specimen collected on the path leading up Pidurutalagalla, about one-third way from the summit, has proved interesting. The rock is dark green, and contains dark red garnets. The minerals are hornblende, pyroxene (chiefly green augite; a little hypersthene), plagioclase (27° 27°, probably labradorite), magnetite, apatite, and zircon. The garnets are not quite evenly distributed. In parts free from garnet, hornblende is more abundant than augite, and these two minerals, with plagioclase and quartz, compose a granular rock. In the neighbourhood of the garnets, augite is far more abundant than hornblende, and magnetite becomes conspicuous. Some parts consist almost entirely of garnet and augite, thus resembling eclogite. Throughout there is a limited amount of a pleochroic (pink to sea-green) pyroxene, probably hypersthene. Only when the garnets are in contact with plagioclase rather than augite are the remarkable coronas seen. These consist of radiating fingers of pyroxene (augite, hypersthene) proceeding from the garnets and standing out against a plagioclase background, the plagioclase showing between crossed nicols the usual twin-lamellation, which is not interferred with by the fingers of pyroxene. The larger magnetics of early consolidation are curiously related to the garnets. They are quite irregular in shape, and occupy bays and depressions in the garnets, but are separated from them by straits of plagioclase constant in width. The garnet appears to have been moulded on the magnetite, and subsequently separated from it by a narrow strip of felspar. From the almost ophitic way in which the plagioclase associated with the coronas and straits occurs, it appears to have been the last mineral to consolidate (Pl. XXXIII, figs, 3 & 4).

Ragedara.— In one specimen from Ragedara somewhat similar

appearances are seen. Some of the garnets are surrounded by irregular graphic intergrowths of augite, biotite, magnetite, and plagioclase. The larger grains of magnetite are either included in the larger garnets or in the augite. In the latter case they are almost always separated from the augite by a narrow rim of garnet.

Kandy.— A very dark pyroxene-granulite from the 'Mica Estate' at Talatuoya shows traces of a similar structure. In this rock the pyroxene, though pleochroic, appears to be monoclinic.

Fig.1:- *Corroded garnet (G), associated with brown mica (M) and surrounded by orthoclase-microperthite (F), in pyroxene-granulite; Ragedara (×57). Sep.595.*

[It is supposed that the three portions of garnet are remnants of a single individual.]

Type 2. Ragedara.— In two specimens of pyroxene-granulite associated with graphite-veins, the felspar is almost entirely ortho-clase and the pyroxene is rather scarce, although the rocks are dark. The minerals are quartz, orthoclase-microperthite, garnet, biotite (scarce), hornblende (replacing pyroxene), sphene, magnetite,

apatite, and zircon. The structure is granulitic. The garnets are transparent and fresh. They show no trace of crystalline form, but have a much corroded appearance; in many cases the grains are deeply hollowed out so as to be crescent-shaped, in others two or three small grains near together appear to be the remnants of a larger grain (see fig.1, p.594). The garnets are more often surrounded by the microperthite than in contact with the abundant quartz. With the corroded garnets and the microperthite is associated the brown mica, which lies in the bays and hollows that appear to have been eaten out of the garnets (Pl. XXXIII, fig.2).

Type 3. Dondra Head (the southernmost point of Ceylon).— Much crystalline rock in large boulder masses is found on the sore. In one specimen the minerals are orthoclase-microperthite, quartz, plagioclase, garnet, hypersthene, hornblende, magnetite, apatite, and zircon. The dark minerals are arranged somewhat in parallel layers, and the quartz-grains are elongated in the same direction. A precisely similar rock is found on the little island near the shore, south of the rest-house of Matara.

Galle.— A very similar rock occurs in the quarry behind Galle town, associated with some decayed and schistose rocks among which a dark biotite-schist is found.

Colombo.— Irregular exposures of dark rock overlain by laterite are found near the site of the new graving-dock. The minerals are quartz (elongated grains), hypersthene, garnet, magnetite, and spinel. The magnetite is fairly abundant, and usually contains irregular patches and strips of resinous-looking, dark green, isotropic spinel. Occasionally very small portions of this green mineral seem to react on polarized light: this phenomenon may be due to local strain, or to the presence of chlorite.

Kandy.— There is an extensive exposure of very various gneissic rocks in the Mahaweli Gorge at Hakinda, near Kandy. A single specimen (on which a species of *Podostemon* was growing) is a fine-grained grey pyroxene-granulite composed of quartz, plagioclase, augite, and some graphite. A distinct parallel arrangement is noticeable, especially marked by the graphite and pyroxene: the appearance is suggestive of flow-structure.

Newera Eliya.— A greenish rock collected between Newera Eliya and the Rambodde Pass is a garnet-free pyroxene-granulite. The minerals are plagioclase (20° 23°, probably labradorite), quartz (elongated), pyroxene, apatite, ilmenite, apatite, and greenish decomposition-products.

Other pyroxene-granulites are described by Melzi [29] from Pidurutalagalla, and Diersche [30] from the quarry between Newera Eliya and Hakgala. They are also recorded by Dr. Diersche from Ragedara, Adam's Peak, and the Dimbula Estate, and by Melzi from bandarawela (garnetiferous pyroxenite).

(2) Normal Granulites.

These and allied rocks are abundant in Ceylon. They are typically white or grey, and contain currant-red garnets; the texture is compact or saccharoidal, and traces of graphic structure are often recognizable. The quartz-grains do not contain lines of fluid-cavities, and the rocks seem igneous rather than metamorphic. The minerals include quartz (elongated grains), felspar (orthoclase- and microcline-microperthite and plagioclase), felspar garnet, with often also biotite, magnetite, ilmenite, apatite, and zircon. The microscopic structure is very granulitic; the minerals being much and confusedly interlocked, and the quartz-grains considerably elongated.

Kandy.— A characteristic white saccharoidal granulite with red garnets occurs in the quarry between the lake and the reservoir. It is largely used for road-metal.

Rocks collected at the far end of the reservoir contain conspicuous large garnets, sometimes in a white quartzose matrix, in other cases forming with biotite a biotite-schist. In the paler specimens the minerals are quartz, garnet, and plagioclase; the last -named contains curious tabular inclusions with their long axes nearly parallel to the twin lamellation : see Diersche [30] pl. vii, fig.3.

Ambakotte (near Kandy).— At the Gangapitiya moonstone-pits the coarsely-banded normal granulites are quarried by means of a short, sloping tunnel. The moonstone (orthoclase without microperthitic structure) occurs in rather large individuals in the granulite. Big pieces suitable for cutting are not very abundant, as the cleavage is well developed and the large individuals break into small fragments. The granulite, as seen under the microscope, consists of quartz (elongated and very irregular), and fine-grained, often micrographic; and very confused intergrowths of orthoclase-microperthite, plagioclase, and quartz. The elongated quartz-grains give to parts of the rock somewhat the appearance of a graphic granite. The appearances presented seem to suggest movement in a nearly consolidated magma, but might conceivably result from the recementing of a crushed quartz-felspar rock. The effect of movement in a viscous magma might be the production of the elongated quartz-grains, and the formation of confused intergrowths from the rest of the magma, the minerals not being able to grow in a well-individualized way after the remixing due to movement in the viscous mass. The orthoclase-grains are sometimes bordered by a peripheral, more transparent quartz-like zone extending irregularly into their interior, and the appearance is often very like that of Lacroix's figure [17] fig.49, p.300. Careful examination between crossed nicols, however, shows that the apparent quartz is here generally plagioclase, very faintly twinned, while in some cases it may be actually quartz. In the latter event the appearance is more conformable with the idea contained in the phrase 'quartz de corrosion' than is the case with the sharply-defined vermicular quartz often occurring in the felspars, where its presence is probably due to the contemporaneous intergrowth of the two minerals.

Galle.— A light-grey rock with scales of graphite, associated with the wollastonite-scapolite-bearing rocks, consists of microcline and some orthoclase-microperthite and quartz (elongated), and also very little sphene, augite, apatite, secondary calcite, and muscovite.

A vein, apparently intrusive in the same series, is very similar, but contains a little plagioclase and hornblende and has no graphite.

South of Galle Harbour a new path has been made from the shore to the plague-camp established on Buena Vista Hill, and near the top fresh specimens are obtainable from the blasted rock. The rock is greenish-grey, and shows a greasy lustre. The rather large felspars resemble those seen in laurvigite, but have no blue sheen. The minerals are orthoclase-microperthite and quartz, with also plagioclase and a very little pyroxene, ilmenite, zircon, and apatite. The structure is not granulitic.

Ambalangoda.— Immediately north of the little river is a small rock-exposure projecting from the sand, containing what was apparently a dark vein about 8 inches wide. A section of the 'junction of rock and vein' shows that the former is a quartz-felspar (chiefly plagioclase) rock with very scarce biotite, magnetite, apatite, and rhombic pyroxene. The quartz-grains are elongated. This granulitic rock appears uncrushed up to the 'junction', where however, there is every appearnce of shearing. The very minute biotites are arranged in a stringy way, so as to include lenticular, areas of quartz and felspar. As one proceeds farther into the 'vein' the shearing becomes less evident, though for nearly $\frac{1}{2}$ inch included in the slide it has a crushed appearnce, which the outer rock has not. The more abundant biotite chiefly distinguishes the dark 'vein' from the outer rock. It seems clear that an old line of shearing has been recemented. It seems clear that an old line of shearing has been recemented. It is noteworthy that this is the only Ceylon rock that I collected wherein clear evidence of mechanical deformation appeared.

Ragedara.— Normal garnetiferous granulites were noticed at the graphite-mines, where they were also found by Dr. Diersche.

Kalawewa.— Following the bund past the P.W.D. bungalow, about $1\frac{1}{2}$ miles from it the road ultimately leaves the great reservoir. A greyish-green rock collected at this point resembles that from Buena Vista, Galle, consisting essentially of orthoclase-microperthite (Pl.XXXIII, fig.5) and quartz, with some hornblende and very little plagioclase, augite (?), magnetite, and zircon. In this and the Buena-Vista rock the microperthitic structure is very clearly seen.

Anuradhapura.— A normal granulite without garnet occurs near the Newerawewa sluice, where the outflow runs under the road. A specimen collected on the Anuradhapura road, nearly 69 miles from Kandy, is a yellow rock with minute garnets : it is a normal granulite.

(3) Diorites.

These were noted at Ambalangoda and Kandy. They are dark rocks with a characteistic 'pepper-and-salt' appearance. The varieties include normal diorites, and quartz and quartz-hypersthene-diorites. The structure is granular ; where foliation is present the quartz-grains may be elongated. The hornblende is often partly derived from the pyroxene.

Ambalangoda.— The dark diorite here is apparently intrusive is gneissose rocks. It is well seen on the shore opposite the rest-house, where it is traversed by fine felspathic grains, and it occurs in a dyke-like manner in the quarry on the first little promontory south of the river, south of the rest-house. The minerals are horn-blende and plagioclase (26° 29°, probably labradorite) with a little biotite, apatite, and zircon. Brown-stained and greenish decomposition-products are occasionally associated wih undecayed fragments of a pyroxene, which they appear to represent. Quartz is absent: the rock is a normal diorite.

Kandy.— A dark-banded rock occurs in the quarry between the lake and the reservoir, associated with the more abundant white normal granulite. The foliation in the diorite is due to alternations of hornblendic with purely quartzose-felspathic bands. The minerals are plagioclase (20° 21°, probably labradorite), hornblende, quartz, scarce biotite, magnetite, apatite, and zircon, and also a little decayed pyroxene. The rock is a quartz-diorite.

A very dark diorite is seen in the quarry on the Pallekelle side of the ferry, below Kandy. The minerals are as in the last rock, but a considerable amount of hypersthene is present; the foliation also is similar. This rock is a quartz-hypersthene-diorite.

(4) Dolerites.

Ohiya.– A specimen from the railway-cutting near the first tunnel above Ohiya Station is a hornblende-biotite-dolerite, containing plagioclase (probably labradorite), hornblende, biotite, augite, hypersthene, magnetite, pyrite, apatite, and zircon. The alteration of the pale augite into dark hornblende, commencing along the cleavage-traces, is well seen.

Horton Plains.— A similar rock, collected about 1000 feet higher, near the Horton Plains rest-house, is a hornblende-hyper-sthene-dolerite, containing plagioclase (23° 24°), hornblende, augite, hypersthene, magnetite, and apatite.

(5) Gannro.

Colombo.— Near the Battenberg Battery a dark rock, weathering into large spheroidal masses, has been recently exposed. It consists essentially of plagioclase (35° 36°, probably labradorite) and pyroxene (sea-green augite and pleochroic hypersthene); hornblende and biotite are abundant; magnetite, apatite, and zircon are also

present. The hornblende is partly derived from the pyroxene. This rock is a hornblende-gabbro. (See Pl.XXXIII, fig.6).

(6) Quartz-Norite.

A beautiful example of quartz-norite occurs not very far from Kurunegala, on the right-hand side of the Ragedara road, forming a rock known as Muttatugala, which has been quarried for road-metal. The rock is coarse-grained, and greenish-grey to brown. The hypersthene, plagioclase, and quartz are easily recognizable macroscopically ; the only other minerals are magnetite and a grain or two of hornblende. The plagioclase (27° 29° probably labradorite) has a blue sheen. The hypersthene occurs in large patches which may exceed 2 inches in diameter, or it may be more regularly distributed ; the peripheral portions of the large patches may often be seen to enclose the felspar in an ophitic manner. Crystalline form is quite absent. In very thin sections the hypersthene is scarcely pleochroic, but in thicker cleavage-fragments a normal pleochroism is strongly marked. My friend, Mr. S. Hastings, has kindly analysed this pyroxene in duplicate, with the following results:—

	I. Per cent.	II. Per cent
SiO ...	50.97	49.93
Al_2 O_3 ...	1.30	.15
Fe O ...	27.21	29.61
Mg O ...	20.87	21.22
	100.35	100.91

Specific gravity = 3.55

The ferrous oxide was weighed as ferric oxide and calculated into ferrous; the alumina is probably overestimated in the first analysis.

(7) Serpentine.

A dark-green decayed rock, found on the spoil-heap at Nilhene graphite-mine, is formed chiefly of serpentinous material. Much of this is derived from enstatite, the characteristic cleavage and straight extinction being occasionally distinguishable in less decayed nuclei. Other portions appear to have been derived from olivine. A few crystals of colourless augite and sphene also are present. It is noteworthy that, out of numerous specimens collected at Nilhene, this alone appears to be of the nature of a massive rock and not of a vein-product.

(8) Limestones.

White crystalline limestones occur in several localities, such as Hakgala, Talatuoya near kandy, and between Matale and Dambulla, on the Anuradhapura road. The stairways as Sigiri are made of white limestone, often containing pale mica and prisms of blue apatite. At Anuradhapura it has been used in some cases for

600 *Mr. A.K. Coomáraswámy on* *[Aug. 1900,*

statues. 'Cipolins' from the neighbourhood of Kandy were recorded by Prof. Lacroix.

Hakgala.— The 'limestone' is found on the Albion Estate, where it has been quarried to a slight extent. The rock is white and crystalline, its texture varying from somewhat coarsely crystalline to saccharoidal. It contains abundance of sky-blue apatite, and much white or pale greenish mica; the latter occurs most frequently in the joints, forming six-sided crystals, and is found also in the rock-mass. Under the microscope the rock is seen to consist largely of colourless monoclinic pyroxene (perhaps sahlite) with also apatite, analcite (?), and a little calcite and muscovite: the whole being very granular, and the constituent minerals often confusedly intergrown.

The pale mica has the optical characters of a biotite. Part of a fresh-looking crystal was supplied to Mr. Shepherd, F.I.C., for analysis, with the following result:—

$Si\ O_2$ ...	40.72
$AL_2\ O_3$ ...	26.38
$Fe_2\ O_3$ ...	trace
$Mg\ O$ ...	14.27
$K_2\ O$ ...	10.36
F ...	.78
Loss on ignition ...	7.63
Organic matter ...	trace
	100.14
O_2 replaceable by F ...	.33
	99.81

Specific gravity = 2.57

Talatuoya (near Kandy):— Much limestone is exposed on the 'Mica Estate'. The rock is white, with occasional prisms of blue apatite, flakes of white mica, or flakes of graphite. The white carbonate is probably dolomite, showing under the microscope little twinning and bright interference-colours.

Limestone exposed on the road between Talatuoya and Kandy is very coarsely crystalline, and without the accessory minerals.

(9) Microcline-Gneiss.

This rock occurs in the country north of Matale,and with other gneisses forms the solitary hills which suggested to Prof. Walther the term domoid gneiss. The rocks are well-banded. Their constituent minerals include microcline (as a rule slightly microperthitic), orthoclase-microcline, quartz, plagioclase, and small quantities of pyroxene, hornblende, biotite, muscovite, pyrite, apatite, and zircon.

Dambulla.— A steep black rock rises not far from the village, with, near its summit, a series of Buddhist rock-temples. A

specimen was found to contain chiefly microcline and quartz, with also a little plagioclase, biotite, muscovite, and calcite.

Similar rocks with a little pyroxene and hornblende occur in the shallow road-metal quarry, by the roadside north of the village. Typical microcline-gneiss occurs near the fork in the road some miles north of Dambulla, and also near the new Sigiri rest-house, where the mineral -banding is very conspicuous.

Other exposures were noted on the Anuradhapura road at distances of $51\frac{1}{2}$, $52\frac{1}{4}$, and $72\frac{1}{4}$ miles from Kandy.

(10) Anorthosite-Gneiss.[7]

Certain gneisses found in Ceylon come under this head, inasmuch as the sole or dominant felspar is plagioclase.

Sigiri.— The imposing Sigiri rock is well known from an archaeological point of view, and has been rendered accessible by mr. Bell. It is a large rounded mass of gneiss, overhanging more or less on every side. The gneiss is conspicuously banded. Dark vertical surface-stains are evident, as is generally the case with the bare rocky slopes that occur in Ceylon. Thin slices show quartz and plagioclase, with more or less biotite, and a little secondary muscovite.

Colombo.— From quarries in the Madampitiya and College roads several specimens were obtained, and six microscope-sections prepared. The minerals are quartz, orthoclase, plagioclase, biotite, hypersthene, magnetite, apatite, and zircon. The term anorthosite-gneiss is applicable to those varieties which consist chiefly of quartz and

[7]The term gneiss is used for convenience, here and elsewhere, though it is perhaps hardly correct to speak of igneous rocks as gneisses.

plagioclase, with a little biotite and hypersthene. In specimens from the College-road quarry minute quantities of spinel are associated with the magnetite, as is the case in the rock from the site of the new graving-dock (p.595).

Kalawẹwa.— A specimen from near the big sluice shows in a microscope-section plagioclase (17° 17°, probably oligoclase), quartz, hornblende, augite, biotite, magnetite, apatite, and sphene.

(11) Granite.

The curious 'animal-shaped' hills in the Kurunegala district were described by Dr. Diersche as granite. He says also that the columns of the Brazen Palace at Anuradhapura are hewn granite.

At Mahara, a large convict-quarry in the Colombo District, a great depth of rock is exposed; it is overlain by laterite. The rock is best described as a rather variable gneissose granite. It is sent in large quantities to Colombo, where it is crushed and mixed with cement, and moulded into large blocks, which are used in the construction of the new breakwater.

602 Mr. A.K. Coomáraswámy on [Aug. 1900,

(12) Pegmatite

Veins of coarse pegmatite traversing other rocks seem rare in Ceylon. They were only seen in a quarry by the roadside between Wattegama and Pamvile.

Much of the rock exposed at Ambalangoda is pegmatitic in the true sense. Thus, in some cases, fairly large idiomorphic individuals of hornblende and orthoclase are embedded in a quartzose matrix. It appears that graphic granite is brought from Ambalangoda, but I did not find this rock, although traces of graphic structure are abundant.

(13) Rocks of Uunusual Composition.

Galle.— The rocks now to be described from Galle are peculiar in their petrological characters and mineral composition. They are specially marked by the presence of scapolite, wollastonite, and a green monoclinic pyroxene (spoken of as green augite, but probably in each case having a composition near that of manganhedenbergite—see analysis, p.604). The conspicuous mineral-banding and the tendency in the wollastonite to mould itself on the other minerals are noteworthy. The peculiar conditions giving rise to rocks of

Vol. 56.] Ceylon Rocks and Graphite. 603

this kind are far from evident; it is important to notice that they closely resemble rocks from Nilhene, which appear to be vein-products.

Fig.2.— *Section near the flaystaff, Galle Fort, showing mineral-banding in the scapolite-and wollastonite-bearing rocks.*

[Below the letter *A* in the figure a 2-foot rule is seen.]

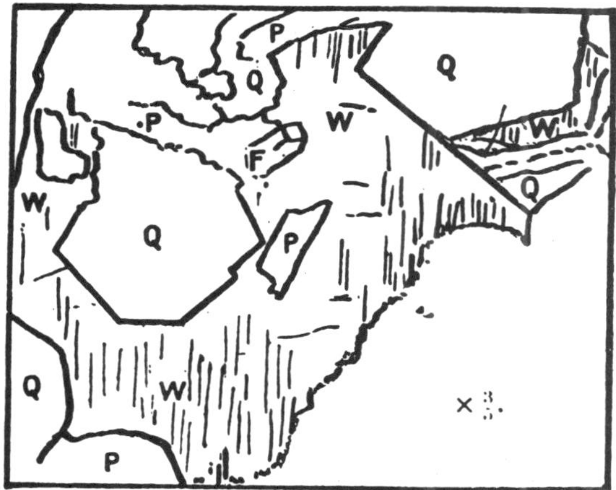

Fig.3:- *Ophitic plate of wollastonite (W), enclosing idiomorphic quartz (Q), pyroxene, (P), and felspar (F).*

A series of rock-exposures is found in the grassy area next the sea-wall of the fort, especially near the flagstaff. The rocks are of exceptional mineral composition, and mineral-banding is very conspicuous (see fig.2, p.602); the appearances resemble those which would result from movement in a non-homogeneous magma. A large specimen from the middle of the exposure seen in the figure, varies from a moderately coarse pegmatite to a finer-grained rock in which the separate minerals are scarcely recognizable macroscopically. The pegmatitic portion shows idiomorphic quartz, green pyroxene, and rarer orthoclase embedded in a large individual of wollastonite (fig.3). The cleavage-faces of the wollastonite have a pearly lustre and a striated surface. It breaks up readily in a fibrous way, and in some of the flakes the emergence of two optic axes may be seen in convergent polarized light, the open axial plane being perpendicular to the elongation of the flakes. The optical character is negative. From the large ophitic individual material for analysis was obtained and supplied to Mr. Shephered, F.I.C. The results of his duplicate analysis are as follows.—

	I	II
	Per cent.	Per cent.
Si O_2	51.28	54.59
Ca O^2	45.55	40.85
Al_2 O_3	2.01	1.77
Fe O	1.34	.70
Loss on ignition	not est.	2.29
	100.18	100.20
Specific gravity	2.76	2.78

These analyses correspond nearly to the formula $CaSiO_3$.

Two or three microscope-sections cut from different parts of this specimen are seen to contain green augite, orthoclase-microperthite, quartz, scapolite, sphene, and iron-ore in different proportions. The scapolite gives negative uniaxial interference-figures, and dipolarizes vividly even in thin sections. Grey fibrous decomposition-products are often present.

A banded, dark rock, collected near by, contains scapolite, augite, quartz, wollastonite, sphene, graphite, pyrite, and calcite. Scapolite is very abundant, and may be recognized by its rectangular cleavage, straight extinction, and negative interference-figure. It is water-clear, but contains some minute lath-shaped opaque inclusions and needles of a transparent mineral arranged with their long axes parallel to the vertical axis of the scapolite. The wollastonite occurs in fairly large individuals, which are evidently moulded on the other minerals. It is colourless, and has a moderately high refractive index, but rather weak double refraction. Its biaxial character can be ascertained: the open axial plane is at right angles to the parallel cleavage-cracks. The wollastonite has been sometimes replaced by a quartz-calcite mosaic which occurs in some small areas between the other minerals, often associated with a residual fragment of wollastonite. Two portions of wollastonite extinguishing simultaneously may be thus separated by a fine-grained quartz-calcite mosaic: in some cases no wollastonite is left, in others no trace of alteration is visible in it.

An even, but not fine-grained, green-and-white rock from the same locality contains the usual pale green 'augite', orthoclase-and-microcline(?)-microperthite, scapolite, sphene, zircon(?), apatite, and graphite. A considerable quantity of this rock was crushed and passed through a mesh with 30 holes to the square inch, but was retained by one with 60. The grains thus obtained were not composite. The pyroxene and sphene were separated by means of cadmium borotungstate, and the pyroxene was then picked out with a camel's-hair brush. More than 3 grammes of perfectly pure light green pyroxene being thus obtained was supplied to Mr. Shephered for analysis, with the following result:—

		Per cent.
$Si\ O_2$	....	50.91
$Ca\ O$	....	24.41
$Mg\ O$	....	.58
$Al_2\ O_3$	....	1.78
$Mn\ O$	....	2.64
$Fe\ O$	....	19.91
		100.23
Specific gravity =		3.377

This composition corresponds almost exactly with that of manganhedenbergite, to which species the pyroxene must be referred. Neglecting the alumina and a little of the silica, the formula $CaFe\ (SiO_3)_2$ is obtained.

Nilhene graphite-mine.— Specimens from this locality are very similar in petrological characters to those described from Galle. They differ chiefly from the Galle rocks in being associated with vein-graphite; in the presence of a micrographic intergrowth of quartz and calcite in some cases; and in the greater abundance of calcite.

A specimen associated with a vein of graphite (see p.610) may be described in detail. The minerals are quartz, calcite, augite, scapolite, and sphene. The augite is of the usual pale green, with extinction-angles ranging up to 44°: the smaller individuals join in a pegmatitic structure with the quartz and calcite; but with larger grains this is only true of their border. The intergrowth of quartz and calcite is such, that in a small area the quartz and calcite severally extinguish uniformly (Pl.XXXIII, fig.1). In other parts of the slide there is merely a quartz-calcite mossaic. The scapolite is recognized by its rectangular cleavage and negative uniaxial interference-figure, etc.

In other specimens the mineral associations are:—

(1) Quartz, calcite, and graphite (scattered);
(2) Augite, scapolite, orthoclase, wollastonite, graphite, sphene, iron-ore and calcite;
(3) Orthoclase, augite, sphene, magnetite, graphite, calcite, quartz ; the quartz and calcite forming a mossaic in small patches, probably replacing wollastonite;
(4) Orthoclase and scapolite;
(5) Quartz, orthoclase, and graphite.

In many of these cases a small quantity of an undetermined mineral is present. It occurs in sheaf-like aggregates or in radiating flakes, in connexion with the nests of stringy and dusty graphite. Cleavage parallel to the length of the flakes is well-marked, less common is one perpendicular thereto. Extinction is parallel to the principal cleavage. The mineral is colourless, and has a low refractive index, bright interference-colours, and strong double refraction. It does not appear to be muscovite.

These rocks, some of which as I said, are associated with vein-graphite, are very probably themselves rather of the nature of vein-products (see p.610).

Ambakotte (near Kandy).— At the Gangapitiya moonstone-pits occur, besides the normal granulite, some curious banded or nodular rocks. Some specimens show straight, or more or less concentric, zones of grey, dark green, and white rock. The grey inner part seems to consist almost entirely of grey granular augite, rarely showing cleavage. The succeeding green band (reaching about $\frac{1}{2}$ inch in width) contains augite, plagioclase, spinel, and an abundant pleochroic undetermined mineral, the two last-mentioned giving the green colour. The undetermined mineral is pale greenish-blue by transmitted ordinary light, and is strongly pleochroic from pale green to sky-blue : multiple twinning is conspicuous ; the appearances suggest a ferromagnesian mineral. The spinel is granular, pale green, and has a very rough surface. Dusky

606 *Mr. A.K. Coomáraswámy on* *[Aug. 1900,*

inclusions may occur in defnite planes. The grains are not attacked by drops of hydrofluoric acid placed on the slide. The light-coloured succeeding portion of the rock consists chiefly of quartz and plagioclase.[8]

(14) Quartz-Calcite Micropegmatite.

A graphic intergrowth of quartz and calcite has been described above (p.605 & Pl. XXXIII, fig.1). I believe that such an intergrowth has not been previously observed.

[8] I hope to be able to obtain more material, and isolate the undetermined mineral. This has not yet been possible, owing to its fine grain and the small amount of material available.

Graphic intergrowths, probably resulting from the simultaneous intergrowth of two or more minerals forming an cutectic mixture[9] are well known to occur in rocks which have consolidated from igneous fusion. It is hardly probable, however, that in the present case the quartz and calcite have consolidated from igneous fusion. It is easier to suppose that we are dealing with a vein-product, the result of a kind of solfataric action, rather than with a true magma-rock. Quartz and calcite are common vein-products, and often occur together in an ordinary way. It seems to me possible that conditions analogous to those found in consolidating igneous magma might be produced in connexion with the formation of vein-products. In this way a micrographic intergrowth of quartz and calcite might be produced.

(15) On Centric Structures in Pyroxene-Granulites.

Intergrowths of pyroxene and plagioclase from Pidurutalagalla, Ragedara, etc. have been described above (pp.593, 594 & Pl. XXXIII, figs. 3 & 4). Similar intergrowths, on a larger scale, of pyroxene, oligoclase, and quartz, and of pyroxene and oligoclase, have been recorded by Lacroix [17] from Salem and Ceylon. He described from Salem a micropegmatite of amphibole and orthoclase forming a corona 1 cm. thick, surrounding garnets of the size of a man's fist.

The radiating pyroxene-plagioclase intergrowths of the Pidurutalagalla rocks may be considered first. Here the 'straits' of felspar, have also to be explained. The following is a suggested theory for this case:— Magnetite and garnet were formed early, the garnet often enclosing the magnetite. Pyroxene, and then felspar followed; in parts of the rock not containing garnets they were able to crystallize without interfering one with the other. Some change in the physical conditions occurring at this stage caused the corrosion of the garnets, and perhaps partly also of the pyroxene, by the viscous felspathic magma. Thus channels were eaten out between the garnet and magnetite, and to some extent between the garnet

and pyroxene. The constitution of the viscous magma would be altered by the absorption of garnet-material, which would also tend to increase the viscosity in the neighbourhood of the garnets. So that in the area between the now corroded garnets there was present a pasty felspathic magma containing in solution the remnants of the pyroxene-elements, with some derived from the corroded garnets. As cooling proceeded some pyroxene was added to the original grains (the appearances tend to suggest this), and the remaining narrow strips of magma consolidated as an intergrowth of pyroxene and felspar, the felspar forming ophitic plates and the garnets serving as a point of attachment for the root-like fingers of pyroxene. Thus, in the end, we have slightly more pyroxene present than if the garnets had not been corroded, for portions of the magma which would

[9]Fr. Becke 'Die Gneiss-Formation des nieder-œsterreichischen Waldviertels' Tscherm. Min. u. Petr. Mitth. vol. iv (1882) p.406; Teall. 'British Petrography' 1888, p.401 ; Bonney, 'On a contact-structure in the Syenite of Bradgate Forest', Quart. Journ. Geol. Soc. vol. xlvii (1891) p.101.

have consolidated as pure felspar may, with the addition of magnesia, iron, and calcium which a pyrope would supply, have actually formed a pyroxene-felspar intergrowth. Intergrowths of plagioclase, pyroxene, biotite, and magnetic in connexion with garnets occurring in a Ragedara rock are somewhat similar; and indications of the same structure are found in a granulite from Talatuoya. These radiating coronas differ from those described by Prof. Lacroix in the root-like form of the pyroxene.

In the pyroxene-granulite of Ragedara, associated with the graphite (see p.594, text-fig. 1, & Pl.XXXIII, fig.2), somewhat similar phenomena seem to have occurred; but here the prevailing felspar is orthoclase, and accordingly the result of the corrosion of the garnets has been the formation of brown mica. Probably the garnets appeared early in the history of the rock, and served as nuclei round which the felspathic material tended to collect : subsequent change in the physical conditions caused the corrosion of the garnets, without sufficient movement taking place in the viscous magma to separate the product of corrosion from the garnets.

The occurrence of centric and granulite structures in pyroxene-granulites was noted by Dr. Becke; and it probably has some special significance in connexion with their origin which is not yet fully understood.

The contemporaneous formation, by the interaction of two minerals, of a third mineral or mineral variety is comparable with the fact that in any ordinary rock in which, for example, hornblende, biotite, and felspar are found, the biotite is often intimately associated with the hornblende. This probably results from a slight mixture of materials at the edge of the newly-formed hornblende-crystals, taking place as the magma consolidates[10].

(16) Microperthite.

The orthoclase of Ceylon rocks is scarcely ever without the fibrous microperthitic structure. In many cases it can be seen, even

608 *Mr. A.K. Coomáraswámy on* *[Aug. 1900,*

with a low power in a thin section, that this appearance is due to the presence of parallel rods and spindles (of plagioclase). If ordinary convergent light is used, and all but the central rays are cut off by a diphragm, the plagioclase stands out very clearly against the orthoclase by reason of its higher refractive index. Between crossed nicols the structure becomes still more evident. The plagioclase-strips are usually spindle-shaped and pointed at both ends, more rarely they are lamellar, or irregular. In hand-specimens of the rocks containing them, these felspars appear greenish-grey with a rather greasy look, resembling the felspars of the well-known rock from Laurvig, but without their blue sheen.

[10][See parsons, 'The development of Brown Mica from Augite by Reaction with Felspathic Material' Geol. Mag. 1900, p.316-July 15th, 1909.]

The structure is best seen in sections parallel with $M(010,\infty P\breve{\infty})$. The orthoclase-cleavage on $P(001, OP)$ is clearly marked, and crosses the les evident cleavage on $100(\infty P\bar{\infty})$ at about 110°. The plagioclase-spindles are seen to have their long axes parallel to the latter cleavage, that is, they run parallel to the vertical axis of the orthoclase[11]. A basal cleavage is observed also in the plagioclase, and it is parallel to, or continuous with, that of the orthoclase (see Pl. XXXIII, fig.5). The spindles may be of uniform size, or there may be many smaller with larger individuals among them. They may be few and scattered, or so numerous as to exced the orthoclase in amount.

The homogeneous-looking orthoclase background gives rather high extinction-angles (up to 8°) from the cleavage parallel with P (0 0 1, OP). The extinction-angle of the plagioclase-spindles is not so easy to determine. Cleavage-flakes or sections for measurement must be very thin and of uniform thickness, otherwise the higher refractive index of the plagioclase renders it difficult to be sure of the exact position of extinction. In a suitable flake from the Kalawewa rock (see p.597), the following data were determined :— The cleavage-flake is parallel with $M(010,\infty P\breve{\infty})$ and is bounded above and below by the basal cleavage, which makes an angle of 108° with the fine striation representing the orthopinacoidal cleavage of the orthoclase, to which the plagioclase-strips are parallel. The extinction-angle of the orthoclase from the trace of OP was 8°, that of the plagioclase from the same trace (that is, on $\infty P\breve{\infty}$) was 14°.

In a specimen from the Buena-Vista rock (p.597) the orthoclase cleavage-angle was 112°, the orthoclase extinction-angle from the trace of OP 7°, and the plagioclase extinction-angle from the same trace 14°. These results agree with those of Dr. Diersche [30] p.250, who says:

'extinction-angles of 13° to 17° in relation to the trace of OP were measured in the lamellae, showing their identity with an albite or labradorite;'

that is, the extinction on $M(010,\infty P\breve{\infty})$ is from 13° to 17°, leaving the ambiguity between albite and labradorite.

He says, however, that in basal sections where the orthoclase has

straight extinction with regard to the edge P/M (that is, the trace of the clinopinacoidal cleavage), the extinction o the lamellae does not differ from that of the orthoclase more than from 2° to 4°. That is, for the plagioclase-lamellae the extinction on P (0 0 1, OP) is from 2° to 4°. This fact, correlated with the extinction on $M(010,\infty P\breve{\infty})$ indicates that the lamellae probably belong to an 'oligoclase-albite' and not a labradorite. Accordingly the plagioclase-spindles are spoken of in this paper as albite-lamellae, though it is very possible that the felspar is not the same in every case.

Microcline, where it occurs, is found to contain similar albite-spindles, forming microcline-microperthite. The microcline is easily distinguished from orthoclase in the usual way.

[11]In the Ragedara pyroxene-granulite the intergrowth seems to be a little more complicated.

IV. THE GRAPHITE : ITS MODE OF OCCURRENCE AND ORIGIN.

Graphite is Ceylon's most important mineral product. Its export began before 1830, and now amounts to about 18,000 tons yearly. The Ceylon graphite is remarkably pure, and it is often found in large masses free from other minerals; specimens weighing more than 5 cwt. have been thus obtained. An excellent account of the Ceylon trade in graphite has been given by Mr. A.M. Ferguson [12]. F. von Sandberger [14] described the Ceylon graphite mineralogically; and it has been used among others by W. Luzi[12] in his chemical researches on the mineral.

While graphite is not unfrequently found in the igneous rocks of Ceylon in the form of small scales and strings, it is as yet only where it occurs as a vein-stone that it has been worked commercially.

Its singular mode of occurrence in branching-veins was first noted by Prof. Walther [18] on the eastern banks of the Kaluganga, about six hours' sail from Kalutara. Here, from the

'light red laterite exposed in the quarry to a depth of 12 metres a system of much-branched veins of black graphite stands out most conspicuously'.

The laterite-matrix was free from graphite, which occurred only in the veins. He supposes that fissures originated in the unweathered gneiss by dislocations, and were filled up by graphite :

'When at a later period the gneiss decomposed into laterite, the carbon-filled veins remained unaltered and occur in the laterite to-day a telling proof that this rock was formed in situ and has not been remanié'.

The occurrence of graphite at Ragedara as a vein-stone was noted by Prof. Zirkel and Dr. Diersche; I have also found it in the Baddegama district.

The vein-graphite shows coarse platy, or stem-or needle-like, forms arranged at right-angles to the edges of the veins. Sometimes other layers of graphite-elements follow, inclined at a slightly different angle, representing probably a second period of deposition. Graphit may occur also in radially arranged groups of crystal-like

forms in quartz-veins. It is curious that the flakes and scales of graphite often show on their surface a system of lines, perhaps representing cleavage-cracks, which intersect at angles approximating to 60°, 90°, and 120°.[13].

The graphite-veins contain numerous inclusions. Dr. Diersche mentions quartz, iron-ore, orthoclase, hornblende, mica, apatite, calcite, and especially rock-fragments, as occurring included in the graphite of Ragedara, where the matrix consists of normal and pyroxene-granulites. The rock-fragments included specimens of these rocks, also

[12]Berichte d. Deutsch. Chem. Geselisch. vol. xxiv (1891) p.4085 & vol.xxv (1892) pp.214, 1378.
[13]Diersche [30] pl.vii, fig.5 & p.277; Rauff [25].

a granite and a quartz-felspar-garnet-rock with muscovite, and very large zircons and apatites. None of these rocks seem very far removed from possible varieties of the matrix. Hence we may hope to gather some evidence concerning the origin of graphite as a vein-stone from the relations existing between the graphite-veins, the contiguous matrix, and the included fragments.

Two specimens from Ragedara show these relations well. The first, a greenish pyroxene-granulite, is surrounded on three sides by a graphite-vein varying in thickness from 2 to 20 mm. The stalky graphite is arranged compactly, perpendicular to the surface of the rock. A thin section has been made, including about $\frac{1}{4}$ inch of graphite and rather more than $\frac{1}{2}$ inch of the rock. Held up to the light, it shows a sharp division between the opaque graphite and the semitransparent rock. Under the microscope, it appears that graphite is absent from the greater part of the rock, but that as one approaches the junction between rock and vein, the former becomes thickly speckled with dusky graphite, and plates and scales of graphite appear inclined at various angles. The rock is perhaps rather less well preserved in this narrow zone next the vein itself, suggesting infiltration.

In another specimen from the same locality the relations between rock and graphite-vein are clearly seen on a larger scale (fig.4, p.611). The photograph represents a smooth surface taken by reflected light, so that the graphite appears white, the rock mottled grey. The vein is of flaky graphite, the veins being as usual perpendicular to the rock-surface; about $\frac{3}{4}$ inch of graphite and $2\frac{1}{2}$ inches of rock are seen. For about $\frac{1}{4}$ inch from the vein the rock is impregnated with scales and flakes of graphite, some of the flakes occupying tiny cracks in the rock. This graphite-bearing zone looks rather less well preserved than the rest of the rock, which is compact and hard and contains no graphite.

Some specimens from Nilhene graphite-mine are slightly different. Here the rock on or in which the graphite is found is probably a vein-product (see p.605). From two specimens (one of which has an external graphite-vein of the usual type, the other having two or three veins running through it) thin slices have been prepared. The rock-portion of the first specimen consists of quartz, calcite, augite, scapolite, and sphene. There is a sharp junction between it and the graphite, none of the latter being present in the

former. The vein itself is almost entirely made up of plates of graphite, which in the slice are generally perpendicular, with occasional narrow interspaces occupied by a zcolite.

In the other specimen, a graphite-vein $\frac{3}{8}$ inch wide is shown, and much graphite is scattered through the rock, so as to form a second, more irregular vein. A thin section shows much scaly graphite in a rock composed of calcite or quartz-calcite mosaic or micropegmatite. Where part of a vein is included in the section, the graphite-scales in it are parallel and closely set. In other parts of the slide the graphite is irregularly arranged. A zeolite is sometimes found between the scales in small quantity.

Fig.4:- *Junction of pyroxene-granulite [below] and graphite-vein [above]; Ragedara* $(\times\frac{6}{5})$

These Nilhene specimens differ from those found at Ragedara, in that the graphite-bearing rock is not an ordinary igneous rock as at that locality, but more probably a vein-product whose formation was contemporaneous with the deposition of the graphite.

The graphite mined at Mawatakelle and Deldahawatte in the same district, and at Godaduwa near Galle, I believe also occurs in veins.

We see, then, that the Ceylon graphite occurs chiefly in this way, and not in beds or lenticular streaks. When, however, it occurs as a rock-forming mineral (as, for example, in a specimen from Hakinda described on p.595), it is present in small specks and scales, scattered through an igneous rock in quantities too small to repay the cost of working. In these cases it may be supposed to have consolidated from igneous fusion, bearing in mind the well-known occurrence of graphite as a furnace-product, and in meteorites.

It seems impossible to regard the vein-graphite as the highly metamorphosed remains of an ancient flora or fauna. The high melting-point of graphite and other considerations make it unlikely that the veins arose as dyke-like intrusions of molten graphite. Prof. Moissan[14] suggests that graphite may have crystallized out of an iron-magma, the iron having been subsequently removed chemically, but here no traces have been found of the immense quantity of iron that would be required for this process. Moreover, the included rock-fragments and the portions of the matrix next the veins would have been considerably altered by the high temperature required. It seems likely that the formation of these graphite-veins has not been associated with very extreme temperature at the point of deposition of the graphite.

Dr. Diersche [30] p.287 suggests that the graphite may have been introduced in liquid hydrocarbons (such as petroleum, naphtha, or asphalt) which were reduced *in situ*. These liquids, rich in carbon, might have been introduced either from above or below.

Slightly different is the sublimation-theory of Prof. Walther [18], who refers to the occurrence of a graphite-like substance which is deposited in the flues of gas-works. He supposes, not that the carbon itself was sublimed, but that it was deposited in the veins from sublimed hydrocarbons.

Two hypotheses thus seem to be plausible, both involving the existence of suitable hydrocarbons; (i) the liquid theory of Diersche; and (ii) the gaseous theory of Walther. Each may be right in different cases, and neither presents any insuperable difficulty. The association of graphite, quartz and calcite at Nilhene is, I think, in favour of the liquid theory; so too, perhaps, the slight impregnation of included rocks noticed at Ragedara.

V. LITERATURE.

1. 1821. DAVY. JOHN. 'Travels in Ceylon' London, 4to.

2. 1847, GYGAX, RUDOLPH. 'On the Mineralogy of Ceylon' Journ. Roy. Asiat, Soc. Ceylon Branch, vol. i. No.3. pp. 1-5.

3. 1818, GYGAX, RUDOLPH. 'Report on the Geology & Mineralogy of the Sabaragamuwa District. [Parl. Reports.]

4. 1848, HENDERSON, Lient. 'On Ceylonite found near Trincomalee' Journ. Roy. Asiat. Soc. Ceylon Branch, vol. ii, No.4, pp.97-99.

5 1849, KELAART, E.F. 'Notes on the Geology of Ceylon; Laterite—Formation -Fluciatile Deposits of Nuwara Eliya', Journ. Roy. Asiat. Soc. Ceylon Branch, vol.ii, No.5, pp.210-19.

[14]Comptes-rendus Acad.Sci.Paris, vol.cxx (1895) p.17

6 1860. TENNENT, Sir JAMES EMERSON, 'Ceylon' 4th ed. London, 2vols, 8vo.

7 1860, RICHTHOFEN, Baron FERDINAND VON. 'Bemerkungen *über* Ceylon' Zeitschr.d. Deutsch. Geol. Gesellsch. vol.xii, pp.523-31.

8 1871, NEVILL, H. 'Notes on the Geological Origin of S.W. Ceylon together with its Relation to the Rest of the Island,' Journ. Roy. Asiat. Soc. Ceylon Branch, vol.v, No.16, pp.11-20 with map.

9 1876, 'Ceylon'. By an Officer of the late Ceylon Rifles. London.

10 1880, DIXON, A.C. 'The Rocks & Minerals of Ceylon' Journ. Roy. Asiat. Soc. Ceylon Branch, vol. vi, No.22, p.39.

11 1881, DIXON A.C. 'Gold in Ceylon' *ibid* vol.vii, No.23, pp.12-14.

12 1885. FERGUSON, A.M. 'Plumbago, with special Reference to the Position occupied by the Mineral in the Commerce of Ceylon, & the question discussed of the Alleged Existence in the Island of the Allied Substance Anthracite' *ibid.* vol. ix, No.31, pp.171-266.

13 1885, TRIMEN, H. 'Remarks on the Composition, Geographical Affinities, & Origin of the Ceylon-Flora' *ibid*, pp.139-59.

14 1887, SANDBERGER, F. VON, 'Beitrag zur Kenntniss des Graphits von Ceylon and senier Begleiter' Neues Jahrb. vol.i. pp.12-16.

15 1887. FERGUSON, A.M. & J. 'Ceylon Handbook & Directory'. Colombo, 8vo.

16 1888. FERGUSON, A.M. 'Gold, Gems & Pearls in Ceylon', London & Colombo.

17 1889. LACROIX, A. 'Contributions á lÉtude des Gneiss á Pyroxéne et des Roches á Wernérite,' Bull. Soc. fran c. Min. vol.xii, chap.vii, pp.282-350.

18 1889. WALTHER JOHANNES. 'Ueber Graphitg*ä*nge in zersetztem Gneiss (Laterit) von Ceylon,' Zeitschr. d. Deutsch. Geol. Gesellsch. vol.xli, pp.359-64.

19 1882, BUSZ, K. 'Mittheilungen aus dem Mineralogischen Museum der Universit*ä*t Bonn' Zeitschr.f*ü*r Kryst.vol.xv, p.616.

20 1890. MODDER, F.H. 'The Animal-shaped Rocks of Kurunegala' Journ. Roy. Asiat. Soc. Ceylon Branch, vol.xi, No.40, pp.377-425.

21 1891. FOOTE, R.B. Translation of [18] *pars*, Rec. Geol. Surv. India, vol.xxiv, pp.42-45.

22 1891. MALLET, F.R. Translation of [17], Rec. Geol. Surv. India, vol.xxiv, pp.155-200 & pl.iii.

23 1892, DICK, A. 'On Geikielite, a New Mineral from Ceylon,' Min. Mag. vol.x, pp.145-47.

24 1892, FLETCHER, L. 'On Baddeleyite (Native Zirconia), a New Mineral from Rakwana, Ceylon,' *ibid*, pp.148-60.

25 1893, RAUFF, HERMANN. 'Ueber angebliche Spongien aus dem Archaicum' Neues Jahrb. vol.ii, pp.57-67.

26 1893. PRIOR, G.T. 'Fergusonite from Ceylon' Min. Mag. vol.x, pp.234-38.

27 1896. ZIRKEL, A. 'Ceylon-Vortrag gehalten im Verein für Erdukunde zu Leipzig am 5ten Februar'.

28 1897. MODDER, F.H. 'A Geological & Mineralogical Sketch of the North-western Province, Ceylon' Journ. Roy. Asiat. Soc. Ceylon Branch, vol.xv, No.48, pp.39-79.

29 1897, MELZI, G. 'Sopra alcune Rocce dell Isola di Ceylan' Rendic. R. Ist Lomb. di Scienze e Lettere, ser.2, vol.xxx,pp.89-102.

30 1898, DIERSCHE, MAX. 'Beitrag zur Kenntniss der Gesteine and Graphitvorkommnisse Ceylons' Jahrb. d. k.-k. Reichsanst. vol.xlviii, pp.231-88 & pl.vii.

VI. SUMMARY.

Ceylon is surrounded by raised beaches, and has been elevated in recent geological times; fluviatile deposits also occur. The gems for which the island is famour are obtained from gravels in the Ratnapura district. With the exception of these quite recent deposits, the island probably consists entirely of ancient crystalline rocks, which may be considered Archaean in the absence of any evidence to the contrary.

614 *Mr. A.K. Coomáraswámy on* *[Aug. 1900,*

Pyroxene-granulites are widespread and characteristic; their mineral composition is rather various. Certain structures in connexion with garnets are very characteristic of some types; they appear to result from the corrosion of the garnets by the magma. Normal granulites are also abundant; they are typically white or grey, and contain currant-red garnets. The granulitic structure and much elongated quartz-grains are characteristic. Microcline-gneiss, sometimes with hornblende, occurs in steep dome-like hills, originating the term domoid gneiss employed by Prof. Walther. Anorthosite - gneiss, gneissic granite, and pegmatite are also found, the latter occurring in veins, but very rarely. Other rocks include dark diorites and hornblende-gabbro. A quartz-norite containing large individuals of hypersthene occurs near Kurunegala. The white crystalline limestones often contain pale mica and blue apatite; in some, colourless pyroxene is very abundant.

Banded scapolite and wollastonite-bearing rocks are found at Galle. Certain rocks similar to these, but apparently vein-products, are also described; these contain in some cases quartz and calcite micrographically intergrown.

Grahpite occurs chiefly in branching-veins in igneous rocks, which at Ragedara are granulites and pyroxene-granulites. The relations to the matrix are described, and are held to favour the idea of the deposition of the mineral as a sublimation-product. (Walther), or from the decomposition of liquid hydrocarbons (Diersche). Chemical analyses of several minerals, including manganhedenbergite, are given.

EXPLANATION OF PLATE XXXIII.

[Photographs taken by ordinary light, except fig.1.]

Fig.1. Quartz-calcite-micropegmatite from Nilhene graphite-mine. x33. Taken between crossed nicols. (See pp.605,606.)

2. Pyroxene-granulite from Ragedara. ×15. This shows garnet, corroded and surrounded by orthoclase-micro-petthite, and intimately associated with brown mica. (See p.594 & text-fig.1).

3. Pyroxene-granulite from Pidurutalagalla (Newera Eliya). ×15. This shows garnets with fringes of pyroxene-plagioclase intergrowth (as in fig.4) on one side; the garnets on their other side are in contact with the augite. A second generation of iron-ore is associated with the pyroxene-fingers. (See pp.593, 606.)

4. Pyroxene-granulite from Pidurutalagalla (Newera Eliya). ×15. The black portions are magnetite; they are separated by narrow strips of plagioclase from the garnet which has fringes of pyroxene-fingers projecting into the plagioclase; in the upper middle part of the figure a whorl of radiating pyroxene-fingers has for its centre a very small (residual) fragment of garnet. The remaining portions are chiefly pyroxene (augite). (See pp.593, 606.)

5. Orthoclase-microperthite from Kalawewa. ×24. The right-hand crystal shows the cleavage parallel to OP in both felspars. (See pp.597, 608).

6. Hornblende-gabbro from the Battenberg Battery, Colombo. ×13. (See p.598.)

Vol. 56.] *Ceylon Rocks and Graphite.* *615*

DISCUSSION.

Prof. BONNEY expressed his sense of the importance of this paper, on which he knew that the Author had expended great labour. It was interesting to find so great a mass of ancient crystalline rocks practically unaffected by pressure. The relations of the garnets, felspar, and pyroxene were most interesting and suggestive, and so, too, was the mode of occurrence of the graphite. He thought the paper also very valuable as illustrating the formation of a root-like graphic structure, which he thought implied crystallization under obstruction, while the ordinary rectilinear 'graphic' structure (to which the name was originally given) implied that one of the two minerals was not resisted by the other.

Dr. J.W. Evans referred to the resemblance between the rocks described in the paper and those of Southern India, of which Ceylon was geologically an integral part. Towards the south of the Indian peninsula the ancient sedimentary rocks disappeared, though the widely-extended granitoid gneiss still continued. The most remarkable fact, however, was the extraordinary development of the charnockite-series of Mr. Holland, which appeared to be identical with the present Author's pyroxene-granulites. These fomred the lofty mountain-masses of the Nilgiris and Annamallais, and were, it seemed, the chief feature of the most elevated regions of Ceylon. Graphite also occurred in Southern India, not only in dykes in Travancore, but in flakes in holocrystalline igneous rocks both in Mysore and in British India.

The PRESIDENT also spoke.

The AUTHOR stated, in reply to Dr. Evans, that small flakes of disseminated graphite occurred in several of the igneous rocks of Ceylon. He distinguished between the graphite occurring thus as a subordinate rock-forming mineral, and the vein-graphite which had been deposited after the consolidation of the rocks in which it was found. In conclusion he heartily thanked the Fellows for the very kind way in which they had received his paper.

MINERAL RESOURCES

by

A.K. COOMÁRASWÁMY, DIRECTOR, MINERAL SURVEY OF CEYLON

GRAPHITE

GRAPHITE (plumbago) is by far the most important of the minerals of Ceylon. It occurs in minute scattered crystals i some of the granulites and crystalline limestones of Ceylon; but it is only where it occurs in beds or veins crossing or running along the foliation planes of the crystalline rocks that its quantity and purity make it of commercial importance. Occurring thus, it is found in beds, veins, or nests, varying from the smallest size to a yard or more in width, and consisting of pure graphite often unmixed with other minerals. The latter however (including quartz, mica, felspar, pyronene, apatite, pyrite, & c.) are often found in association with the graphite veins.

The mode of origin of the veins of the graphite is not certainly known, though it is quite clear that they do not represent the metamorphosed remains of ancient vegetable deposits. The graphite has probably been deposited from some carboniferous liquid or gas, whose introduction was posterior to the consolidation of the rocks in which it occurs. The small flakes of graphite occurring in some of the granulites and crystalline limestones seem, however, to be original.

Graphite is principally mined in the Southern, Western, Sabaragamuwa, and North-Western Provinces, at distances from the sea ranging from 5 to 50 miles, although the other Provinces are worked to a less degree.

---

(2)

The greater part of the graphite comes from pits of mines, the largest of which are sunk to a depth of as much as 400 or 500 feet and in some of which steam pumps are employed to remove flood water or that arising from springs. There is usually a main shaft (see illustration opposite) with a windless for banding the graphite in a sort of tub, while the workmen ascend and descend by means of rough wooden ladders tied with jungle ropes, and rendered exceedingly slippery by the graphite dust and water. The graphite when brought to the surface is partially cleaned and freed from minerals or pieces of rock with which it is associated, and sent on to Colombo, where it is more carefully cleaned or "cured", and sorted into the various qualities which are placed in casks for export.

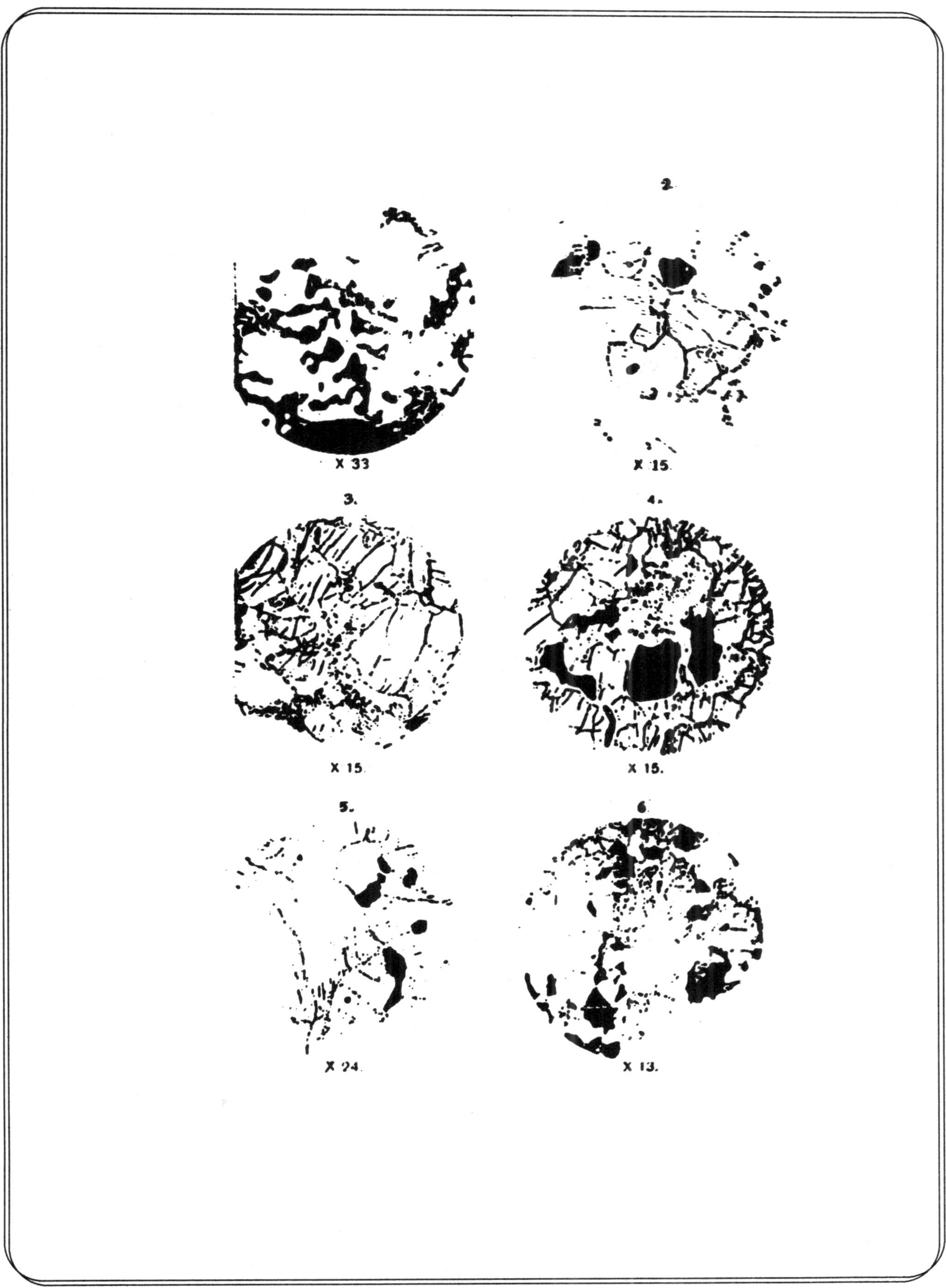
2.
X 33
X 15.
3.
4.
X 15.
X 15.
5.
6.
X 24.
X 13.

Beside the larger mines, which are more or less constantly in work and of which there are about twenty-five in which machinery is employed, there are innumerable small holes and shafts which have been opened by villagers and others who have discovered traces of graphite in the course of their agricultural operations. At the present time there are any number of abandoned pits of this sort; in many cases the quantity of graphite obtainable proved remunerative only while the inflated prices of a few years ago prevailed. At that time, when even a pound of graphite had a value, every small occurrence was sought out and worked. There is, however, no sign of any exhaustion of the supply of graphite which the Island can produce upon demand.

An export duty of Rs.5 per ton is levied on all graphite shipped. From very small beginnings, less than a thousand hundred weights fifty-eight years ago, the amount exported has increased, until in 1899 it amounted to 616,385 cwt, since when, owing to the lessened demand, the amount exported has decreased to 495,508 cwt. (valued at Rs.10,516,360 in 1902, yielding in the latter year a royalty to Government of Rs.125,946; the amount exported has, however, increased annually since 1900.

The price of the finest graphite varies at present from Rs.400 to Rs.550 per ton, and of the lower qualities in the form of chips and dust from Rs.100 to Rs.200 per ton. On the exported graphite, America takes the greater part, the United Kingdom, Germany, and Belgium coming next in order.

The digging for and local trade in graphite remain chiefly in the hands of the Sinhalese; it has, however, been mined on a number of tea estates and proved remunerative. It has been estimated that the various processes of mining lifting, carting, sorting, packing, and shipping graphite, as well as the making of the casks in which it is exported

---

(3)

---

provide employment for about 50,000 men, women, and children.

Its uses are manifold; about one-third of the export is used in the manufacture of crucibles, one-third for stove polish ("black lead"), and the remainder for lubricating purposes, electroplating, and in the manufacture of black paint and of lead pencils.

MICA

Mica- Sometimes erroneously called talc-is found in large quantities in the central districts, where it not infrequently shows on the surface and generally exists in the neighbourhood of the beds of crystalline limestone. Almost the only mica hitherto found in Ceylon in pieces large and clean enough to have commercial value belongs, as in Canada, to phlogopite; the greater part of the Indian mica, however bleongs to muscovite. The Ceylon mica is accordingly coloured, amber and bottle-green colours being most usual Colourless phlogopite is also found occasionally.

Mica is not mined for in a systematic manner, but is collected by hand from pits dug where there are surface signs of a vein; the veins are not as a rule more than one or two feet in width, and are composed of closely packed hexagonal crystals of mica of variable size, usually free from and admixture of other minerals. It is chiefly found in the Uva and Central Provinces. There is some local demand for mica in connection with the Kandyan art industries. It was largely used in the time of the Kandy kings for decorative purposes. Locally procured mica was also used in connection with the electric lighting of Kandy. The quantity exported has shown a steady decrease from an amount valued as Rs.43,637 in 1896 to an amount valued at only Rs.375 in 1902. There seems no reason why a more extensive trade should not be carried on.

Its splendid insulating properties, combined with its ability to withstand the highest degree of heat and to resist damp, and the fact that it is incompressible, tough, and non-combustible, and does not absorb oil or moisture, render it extremely valuable in the manufacture of electrical machinery. It is practically the only material now used for commutator bars, for which it is eminently suitable on account of its insulating properties, great durability, and perfect cleavage, in consequence of which it can be procured of any desired thinness.

It is used frequently to separate the iron discs that go to make up the iron core of an armature; the completed core is again very generally insulated with this material to keep the armature conductors electrically separated from the iron

---

(6)

---

stones, or asteria, which when cut en cubachan in a particular direction exhibit a six-rayed star of light. This effect is caused by the presence of minute inclusions arranged in planes corresponding to the stages in the growth of the hexagonal crystal, producing an appearnce known as "silk". In rare cases a twelve-rayed star has been observed. Many fine specimens show the star in ordinary light, but if diffused, the light decreases the brilliancy of the star by reflection; the star is best seen in sunlight or artificial light. Ordinary specimens are of little value, but a lime stone commands a high price. Star stones are almost peculiar to Ceylon.

The varieties of *chrysoberyl* are very interesting. The *cat's-eye* is highly valued, and fine specimens have realized large sums, but it is affected by the caprice of fashion, not commanding a general admiration, as do the sapphire and the ruby; the result is that in some years its price is increased by a demand which in others as suddenly falls. There are inferior kinds of stones resembling cat's-eyes, such as the quartz cat's-eye, and crocidolite which is now stained to resemble the chrysoberyl or true cat's-eye; but in no case do these compare with the real cat's-eye, which is said to be peculiar to Ceylon. Although found in several districts, the finest have been produced from the gem pits of Morawak korale.

In the same district, and it is said almost exclusively there, is found the beautiful gem called *alexandrite*. This mineral was formerly found only in the northern part of the Russian Empire and took its name from the Imperial family. The characteristic of this gem when really fine is its rich vivid green hue by day (much darker than the emerald, and slightly bronzed), which by artificial light is completely changed to a deep red. Like the cat's-eye, this gem occasionally commands a high price in the European markets, and it sometimes sought after by Americans, and Russians who are led to suppose that the stones are of Russian origin.

The stone known as *zircon* is classified under various names, according to slight variations of colour or the imagination of the dealer who introduces it to the market. Its usual colours are various shades of brownish and yellowish red, showing in fine specimens a tiery hue which the ancients were wont to credit with supernatural poweers. Man yother qualities it was supposed to possess; amongst others, the power of composing the werer to sleep and protecting him from unseen enemies. Another kind of zircon is almost colourless; it is a whitish crystal with a faint smokiness, and it is often spoken of as Matara diamond. It has of course no connection with the real diamond.

---

(7)

Spools are often of a bluish colour, or red or green, usually with a touch of smokiness. The red varieties are often known as false or Balas ruby. Really good specimens of the latter variety may be very beautiful and worth as much as Rs.100 or more, but most of the spinels obtained are very cheap.

The pale green *beryls* are found in large flawless crystals and sold under the name of *aquamarine*; it is only very occasionally that Ceylon beryls possess the true *emerald* colour.

To the mineralogist the gems are of most interest in them uncut state and in connection with their mode of occurrance in the rock. Unfortunately most of the interesting gems of Ceylon have not yet been found *in situ*, but only as more or less water-worn pebbles in the river gravels of the Balangoda, Rakwana, and Ratnapura districts. Several new minerals have been found in the heavy refuse from gem washings during the last fifteen years, and it is probable that others remain to be discovered.

Recent Marine Clays at Kuchaveli, Ceylon :

Aananda K. Coomaraswamy

Note on some Post Tertiary Mollusca from Ceylon :
R. Bullen Newton

Geol. Mag., Dec. v. vol. II., No.XI., November 1905.
Spolia Zeylanica, 1906

A recent marine clay is exposed near Nachchiarmalai on the Matti-aru (near Kuchaveli, 22 miles north of Trincomalee) about

$1\frac{1}{2}$ mile from the coast, at about sea level. The section shows about 9 feet of alluvium resting on the clay bed which is about $2\frac{1}{2}$ feet thick, the lower part being under water. The bed is full of marine shells, and contains also concretionary nodules with shells, serpulae, and crab remains. All these also occur commonly washed out and scattered over the sand banks of the Matti-aru. The common crab, used medicinally, is ***Macropthalmus Latreillen*** (Desmarest), the "medicine" crab of the Chinese pharmacopœia ; it has been found in the Post-tertiary clays of Southern China. A single chelate pincer belonged to *Scylla serrata* (De Haan), the great Indian swimming crab. The following shells together with annelid tubes, Balani, fossil wood, and the crabs complete the list of organisms collected:—

Telescopium telescopium (Linæus).
Potamides fluviatilis (Potiez and Michund).
Purpura carinifera (Lamarck).
Nassa ornata (Kiener).
Melongena pugilina (Bour).
Natria, sp.
Scapharca rhombea (Bour).
Arca, sp.
Placuna placenta (Linnaeus). This is the "widow-pane oyster," and it is exceedingly abundant in the bed.
Tapes undulata (Bour).
Tapes textrix (Chemintz).
Diplodonta, cf. oblonga (Hanley)
Tellina, sp.
Dosinia salebrosa (Roemer).
Chione allied imbricata (Sowerby).

Three points in Mr. Bullen Newton's paper must be noticed:—

(1) As stated in my paper, nodules occur in the clay bed, and one crab-nodule was found *in situ*; there is therefore no foundation for his suggestion that the fossils in the nodules may be younger than those in the clay.

(2) It does not appear how the identity of littoral fauna can prove the recent connection of Ceylon with India.

(3) The particular specimens examined were not presented as stated.

ANANDA K. COOMARASWAMY.

CEYLON.

REPORT ON THORIANITE AND THORITE

By Mr. ANANDA K. COOMARASWAMY, B.Sc.,
Director, Mineralogical Survey, Ceylon

The newly-discovered mineral thorianite, together with a mineral regarded as thorite, and a number of other heavy minerals, is found near Kondurugala, Bambarabotuwa, Province of Sabaragamuwa. The principal deposit occurs in and near the bed of the upper part of the Kuda Pandi-oya a small stream which at first occupying a small north-west and south-east strike-valley north of the Hopewell-Hapugastenna bridle path, turns through nearly a right angle and joins the Maha Pandi-oya a little below the same path. For the localities referred to the accompanying map should be consulted. In the bed of the Kuda Pandi-oya and in a small "deniya" (swamp) just below the swamp, the thorianite is to be obtained in a considerable abundance. It could not be discovered *in situ*, but it can hardly be doubted that it is derived from some rock outcropping not far distant from the highest part of the little stream; the stream is so small that after a few days of drought (not very usual in this west district) no running water is met with above the camp. The matrix is no doubt a rock of granitic type similar to those containing zircon, allanite, & c., which have been met with elsewhere in the balangoda district, intrusive in the charnockite series, and classed as belonging to the balangoda group.[11] Of these rocks the largest exposure (a narrow lenticular mass two miles or more in length) is that of zircon granite on Massena estate (six miles);[12] the rock consists of felspars, quartz, biotite, zircon, and ilmenite; smaller exposures of zircon granite are found on Herimitigala (eight miles) and Hopewell estates (fifteen miles); at Denegamabridge on the road between Balangoda and Belihul-oya, and near the 91st milepost on the same road. There are also similar granites without accesory minerals exposed, *e.g.* on the main road about a mile below Balangoda. The allanite-granite, or pegmatite, is best seen on Lower Denegama estate, where it occurs as a dyke three or four feet thick, forming a conspicuous ledge in the bed of the stream which runs through that part of the estate. The rock is composed of felspars (chiefly orthoclase), quartz, biotite, and allanite. The allanite forms thin tabuylar idiomorphic and larger irregular crystals, the biggest having a greatest diameter of three inches. Almost identical rocks occur (1) in the Weweldola on Dikmukulana estate (eleven miles) and (2) at Weligopola (nine miles). It is important to observe that a pegmatite rock, composed mainly of pink orthoclase, quartz, and biotite, with accessory apatite, tourmaline, ilmenite, & c., has been observed on Ambalawa estate, Gampola ("Spolia Zeylanica," vol. I., part IV., 1904), containing in addition to the above-named minerals a few cubic crystals of a black mineral at first regarded as uraninite, but which is almost certainly thorianite.

[11]See Contribution to the Geology of Ceylon. III., The Balangoda Group, by A.K. Coomaraswamy - *Geological Magazine, 1904, pp.418-422.*

[12]The distances are quoted from Balangoda

To return to the Kondurugala locality, the thorianite is probably derived from a granitic rock with no very large outcrop, but in which it occurs in considerable abundance; perhaps together with the zircon, thorite, and ilmenite, which quite possibly, however, occur separately in other rocks of a similar character. The outcrop of this rock could not be discovered owing to the dense jungle and thick soil and landslips; rock (mainly decomposed granulite) is indeed exposed at several points in the beds of the Kuda Pandi-oya and of the small streams joining it; but any search for the outcrop of a particular rock in the adjoining jungle could only be expected to succeed at the cost of a large expenditure of time and money, and might even then result in failure.

The thorianite occurs in the form of very heavy black cubic crystals more or less waterworn; associated with it is a large quantity of zircon in good crystals not much waterworn, and rolled pebbles of ilmenite; there is also a much smaller quantity of a brownish-yellow to orange-coloured mineral identified as thorite, and occurring in crystals usually very much worn. This mineral (at first identified as monazite) has a chemical composition and specific gravity corresponding to those of thorite; but the shape of the comparatively few crystals whose form is well preserved is quite unlike that of a tetragonal mineral, and I do not at present feel satisfied that the mineral is really thorite. All these minerals are also found in broken and rolled fragments of a subrecent ferruginous conglomerate which has formed in the bed of the stream and been subsequently broken up. This conglomerate contains rounded pebbles of rock, heavy minerals, & c., held together by a ferruginous cement. Not very much of this material is found, the greater part of the heavy mineral occurring in loose crystals.

Prospect of further discoveries:— The distribution of particular minerals in gem-bearing gravels in Sabaragamuwa is very local. As might be expected, it is generally the case that any one mineral is found in abundance in the bed and valley of a very small stream, and it often happens that the mineral is absent and another present in the bed of other small streams quite near. In the more extensive deposits of large rivers at a greater distance from the hills all the minerals whose sources are tapped by the various branches of the river may be found, but no particular one in any large quantity; the heaviest minerals moreover are not carried far and remain collected in deniyas (swamps) and potholes near their place of origin.

---

(2)

There is a large area, including at least the whole of Sabaragamauwa and parts of the Central, Western, and Southern Provinces, wherein this or other rare heavy minerals may be looked for. There is a possibility of their occurrence almost anywhere, but they are not likely to be detected except in the course of gemming operations; where gemming is carried on, therefore, it is worth while to make a careful inspection of the nambu, or heavy residue which remains at the bottom of the gemming basket, as this residue consists, if the washing has been thoroughly performed, almost entirely of the relatively heavy minerals ; it is indeed on this that the whole process of gemming is based, the lighter minerals being washed away and the heavier gems remaining. The rare heavy minerals such as thorianite, however, are mostly a good deal heavier than the gems themselves, and the nambu is of no value to the gemmer after it has been picked over for possible gems; anyone therefore anxious to prospect for thorianite should proceed as if gemming, but be careful to examine the heavy residues not only from a district as a whole, but from as many as possible of the small (even the smallest) streams in it.

It is unlikely that any very extensive deposit of any of these rare heavy minerals will be found, but they may be expected to occur at various points in moderate amounts. In the Kuda Pandi-oya valley a total of five tons might perhaps be obtained; from the Alupola-dola or the Kuda-oya I doubt if half a ton could be profitably extracted, but these estimates are quite uncertain. The Gampola occurrence is quite without commercial value. The total amount actually obtained in Bambarabotuwa so far does not exceed fifteen hundred weights. In the immediate neighbourhood of Kondurugala the Walaweduwa jungle seems the most favourable district for investigation. An examination of the Ratganga valley for the west showed that thorianite is absent there, and that even zircon is very scarce ; but (on the north-east) one or two heavy minerals not yet determined were actually obtained in the southern part of the Walaweduwa Crown forest reserve, and further work on that side of Kondurugala might be useful. It must be mentioned, however, that the district is a very wet one, the jungles swarming with leeches; it is also very inaccessible, provisions and camping effects having to be carried fully twenty miles from Balangoda; camping in these jungles is absolutely useless except in fine weather; and in the absence of a detailed map observations cannot be very accurately set down.

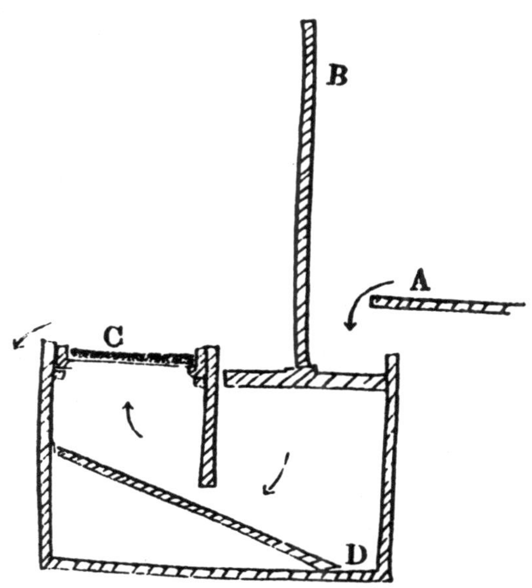

Fig.1:- Section of "jigger".

A - Spout with small stream of water
B - Piston working up and down
C - Layer of thorianite on wire gauze tray
D - Small stuff slipped through the meshes is collected here

An examination of some other districts during the present year has been much hindered by the wet season. Work at Kolonna, Madampe, and Pelmadulla has not led to the discovery of any rare heavy minerals, but they may quite possibly be found at Pelmadulla or elsewhere, as the most elaborate and detailed work extending over years would be required to *prove the absence* of such minerals in any district.

The physical properties of thorianite are briefly as follows:— Specific gravity 8.98 - 9.32 (Mr. Holland says he obtained one crystal of specific gravity approaching 10); hardness 5.5-6; colour black, streak greenish gray; in thin sections translucent, with strong red-brown colour; crystallizes in the cubic system; isotropic; usual form, a simple cube, but penetration twins (tw.pl.o.) (like those of fluorite) common; fracture uneven to conchoidal; brittle; lustre resinous to submetallic; insoluble in acids ; with borax or salt of phosphorus gives reactions for uranium; radio-activity about twice that of pitchblende. The chemical composition is given in Professor Dunstan's report.

A partial analysis made in M. Curie's laboratory in Paris gave 79 per cent. oxides of rare earths (principally thoria) and 14 per cent of uranium oxide. A rough preliminary analysis made by Professor Sir W. Ramsay is as follows:— Oxides like thorium oxide 76.4; uranium oxide 14.9; iron oxide 6.1; lead, arsenic, & c., 2.0; insoluble (not $Si\ O_2$) 0.7. More complete investigations are being carried on at the Imperial Institute and by Professor Sir W. Ramsay.

---

(3)

The specimen of thorite analyzed was one of two which were brought to Mr. Holland before he had discovered the source of the supposed uranite; they were stated to come from Bambarabotuwa, and appear to be similar to the material actually obtained from the Kuda Pandi-oya; of this actual material, however, no analysis is yet available.

How the thorianite is obtained:— The thorianite is so abundant in the bed of the Kuda Pandi-oya that several pounds of nearly pure material can be scooped out of some potholes after a shower. In less favourably situated potholes it accumulates in association with other sand and gravel, and this mixture must be scooped out and washed in the gemming basket. It also occurs in the gravel of a small deniya (swamp) below the camp, and can be extracted by gemming in the usual way. Pits can be sunk and the gem-bearing gravel washed. The whole of this district has been thoroughly worked by illicit gemmers. The only gems obtainable in the Kuda Pandi-oya are golden or brownish-yellow zircons, called by the gemmers *toramalli.* It is difficult to completely separate the thorianite from other minerals in the gemming basket, so that Mr. Holland has employed a jigger (fig.1) for the final separation. The jigger consists of a wooden box partially divided into two compartments, one carrying a wire tray, the other containing a wooden piston worked up and down by hand. A small trickle of water is required to replace the overlfow. The piston is worked up and down and a current of water is forced through the meshes of the wire tray which has been covered with a layer of thorianite mixed with zircon, & c. A thin layer of pure and large material should be laid first on the tray and the material to be cleaned laid over this; the small crystals are thus prevented from sinking through the meshes. After a short time the lighter minerals accumulate on the surface of the layer, and may be strained off with a spoon. Owing to its great weight the material has to be strongly packed in small boxes before it can be safely despatched to Europe.

History of the discovery".— The thorianite was first obtained by Mr. W.D. Holland, who has for many years been much interested in the geology of the Bambarabotuwa district. About five hundred weights was obtained before the place of origin could be discovered. This material was sold to Sir W. Ramsay, but samples were sent also to Sir W. Crookes, Professor Sir W. Ramsay, Professor Curie, and others. Samples were also forwarded by the Mineralogical Survey to the Imperial Institute and provided material for the analysis given in Professor Dunstan's report. The thorianite was at first regarded as uraninite. Meanwhile a very small quantity of the same mineral was found in a pegmatite vein on Ambalawa estate, Gampola, and described ("Spolia Zeylanica"), and of these one was analyzed at the Imperial Institute, the other is now in the Colombo Museum.

There is a piece of massive thorianite or uraninite in the Museum labelled from Kogwell and presented by Mr. Jacob de Mel. I came also across a single crystal in looking through some gem refuse in a Colombo jeweller's shop ; so that there is a fair chance of further discoveries, and it is almost surprising that the mineral has not been found before.

For three weeks in March I occupied a temporary camp beside the Kuda Pandi-oya. I soon found that the Kuda Pandi-oya lay outside the boundaries of what appeared to be Crown land; Mr. Holland had leased 100 acres of Crown land on the supposition that the Kuda Pandi-oya was enclosed in the area leased. I was able to discover two other localities for thorianite, however, one in Alupola within the land actually leased, the other below the path in the Kuda-oya bed. Not very much is obtainable here however. A royalty of 10 per cent has been claimed on the material obtained from the leased land, and a portion of the amount (55 pounds) so obtained forwarded to the Imperial Institute for further investigation. The remainder is in the hands of the Colonial Secretary. Of the supposed thorite, not more than 20 pounds altogether has been got. No work is being done at present; but a search for further deposits is being carried on by the Mineralogical Survey and by private individuals.

A.K. COOMARASWAMY.

Kandy, August 3, 1904.

Postscript:— Since August a small deposit of the mineral provisionally identified as thorite was discovered at Durayakanda, Gilimale South, about six miles from Ratnapura. The occurrence is not likely to prove of commercial value; it will be more fully described in my Tenth Interim Report and in my Administration Report for this year.

A further amount of 1,200 lb. of thorianite has been obtained from Bambarabotuwa since the above report was written.

Kandy, November 1, 1904. A.K.C.

(4)

Map of part of Bambarabotuwa, showing distribution of thorianite.

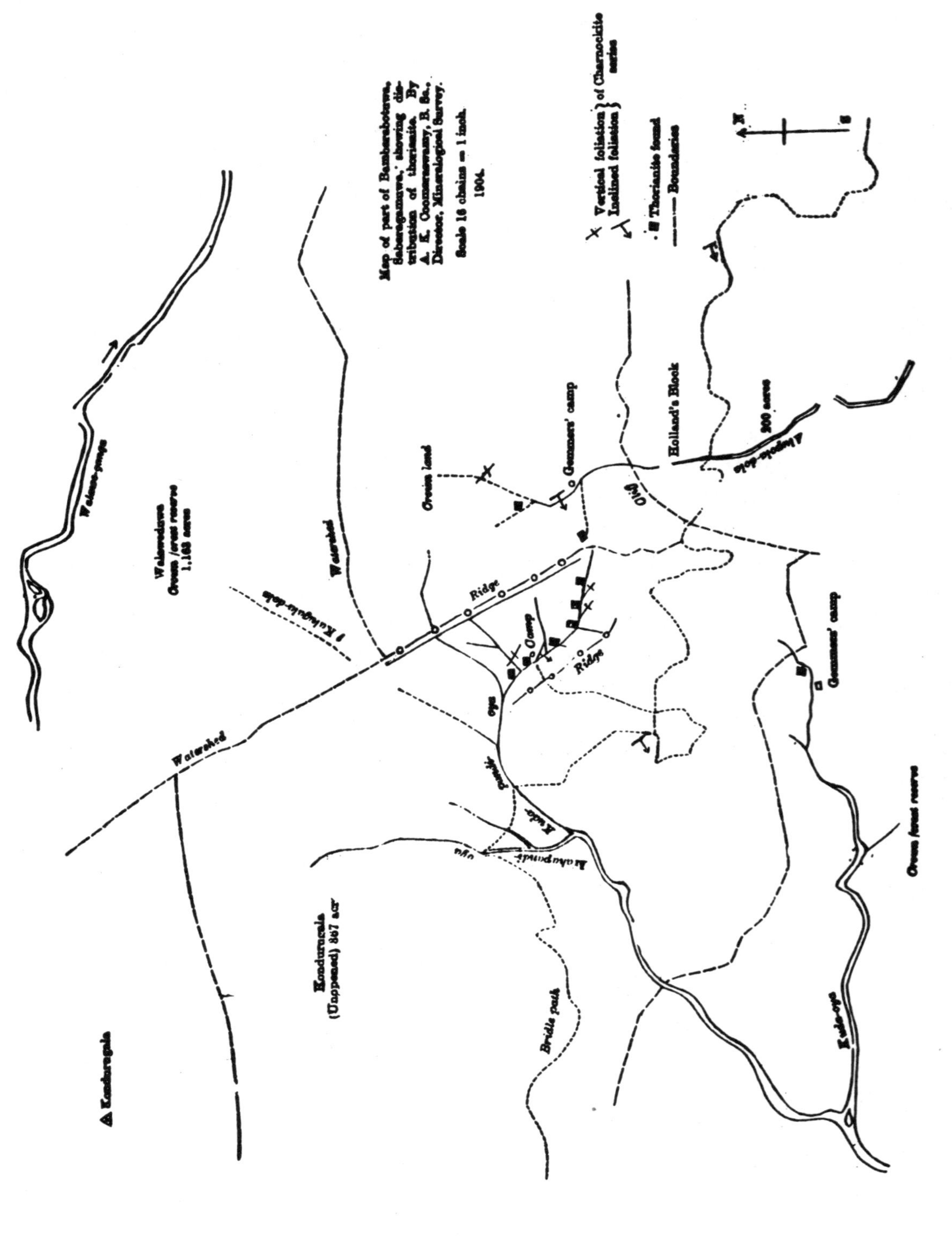

[*Extracted* from the *Geological Magazine*, Decade IV, Vol. IX, No. 457, p.305, July, 1902.]

OCCURRENCE of RÁDIOLARIA in GONDWÁNA BEDS near MADRAS

By A.K. COOMÁRASWÁMY, B.Sc., F.L.S., F.G.S.

[PLATE XVII, Figs.1-6]

THE Upper Gondwána (Rajmahál Series) beds near Madras are divided into two groups, the lower of which has been named from Sripermatúr, a twon 25 miles west-south-west of Madras, and a well-known locality for fossil plants. The group is composed of white and pale buff-coloured shales containing plants, at least 10 feet in thickness, resting upon sandstones, grits, and micaceous sandy shales, at least 15 feet in thickness, whose base is not seen, but which probably rest on the gneiss. Silicified wood occurs in some of the gritty beds. The other fossils occur in the shales, and consist of plant and animal remains. The latter are poor, and comprise two or three indeterminable species of ammonites and several lamellibranch bivalves.

Last year, while staying at Madras I paid a short visit to Sripermatúr, accompanied by Mr. Ranga Chari, M.A., of the Madras University. I did not obtain many plant-remains, but was struck by the very porcellanic character of the shales containing them. These are thin-bedded, white to cream or pale coffee-coloured, and often have a smooth conchoidal fracture. A few specimens were collected from a wayside exposure, on the track between Sripermatúr and Vellum. In the thin slices cut from these specimens Radiolaria are to be seen, though for the most part in a very poor state of preservation. These organisms, together with tiny fragments of clastic quartz, and rarely a few carbonaceous specks, are embedded in a fine siliceous paste which remains dark between crossed nicols. Quite locally chalcedonic patches show a fibrous structure. The fragments of elastic quartz are rarely as large as the Radiolaria,

and graduate downwards to the finest dust. Very minute bedding is clearly seen in the slides. The finer material seems to contain, or to be aggregated into, small globular masses, some of which are rather conspicuous, but very possibly not of organic origin.

Dr. Hinde has very kindly examined the slides, and says "there can be no doubt that the organisms are radiolaria'. "Probably several forms are present, most of them rounded or lens-shaped discs with minute reticulate structure and without radial spines; others are oval with spines, and one form is like a sugar-loaf in section". In some cases there is a body with reticulate strucutre within a clear circular area; this may indicate the presence of radiolarian forms with the inner test better preserved than the outer.

There are also in the slides some groups of small rounded bodies, of inconstant dimensions, which, as stated above, are probably not of organic origin. Dr. Hinde thinks that the disc or lens-shaped forms with reticulate structure (Figs. 5,6), which are by far the most abundant forms present, belong to the genus *Spongodiscus* in the Discoidea section. The conical perforate form (Fig.4), which are by far the most probably belongs to the genus *Dictyomitra* in the Cyrtoidea division. The spined forms (Figs.1-3) are too faintly showsn for determination, but may belong to the Discoidea section.

The association of these radiolarian forms with plant-remains (the shales are known as plant-beds, though the plant-remains are not very abundant; I found no mollusca myself) is of great interest, as the shales must have been deposited in *comparatively* shallow water, a state of things also indicated by the detrital character of the other beds included in the Sripermatúr group. The shales cannot, however, have been formed very near the shore. Mr. Foote says that they "show every evidence of having been deposited in perfectly tranquil water at sufficient depths probably to be beyond the agitation of the waves. The fragmentary nature of the fronds (Mem. Geol. Surv. India, vol.x, p.64, 1873). The most porcellanic shales seem to contain the best preserved plants.

The Radiolaria must have been present in sufficient quantity to give a very siliceous character to the rock, yet the use of the term 'radiolarian chert' would hardly be justified. I might add that Radiolaria are to be found at the present day in shallow water, though not in such abundance as in the open ocean.

Finally, I desire to thank Dr. Hinde for his very kind assistance and advice.

REFERENCES

Mem. Geol. Surv. India, vol. x, p.64, 1873.
Paleont. Indica, ser. IX, p.236, 1875.
Geology of India, 2nd ed., pp.182, 208, 1893.

EXPLANATION OF PLATE XVII.

RADIOLARIA FROM UPPER GONDWÁNA BEDS, SRIPERMATÚR.

Figs.1-3: Spined forms, possibly belonging to the section Discoidea; outlines only distinguishable; ×125.

Fig.4: Conical reticulate form probably referable to the genus *Dictyomitra*, ×420.

Fig.5, 6- Lenticular reticulate from perhaps belonging to the genus *Spongodiscus*, ×200.

PLATE XVII, Figs.1-6.

Radiolaria from Upper Gondwána Beds, Sripermatúr.

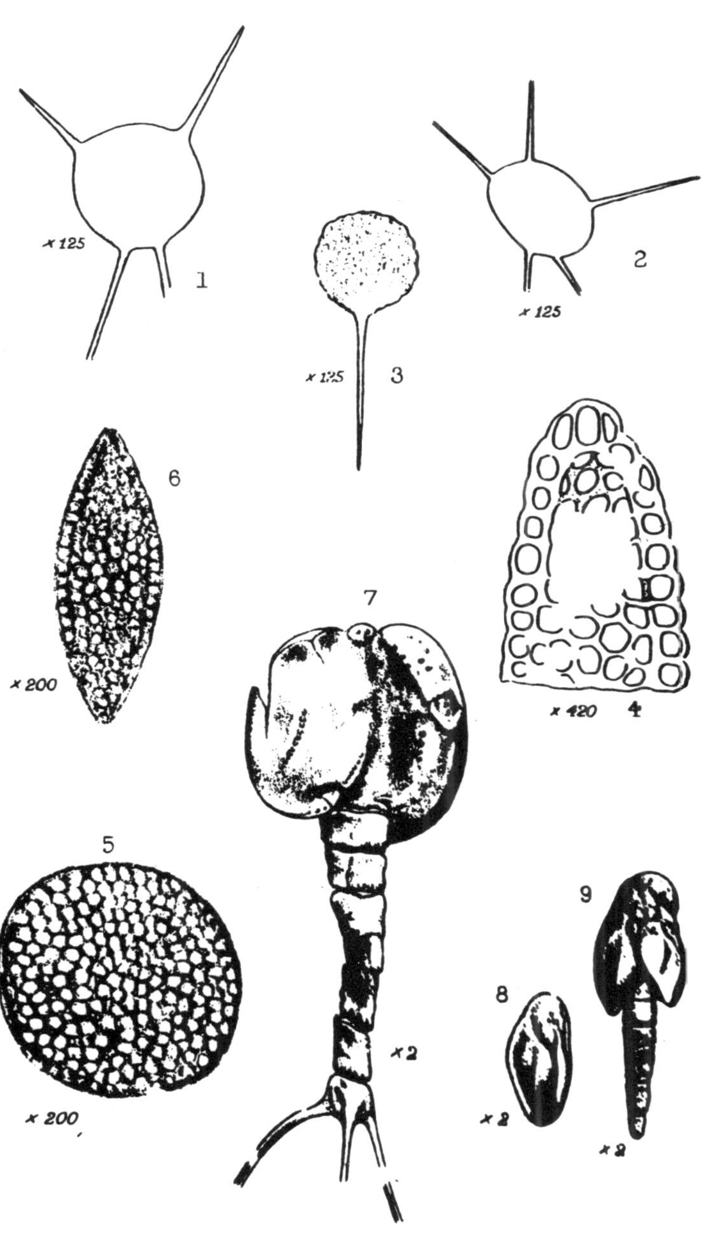

OBSERVATIONS on the TIREE MARBLE, with NOTES on others from IONA.

By ANANDA K. COOMÁRASWÁMY, Esq., B.Sc., F.L.S., F.G.S.

(Read December 17th, 1902)

Q.J. of the Geo. Soc., London,**59** (1903) 91 – 104.

[PLATES VI & VII]

CONTENTS

I. INTRODUCTION.

ALTHOUGH so well known, and represented in many mineralogical collections, the Tiree Marble has never been described in detail. Macculloch[3] says that the Tiree flesh-coloured marble

'resembles the greater number of the primary limestones found in gneiss and mica-slate, and may be considered as a large nodule... The nodule of limestone... appears to be an irregular mass of about 100 feet in diameter, and is surrounded on all sides by gneiss'. [The marble] 'occasionally contains imbedded lumps of granite or gneiss, similar to those which occur in the limestone of Glen Tilt. These are always visible at the surface, from their superior power of resisting the action of the atmosphere.'

He also describes the exposures of limestone in a field south-east of Balephetrish Hill. He says that

'it is possible that the masses of limestone thus found in gneiss have once been stratified, and that they have suffered some posterior changes by which the apperances of this disposition have been obliterated'.

[3] 'Western Islands of Scotland' vol.i (1819) pp.48 & 49.

Prof. Bonney[4] has given a valuable description of the dynamic phenomena so well illustrated in thin sections of the Tiree Marble.

Prof. Cole & Prof. Sollas[5] may originally have been a wind-blown coral-sand rock, with abundant rounded crystals of detrital augite, etc., such as was found on the east side of Mér, one of the Murry Islands. This theory appears untenable; for it is impossible to imagine a volcanic origin for the majority of the accessory minerals, which, with the exception of felspar, are, on the contrary, the typical accessory and contact-minerals associated with crystalline limestones in all parts of the world.

This paper embodies observations made by me in the summer of the present year (1902).

II. GENERAL DESCRIPTION.

The exposures of marble and gneiss near Balephetrish are shown in the accompanying maps (figs. 1 & 2). Unfortunately, the relations are much obscured by more recent deposits. It is clear, however, that the gneiss has a general south-westerly and north-easterly trend and that the masses of limestone occur as lenticles of various size in the gneiss, exhibiting a similar foliation. The latter in the gneiss is more or less interrupted by the large limestone-augen round which it sweeps, and the result is great local irregularity in its direction; the foliation of gneiss and limestone is, however, always parallel.

On the shore west of Balephetrish the two rocks are most intimately associated, some quite small inclusions and narrow streaks of marble, besides larger masses, occurring in the gneiss.

The marble is by no means of uniform character; it will be simplest to give brief descriptions of the main types notice.

(1) Pink marble of Balephetrish Quarry:— This is the well-known Tiree Marble, so common in collections,. The rock consists of minutely granular calcite, with some subrodinate dolomite; embedded in this compact carbonate-matrix are abundant crystals of dark-green hornblende (often idiomorphic), and coccolite, with scarcer greenish-brown mica, and minute brilliant grains of pale hair-brown sphene. In other specimens coccolite is the predominant or only silicate. Minute crystals of scapolite were also found (in residues) by the later Prof. Heddle.

[4]'The Effects of Pressure on Crystalline Limestones' Geol. Mag. 1889, p.483

[5]'The Origin of certain Marbles: a Suggestion'. Sci. Proc. Roy. Dublin Soc. vol. vii (1891) p.124; see also Haddon, Sollas, & Cole, 'On the Geology of Torres Straits,' Trans. Roy. Irish Acad. vol xxx (1892-96) pp. 436,470.

Fig.1:- *Geological Map of a portion of TIREE*
by A.K. Coomáraswámy, B.Sc.

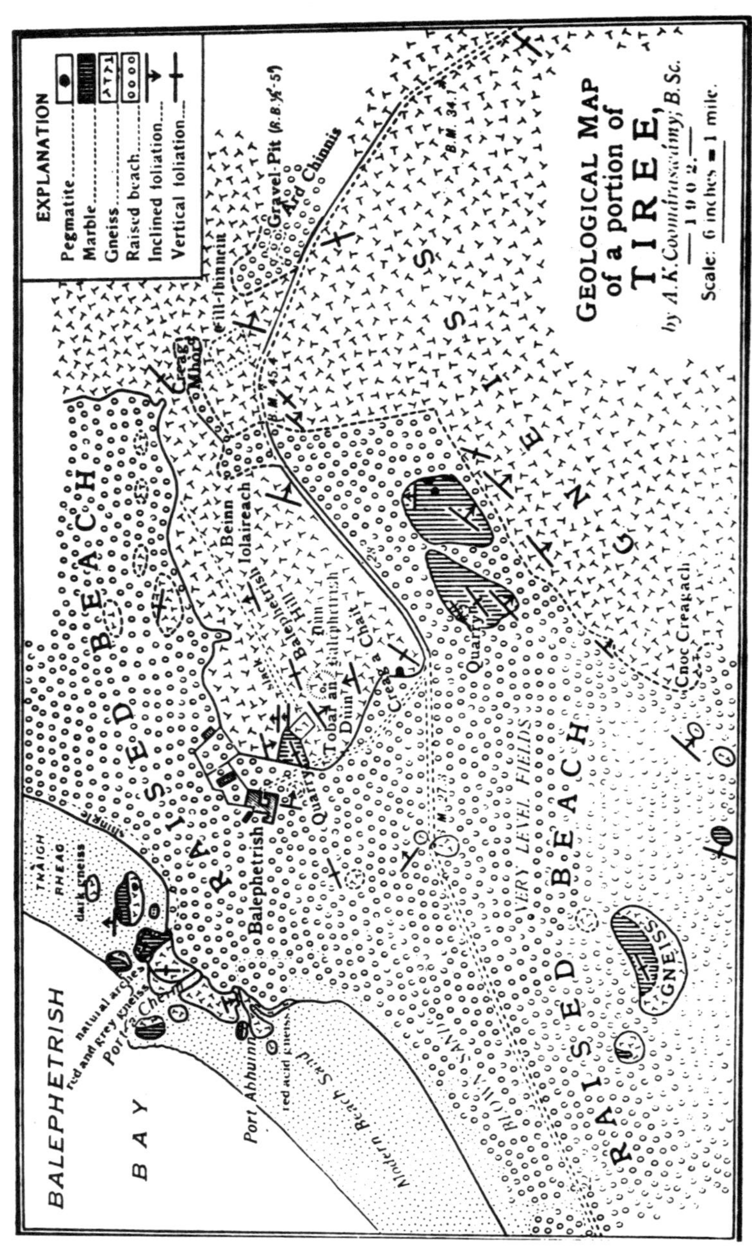

Fig.2:- *Enlargement of a portion of fig.1*

Fig.3:- *Limestone, with inconspicuous foliation, and abundant forsterite weathered out on the surface. Fields south-east of Balephetrish Hill. (About $\frac{2}{3}$ nat. size.)*

[Sahlite, tremolite, mica, and spinel also occur very sparingly, on this specimen.]

(2) Pink marble from Port Abhuinn (see map, fig.2 p.93):— This rock much resembles the marble from Balephetrish Quarry. It consists of calcite, generally very fine-grained, the result of shearing and strain, the effects of which are well displayed; a few larger individuals are stretched out, bent, and twinned (Pl.VII, fig.2); occasional grains have been preserved uncrushed where protected by silicafes, in the manner described by Prof. Bonney. Coccolite, green in hand-specimens, very pale in sections, is abundant; rounded grains of orthoclase are common; a decomposed mineral, which is probably scapolite, occurs; and there are tiny grains of sphene sparsely scattered through the rock. The same rock, thoroughly mylonized, and containing blocks of gneiss in the crush-breccia (fig.5, p.99), is of a grey colour, and has a smooth, flinty, conchoidal fracture.

(3) Grey marble from south side of field, south-east of Balephetrish Hill (fig.3):— Here the carbonate-matrix is less broken down than in the other types described; the cleavage-faces of carbonate-crystals can be clearly seen on a freshly-broken surface. Microscopic examination shows that much of the carbonate-matrix is, however, as usual, very fine-grained and compact; the larger grains are

usually dolomite, the finer material stains like calcite. Other minerals include:— (a) Very abundant forsterite; the crystals weather-out as brownish lumps on the surface of the limestone; under the microscope complete or partial serpentinization is characteristically displayed. (b) Scarce colourless mica, tremolite, and sahlite. (c) A few scattered grains of greenish-blue spinel; this mineral was not observed at any other point, and even here occurs in very limited amount.

(4) A white marble from the small quarry in the same field (fig.4):— This rock is generally similar to the last-described, but is finer-grained and more compact, and conspicuously foliated. The silicates are forsterite and sahlite, which appear as ovoid augen, round which the foliation sweeps (fig.4) giving quite a gneissose aspect to the rock.

Various inclusions are characteristic of the Balephetrish marbles. They may be classed as (1) gneiss-inclusions, and (2) mineral-aggregates.

(1) Gneiss-inclusions.— These occur in the pink marble at Balephetrish, and in similar marbles on the shore west and north-west of Balephetrish. The minerals characteristic of these inclusions include quartz, felspars, hornblende, augite, scapolite, and sphene.

The presence of some of these inclusions is evidently the result of crush-brecciation (fig.5, p.99); others seem to be more or less modified portions of intruded rock, and recall similar inclusions in the crystalline limestones of Ceylon.

Fig.4:- *Limestone similar to that of fig.3, but of finer grain and well-foliated. From a small quarry in the same field. (About $\frac{4}{5}$ nat. size*

[The small augen consist usually of forsterite.]

There occur also small pegmatite-patches in the eastern part of the exposures of marble in the field of south-east of Balephetrish Hill.

(2) Mineral-aggregates.— Small and large augen of pure white sahlite (the largest measuring 6 feet by 4) occur in the white marble of the field south-east of Balephetrish Hill; the foliation of the marble sweeps around these on a large scale, just as on a smaller scale it bends round the little augen consisting of single grains of diopside or forsterite.

Near the spot marked 'Natural Arch' numerous dark mineral-aggregates occur in the pink marble, varying in size from a few inches to a foot or two in diameter, and of quite irregular lumpy shape. These weather out on the surface, being often attached by quite a narrow neck. They consist of dark-green cocolite, scapolite, sphene, and blue apatite, with also calcite and colourless or brown mica (Pl. VII, fig.3). The first four minerals are especially characteristic; the rock strongly recalls certain augite-scapolite-sphene rocks from Ceylon.[6] The grains of apatite are rather conspicuous, owing to their bright sky-blue colour; this occurrence of blue apatite again reminds one of Ceylon.

[6]A.K. Coomáraswámay, Quart Journ. Geol.Soc., vol.lviii (1902) pp.497 & 680.

III. CONTACT-PHENOMENA.

Contact-phenomena are not especially well displayed, partly owing to the earth-movements which have affected the rocks since their formation, partly owing to the fact that contact-sections are rather infrequently exposed.

A contact of limestone with gneiss of acid type is seen at the southern edge of the quarry at Balephetrish. The gneiss is composed of quartz and felspar, exhibiting strain-shadows and partial granulitization; a few rather large grains of zircon occur in addition. The marble is one of the finest-grained types, with appearances of 'streaming', and with numerous grains of coccolite and also decomposed scapolite (?) and sphene. A dark zone, about half an inch wide, separates typical marble from typical gneiss. In this zone, on the side next the gneiss, the minerals hornblende and felspar (triclinic), with a little accessory golden-brown mica, compose the rock, while greenish pyroxene is rare. On the side next the marble the zone contains pale-green monoclinic pyroxene (that is, cocolite), scapolite with rather abundant sphene, and accessory iron-ores. The change from hornblende-felspar-rock (modified gneiss) to pyroxene-scapolite-rock (altered limestone) takes place rather abruptly near the middle of the half-inch dark zone; but the

minerals are completely interlocked, and there is no actual line of junction seen under the microscope, although an abrupt change is evident. The rock has suffered from frequent minute faulting and slight shearing, the effects of which are quite distinct in the gneiss and contact-zone, but are lost as they pass into the compact marble. The original relations are not obscured. It is rather curious that the only rock here seen in actual contact with the marble should be this quartzo-felspathic gneiss or granulite, while the prevailing rocks at Balephetrish are of a dark hornblende-plagioclase type.

At many points on the shore contacts of marble and gneiss (usually of acid type) are to be seen. In some case, as, for example, near the 'Natural Arches,' a zone of grey rock composed of monoclinic pyroxene separates typical limestone from gneiss; in other cases, no very striking contact-phenomena attract attention.

Enclosures of gneiss in marble appear often to have been much modified by the absorption of lime, with the development of pyroxene, scapolite, and sphene; in one case (Port-Abhuinn marble) a large inclusion of gneiss was separated from the typical marble by a narrow zone of pyroxene-scapolite-sphene-rock, with apatite and calcite, essentially resembling the rock of a narrow contact-zone described above, and of the mineral-aggregates occurring in the marble at the 'Natural Arches.'

The presence of accessory minerals in the limestone is in itself to be regarded as evidence of contact - metamorphism. It is shown below that the rounded character of some of these affords no evidence of detrital origin, such as has been suggested[7]; grains

[7]G.A.J.Cole & W.J.Sollas, Sci.Proc.Roy.Dublin Soc.vol.vii (1891) p.14; A.Harker, 'Petrology for Students' 2nd ed. (1897) p.317.

of rounded coccolite and sphene are associated with idiomorphic hornblende. I have no doubt that all these minerals have crystallized *in situ*; probably under conditions analogous to those of minerals crystallizing in a cooling magma, when, moreover, the limestone perhaps existed in a state akin to fusion.

IV. DYNAMIC PHENOMENA.

In some varieties of limestone foliation is inconspicuous, and the silicates are frequently idiomorphic. In others it is very distinct; and the included minerals are rounded, and form little augen round which it sweeps (fig.4, p.95). The rounded outline of the silicate-minerals is frequently, however, an original character; thus the grains of coccolite and sphene in the pink Balephetrish marble have smooth rounded forms, but are associated with sharply idiomorphic hornblende, showing that the rounded form is for these minerals a characteristic habit rather than a secondary phenomenon.

The effects of pressure are most conspicuous in the carbonates. These are usually present as the fine-grained, granular, compact matrix, in which various silicates are embedded. The various thin sections examined show most clearly that the minutely granulated condition is a result of the complete breakdown of larger grains,

fragments of which are sometimes preserved (see Pl.VII, fig.4) and that the limestone as a whole must once have possessed a more coarsely crystalline character, as was pointed out by Prof. Bonney several years ago.[8] These phenomena are illustrated in Pl.VI, fig.1 and Pl.VII, figs.1 & 2. Other examples of marble showing cataclastic structure are described and illustrated by Messrs. Adams & Nicolson,[9] who have reproduced experimentally the same phenomena. If we accept the results of their experiments, we must conclude that the deformation of the Tirce Marble is the result of pressure, acting at a temperature not above 300° C. We are, then, observing the effects of dynamometamorphism, free from complications due to marked accompanying thermal metamorphism. This must apply also to the gneisses with which the limestone is associated; a detailed examination of these has not been made; they have certainly suffered from minute faulting and local shearing, but to a small extent, compared with the gneissose rocks of lona, or, for example, of the Loch-Maree district. The gneisses are of a general Lewisian type, and include dark hornblendic rocks composing Balephetrish Hill, and also quartzo-felspathic gneisses which are more abundant on the shore. Coarse rocks of granuloid or pegmatitic aspect, and with inconspicuous foliation, occur near the road farther east.

An interesting crush-breccia (fig.5, p.99) occurs at the junction of limestone and gneiss, at Port Abhuinn, on the shore west of Balephetrish. Here we find a foot or two of completely mylonized limestone, having a smooth flinty conchoidal fracture, containing numerous blocks of gneiss, which have also suffered from the pressures. The calcareous

[8] 'The Effects of Pressure on Crystalline Limestones' Geol.Mag.1889,p.483

[9] 'An Exp[erimental Investigation into the Flow of Marble' Phil. Trans. Roy. Soc. vol.cxcv(1901) A p.387 & pls.xxii.xxv. Compare especially pl.xxv, figs. 3& 4 with my figures.

matrix has a delicate 'flow-structure', resembling that of a rhyolite. Something of the same kind has taken place quite locally near the spot marked 'Natural Arch' on the shore. There are gneissose streaks and patches in the limestone, and the two rocks are sheared together. The marble is a fine-grained calcite-rock, with diopside-grains more or less twinned and strained and bent. The sheared gneiss consists of a fine-grained foliated matrix of felspar and calcite-dust, with rounded scapolite and triclinic felspars with undulose extinction, broken porphyritic augite, and occaisonal hornblende and sphene. A dark intervening zone contains augen of felspar, hornblende, and sphene and is perhaps the remnant of a contact-zone. It is more usual, however, for gneiss-inclusion and mineral-aggregates (Pl.VII, figs.3 & 4) in the limestone to have been more or less protected from extreme effects of pressure.

V. NOTES ON THE MINERALS.

The following minerals occur in the limestones or in the mineral-aggregates found in them :— Calcite, dolomite, pyroxene, amphilbole, forsterite, scapolite, mica, sphene, apatite, orthoclase, spinel, and serpentine.

Carbonates:— In most cases calcite is the predominant carbonate; it is rare for dolomite to be abundant. The deformation and granulitization of the carbonates are referred to above. Heddel[10] gave the following percentage analysis of Tiree Marble:—

$$CaCO_3 = 95.94; MgCO_3 = 1.78; FeCO_3 = 0.576; MnCO_3 = 1.028$$

Pyroxene:— The Tiree Marble has always been noted for its 'cocolite'. This mineral (which is not more characteristic of the pink Balephetrish marble than is amphibole) appears to vary in colour from light to dark green, nearly black, the latter colour being characteristic of the mineral-aggregates, composed of pyroxene, scapolite, sphene, and blue apatite. I have in this paper spoken of the coloured pyroxene associated with the marble as coccolite, and called the white pyroxene sahlite, as this name has been employed by previous authors. Masses of sahlite formlarge and small augen in the white marble in the field south-east of Balephetrish Hill. An analysis of sahlite by Heddle[11] is here quoted :

$$SiO_2 = 50.54; Al_2O_4 = 4.68; Fe_2O = 4.14; FeO = 0.04; MnO = 0.69.$$

$$CaO = 23.59; MgO = 14.4; K_sO = 0.31; Na_2O = 0.62; H_2O = 1.48$$

$$Total = 100.51.$$

[10]'Mineralogy of Scotland' vol.i (1901) p.137 [Mr. E.C.C. Stanford gave analyses of Tirec Marble, Rep.Brit.Assoc.1870 (Liverpool) Trans.Sect.p.65-ED.]

[11]Trans.Roy. Soc. Edin. vol.xxviii (1879) p.460.

Fig.5:- *Crush-conglomerate on the shore of Port Abhuinn: mylonized limestone with gneiss-inclusions.*

Heddle appears to classify all the pyroxene as sahlite, as he says

that the sahlite may be dark-green, light-green, or grey, and does not mention coccolite as actually occurring in the limestone at all. A more detailed chemical examination of these pyroxenes with reference to colour seems desirable.

The pyroxenes are often twinned, and even distorted, in those varieties of limestone which have suffered most severely from the pressures (Pl.VI, fig.2).

Amphibole:— Dark-green amphiboles are very characteristic of the pink Balephetrish marble, and are often of large size, and then sometimes intergrown with carbonates. Individuals are frequently idiomorphic (Pl. VII, fig.6). The pleochroism is (a) pale greenish-straw, (b) olive-green, (c) bluish-green; the extinction-angle on flakes= about 23°. Colourless tremolite occurs in other varieties of limestone.

Forsterite:— This mineral (not hitherto recorded from Tiree) occurs abundantly in some varieties of limestone, especially in the white limestone exposed in a field south-east of Balephetrish Hill. Crystals are never idiomorphic, but elongated individuals extinguish in the direction of their length, and examination in convergent polarized light shows that the axial plane is at right angles to the direction of elongation. Partial or complete serpentinization, commencing along the irregular cracks, is typically exhibited.

The forsterite from a specimen (fig.3, p.94 & Pl. VII, fig.1) of marble from the south side of the field has been analysed by Mr. W.C. Hancock, B.A. Weathered, but carefully selected crystals were used for this analysis, which confirmed the identification of forsterite. The specific gravity was 2.80.

Scapolite:— This mineral was found by Heddle in the residues from the pink Balephetrish marble, in rare minute crystals, with the forms a(100), m(110), r(111), associated with sahlite and white biotite. His analysis is as follows[12]:—

$$SiO_2 = 48.923; Al_2O_4 = 22.098; Fe_2O_3 = 3.159; FeO = 1.508; MnO = 0.538;$$

$$CaO = 7.753; MgO = 2.769; K_2O = 6.058; Na_2O = 1.679; H_2O = 5.691.$$

I have found it much more abundant, in larger, not idiomorphic grains in mineral-aggregates (associated with coccolite, sphene, and apatite), in modified gneiss-inclusions, and in contact-zones.

Sphene:— Small, lustrous, blunt-angled crystals, of a watery-brown colour, occur sparingly in the pink marbles. Larger grains occur more abundantly in the coccolite-scapolite-sphene aggregates, and in some varieties of contact-rock or modified gneiss-inclusions. The grains have sometimes quite a pinkish colour; they were mistaken by Jameson[13] for garnet.

Mica:— Rather decomposed, colourless flakes of mica occur in some varieties of limestone and in mineral-aggregates. No tyical phlogopite occurs. Dark-brown mica is occasionally seen in mineral-aggregates.

Apatite:— Grains of sky-blue apatite (not hitherto recorded from Tiree) are found rather plentifully in the coccolite-scapolite-sphene

aggregates occurring in the limestone near the 'Natural Arches' on the shore. The grains are often elongated in the direction of the vertical axis, but are not idiomorphic; in this sections they are practically colourless. The grains are only faintly pleochroic, showing a birght sky-blue for rays vibrating perpendicular to the vertical axis, and a paler greenish-blue for rays vibrating parallel thereto. Individual grains vary in colour from bright sky-blue to a duller bluish-green. Possibly this patite is the mineral spoken of by Heddle[14] as a 'plae-blue watery variety' of sahlite. Mr. W.c. Hancock, B.A., has carefully analysed the blue apatite, with the following result:—

[12] 'Mineralogy of Scotland,' vol. ii (1901) p.53.

[13] 'Mineralogy of the Scottish Isles' vol.ii (1800) p.32.

[14] Trans.Roy.Soc.Edin. vol. xxviii (1879) p.459.

$Cl = 1.85; SiO_2 = 1.5; CaO = 53.92; P_2O_3 = 39.55;$ and H_2O(hygroscopic) $= 3.16$.

Total = 99.98.Specific gravity = 3.20.

This analysis shows that the blue apatite of Tiree is one of the comparatively rare chlor-apatites, and much resembleds analysis of chlor-apatite given by Voelcker.[15]

Spinel:— A few grains of bluish-green spinel (not hitherto recorded from Tiree) are scattered very sparingly in a variety of marble (No.3, p.94) exposed in the field south-east of Balephetrish Hill. The accompanying silicates are:— abundant forsterite, and scarce mica, tremolite, and sahlite. The spinel is quite isotropic; and in thin chips nearly colourless. Mr. G.T. Prior, M.A., F.G.S., has kindly determined the specific gravity as 3.635. The spinel is probably referable to pleonaste.

VI. IONA.

Marble is exposed on the shore south of Dun Dugaidh; a long 'slack' leads down to the – from the sea – rather inasccessible beach known as the 'Marble Quarry.' Much white marble occurs here: portions are greenish or black. It has a silky, almost schistose aspect; it seems to occur as a band or lenticle in the gneissose rocks. The colourless varieties consist of calcite, with some serpentine and tremolite, the last-named in small needles often felted together. In the dark patches serpentine is more abundant, and may practically compose the rock. The rocks inmmediately associated with the marble include actinolite-felspar schists with epidote andpyrite (north wall of 'slack'); hard white rocks, excessively fine-grained, resembling the marble but composed of felspar and muscovite (?) with zoisite. Associated with these was a band of mica-schist-like rock, composed of finely granular quartz, mica, and felspar, with actinolite and pyrite, and exhibiting well-marked foliation. This rock suggests an altered sediment. There are, moreover, bands and streaks of dark rock, hard and heavy, in the marble itself; these are actinolite-felspar-schists, with much pyrite and some epidote. The orientation of the needles of actinolite is very marked, the ether-axis being parallel to the foliation. All these rocks appear to have been entirely reconstructed.

A small exposure of higly sheared silky limestone occurs near the commencement of the path at the beginning of the grassy flat (raised beach) above St. Columba's Bay (Port-a-Churiach); it forms a band in the gnesiss, with corresponding foliation. It contained a few small sillicate-augen; one consisting of colourless monoclinic pyroxene, with secondary tremolite and calcite; another of felspar and calcite, witih zoisite and amphibole. The felspar ismostly triclinic; there is some orthoclae and orthoclase-microperthite; in the latter the inclusions of triclinic felspar have sometimes given rise to secondary clacite, producing a (secondary) banded intergrowth of calcite and orthoclae.

Finally, there is a band of limestone or ophicacite (locally caleed sunstone) in the gneiss about a quarter of a mile north-east of Port Bán, on the western coast of Iona. The

[15] Rep.Brit.Assoc.1857 (Dublin) Trans.Sect.p.59.

rock is less curshed than those above described, and consists of grnaulated calcite with equally abundant flakes of colourless mica (often bent) and rounded grains of serpentine, the last-named very possibly a pseudomorph after forsterite, and also grains of nearly colourless sphene. A single large crystal of decomposed pyroxene was also observed.

The limestones just descibed from Iona are included in the gneiss. The latter is of vaired character, though hornblendic types are perhaps most usual; it has nearly everywhere suffered much pressure.

Sedimentary rocks suggestive of Torrison Sandstone occur along the eastern shore of Iona, especially in the northern half of the island; they include felspathic grits and finer salty or ahly bands. They become very much sheared, especially as the junction with gneiss is approached; there would be much difficulty in drawing a satisfactory boundary. The limesones above described are not associated with these sediments, but entirely included in the gneisses, and seem to correspond to rocks such as the Tiree Marble, but which have suffered much more fromthe effects of pressure, so that we cannot study the original relations or even point to original minerals. The ophicalcite from near Port Bán, however, more nearly resembles the Tiree Marble; differing chiefly in the absence of any portion of the mineral now replaced by serpentine.

VII. Summary and Conclusions.

Tiree.

(a) The various patches of marble occur as lenticular masses in the gneiss. There is no evidence as to the origin of the limestone.
(b) The development of hte accessory mineralsandprobably alsothe original more coarsely crystalline character of the marble, are the result of contact - metamorphism at high temperature.
(c) The hornblendic and acid orthogneissses associated with the marbles were the agents in this thermometamorphic process.

(d) The present foliation fo the marble and possibly in the part of the gneiss, together with the various dynamic phenomena displayed by the marble (and gneiss), is the result of subsequent dynamic metamorphisms unaccompanied by any marked thermal metamorphism.

Iona.

The marbles of Iona were very probably originally similar to those of Tiree, but have, in most cases, suffered more severely from dynamic metamorphism.

Quart. Journ. Geol. Soc. Vol. LIX, Pl. VI

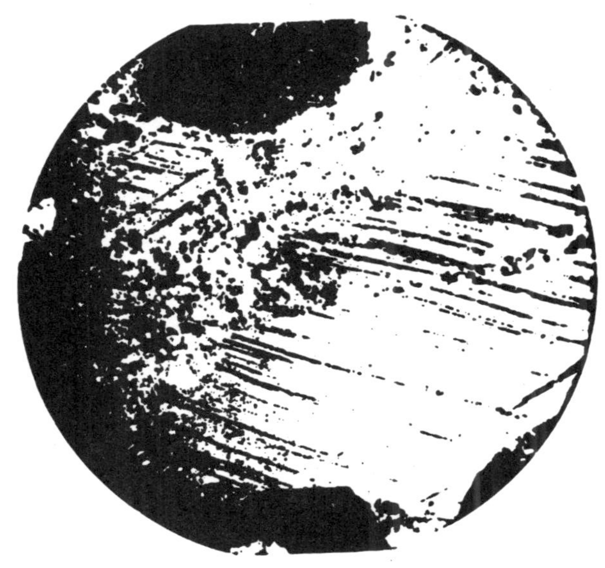

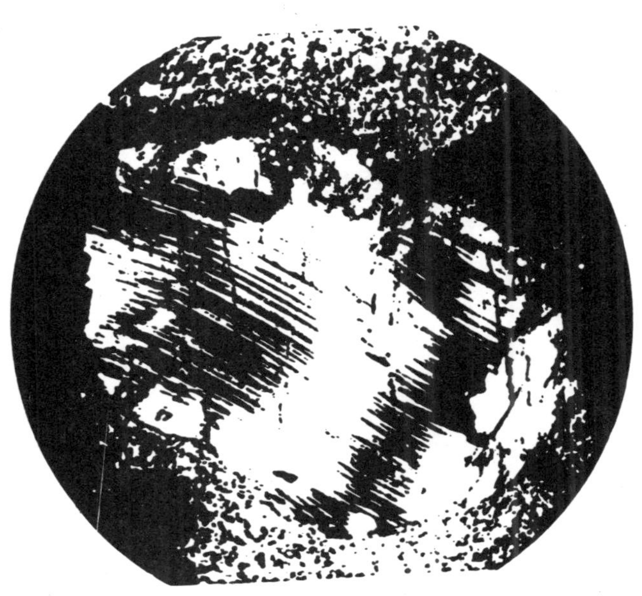

"Tiree Marble"

Quart. Journ. Geol. Soc. Vol. LIX, Pl. VII

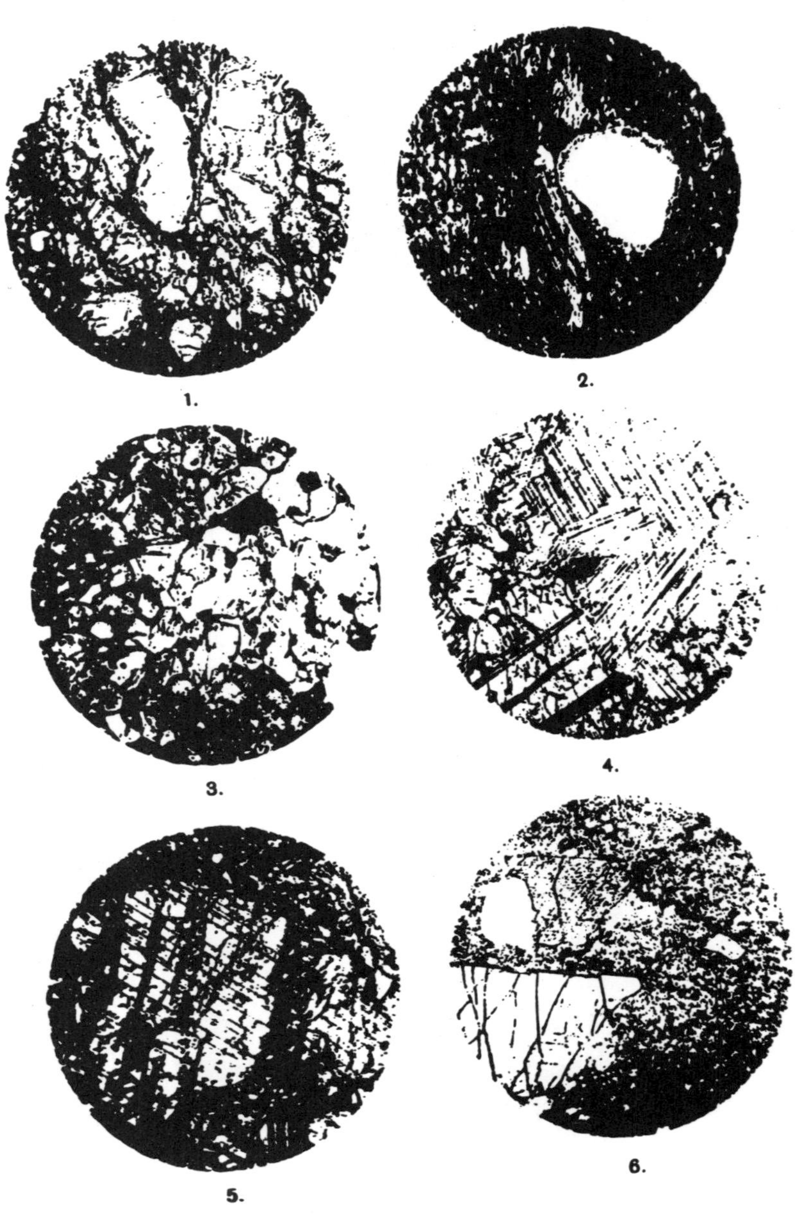

EXPLANATION OF PLATES VI & VII.

Plate VI.

Fig. 1. Incipient cataclastic structure in carbonates. Fields south-east of Balephetrish Hill. × 33, crossed nicols. (See p.98) Slide 1027, Author's collection.
2. Twinned and bent crystal of sahlite, in the mylonized limestone of the crush-conglomerate, Port Abhuinn. × 26, crossed nicols. (See p. 100) Slide 1033, Author's collection.

Plate VII.

Fig. 1. White limestone with forsterite (and sahlite); the lines of crushing are well seen in the carbonate-matrix. From small quarry in field southeast of Balephetrish Hill. × about 9. (See pp. 98 & 100) Slide 1022, Author's collection.
2. Cataclastic structure in pink marble; bending, shearing, and -up of clacite-grains. The large transparent grain is orthoclase. Port-Abhuinn marble. × 12. (See pp. 91 & 100) Slide 1031, Author's collection.
3. Aggregate composed of coccolite and scapolite with sphene, (blue) apatite, calcite, andmica. Natural Arch. shore, Balephetrish. ×15. (see p.96) Slide 1010, Author's colelction.
4. Silicate-aggregate, with abundant calcite and dolomite; the carbonates uncrushed. Same locality. × about 13. (See p.99) Slide 1025, Author's collection. The nearly central small dark patch is an appearance due to local incipient breakdown of a clacilte-grain.
5. Sahite in the same slide as fig. 1; basal cleavage well developed. × about 8.
6. Idiomorphic amphibole, in fine-grained pink marble, Balephetrish Quarry. ×*i*about 11. (See p. 100) Slide 1019, Author's collection.

Discussion.

The President spoke of the fine series of microscope-sections exhibited by the Author, and of the evidences of movement and of mylonizatiooon which they presented. The proofs of contact-alteration of the calcarcous rock by the gneiss seemed clear, and the association of minerals was reminiscent of that shown in the case of the Ledbeg Marble. Was it impossible that the Tiree lencticles, which had the smae trned as the Ledbeg rock, were relics of a south-westerly extension of the same calcarcous formation ? The consclusion that the Tiree Marble was a limestone which had been

first intruded upon, enveloped, andmetamorphosed by igneous rock, and that afterwards the resulting complex had been subjected to dynamicmovements bringing about mylonization, was simple, and appeared to agree with most of the phenomena cited. It would be interesting, however, to know whether there was anything apparent to disprove the theory that the injection and the main dynamic action were simultaneous; or, in view of the experiments of Adams & Nicolson,that the crystals of clacite, etc. might not have developed and enlarged even under pressure;or that the original limestone might not have undergone dynamic action before the injection; or that the entire complex itself might not have been squeezed and yielded more than once afterwards.

Prof. Bonney said that hisknowledge of the Tiree Marble was founded on hand-speciments and rock-slices only, as he had never visited the place. Still, in theAlps, he had somany opportunities of studying the effects of pressure on crystalline limestones, that he could venture to say that the suggestion thrown out by the President as to the coarses calcite-grains in the marble being recrystallized, was no more probable than that the felspar-grains in the 'mylonites' of Glen Laggan were recrystallized. The coarse calcite might be sometimes seen to have been protected between two grains of sahlite, like a bay between headlands. He called attentionto some analogies between the Tiree rockand the well-known crystalline limestones of Côte St. Pierre (Canada), and said thathe did not think any close connexion existed between it and the Ledbeg Marble. The paper to which the Society had listened appeared to him to make a valuable addition to knowledge.

Prof. Sollas thought that the Ledbeg limestone owed its charcter to contact - metamorphism, that of Tiree to this action and something more; for it appeared to be a very mixed product, and some of its minerals might have been originally deposited along with calcarcous seidment, jsut as olivine-sands are now being mingled with clcarcous sands on the shores of Sawaii. Some ofthe felspar of the Tiree Marble presented evidence of secodary growth within a 'calciphyre' was shown by Issel's observations on radiolaria included within albite-crystals, which occurred as constituents of the so-called 'calciphyre' of Rovegna in the Val di Trebbia.

The Author said that he saw no reason for connecting the Ledbeg and Tiree Marbles, the former being post-Torridonian, the latter in all probability pre-Torridonian. It was quite clear that recrystallization had not taken place since the production of cataclastic phenomea. Contact-phenomena were unmistakable in places. He thought that there was little to suggest a detrital origin for the accessory minerals. With the exception of the, not very abundant, felspar, it would be very hard to imagine a volcanic origin for these.

Part II

Mineralogical Survey, Ceylon Administration Reports

CEYLON.

ADMINISTRATION REPORTS, 1903.

MINERAL SURVEY.

REPORT of Dr. A. K. COOMARASWAMY, F.L.S., F.G.S., Director.

Geological Survey & Mines Bureau
1995

ADMINISTRATION REPORT[1]

A separate guide to the geological collections is to be prepared has been undertaken includes the identifications of specimens submitted by Government the preparation of an article to be contributed to the new edition of a general guide Colombo Museum to appear in 1904, and of an article on the mineral resources of Ceylon to be to the handbook for the Ceylon Court, St. Louis Exhibition, and assistance in the collection specimens therefor.

The staff consists of a Director, A.K. Coomaraswamy, B.Sc., F.L.S., F.G.S., and an Assistant or James Parsons, B.Sc., F.G.S., who arrived in Ceylon on March 7. An office peon, two a "Collector," and eight coolies are also usually employed. The greater part of the time of is spent in the field. Bi-monthly reports, accompanied by specimens, are sent to the Director Imperial Institute, where a chemist is retained in connection with the survey.

It is estimated that an area of 1,000 square miles can be superficially surveyed in the course of A smaller area has been examined in 1903, owing to the late arrival of the staff from England to the time occupied in making preliminary arrangements The area examined in 1903 amounted 512 square miles; but the work of the survey can only partially be judged by area, as in the event of the examination of important economic minerals being necessary a prolonged stay in one area may be required.

The minerals examined during the year include mica, graphite, iron ore, moonstone, garnet, and various minerals found in gem washings. Observations on the general geology of Ceylon and or the mode of occurrence and distribution of the minerals referred to are given below:-

1. *General Geological Considerations* :- The greater part of Ceylon consists of ancient crystalline rocks, especially characterized by well-marked mineral banding (foliation) and variability in mineral composition. These are for the most part to be regarded as deep-seated rocks of igneous origin, although it is just possible that amongst them some exist whose origin was sedimentary, but which are now so far metamorphosed and incorporated with those of igneous as to be almost inseparable thereform.

These crystalline rocks are known as granulites, or, using the term in a wide sense, as gneisses, and belong to the same group of rocks as those which are known in Southern India as the Charnockite Series.*. The rocks themselves comprise granular quartz rocks, leptynites, charnockite proper, pyroxene granulites, norites, diorites, gabbros, & c., and more basic rocks composed mainly of pyroxene, amphibole, and garnet, as well as many rock varieties of unusual mineral composition; there exist all variations, from extremely acid to very basic types. Coarse intrusive pegmatites are abundant in some areas.

[1]Figures 1,2 and 3 are missing

The crystalline limestones are interbedded with the other crystalline rocks, and are of importance as a cource of lime for building purposes. Secondary calcareous deposits (tufa) in small quantities are of common occurrence. There are also rocks composed essentially of one or more of the minerals pyroxene, mica, amphibole, scapolite, sphene, spinel, graphite, which occur at the junctions of limestone and granulite, as inclusions in the limestones and in separate beds amongst the other rocks. Rocks of the Galle group resemble these in some respects, but are further characterized by the presence of wollastonite, a mineral not yet found elsewhere in Ceylon.

Of rocks derived from the decomposition of the granulites, laterite (cabook) is the most important; there are also various clays, including beds of kaolin, rarely of a valuable character. More hydrous aluminium silicates, such as halloysite, have also been found to be abundant. Of typical dyke rocks, but few occur in Ceylon, and none have been seen in the course of the work so far accomplished. Other volcanic rocks are not met with in Ceylon.

Of sedimentary deposits, there are river gravels and sand, some of which yield gold in small quantities ; alluvial clays useful for making pottery, bricks, and tiles ; calcareous raised beaches on the coast; and marine-blown sand.

. *2. Economic Minerals :-* Turning to the minerals of economic importance, whose been investigated in the areas examined, mica may first be mentioned.

(Figures 1,2 and 3 are missing)

L2 CEYLON ADMINISTRASTION REPORTS FOR 1903. PART IV.

Mica parallel to foliation and exposed on a dip slope is most liable to deformation and flaws on account of surface slip in the soil. It may, however, be little affected by water. Mica is often discovered in this situation on account of the large exposure in the surface and subsoil. Veins in or parallel to strike or dip joints are on account of their usual high angle apt to form passages for surface water, and the mica suffers by hydration.

Production and Methods of working :- The quantity of mica exported from Ceylon has sunk rapidly from an amount valued at Rs.43,637 in 1896 to an amount valued at only Rs.375 in 1902. Very little mica is now worked, although locally procured mica was used in the electric lighting of Kandy a few years ago. The mica pits are rarely very extensive; rectangular pits, rarely exceeding 10-15 feet in depth and one or two yards in width, are usually sunk along the outcrop of a vein. Work is often abandoned when much blasting becomes necessary, or when timbering is required. The mica is extracted in "books" or crystals, and the small and flawed ones being rejected, the others are split into thin laminae by having the flawed or weathered edges cut a way by a large pair of scissors. The mica is packed in "boxes" containing about 80lb. The right to work the pit is often rented from the owners of; the land, a royally of Rs.5 a box being a very usual remuneration. Time to time been more or less extensively worked on various tea estates, especially Haputale districts. The most important working, at Haldummulla, will be inspected next year.

List of the More Important Localities where Mica has been dug in parts of the Central and Uva Provinces

Central Province.

Talatu-oya, six miles east of Kandy, Central Province - A map is given showing the localities where mica is found in the Uduwela valley. The industry was formerly extensive, and dates back to the time of the Kandyan Kings, who made use of mica for ornamental purposes. The mica belongs to phlogopite, and occurs in the manner described above. Many of the veins are junction veins, being closely associated with the junctions of limestone and granulite. The only pits in work at the present time are two, which are being worked for the Mineral Survey in order to obtain commercial specimens and to gain an acquaintance with the methods of work and necessary expenses.

Hanguranketa - A number of localities where mica has been found in the neighbourhood of Hanguranketa have been examined by the survey, but hardly any of these afford mica of commercial value. The best were at Illagolla (Gangapalata Korale), where crystals of considerable size were met with in Crown land ; the mica is however, much bent, and the inaccessibility of the spot renders it unlikely that further excavation would be remunerative.

Teldeniya - Some large mica has been obtained in one or two localities near Teldeniya, particularly on Kandewatte estate, where "bottle green" mica occurs in a band of pyroxene-mica rock, which can be easily followed through the estate.

Province of Uva.

Badulla - A number of localities in the neighbourhood of Badulla have yielded mica on a commercial scale. The most important of these are found along a belt of country between Dikwella and Udakahowile, in association with a band of crystalline limestone running in the same direction. A sketch map of these localities is given.

On Elmshurst estate (Alugolle) mica is locally abundant, but occurs in the subsoil on dip slopes, and is much flawed. Crystals 14 in. by 10 in. in diameter were found, trimming to 5 in by 5 in. to 7 in. by 5 in. This is perhaps the best locality near Badulla.

At Udakahowile three pits were sunk, and a good deal of valuable mica obtained. One of these has the form of a long trench, and evidently follows the course of a vertical vein of mica, of ; which however no trace now remains.

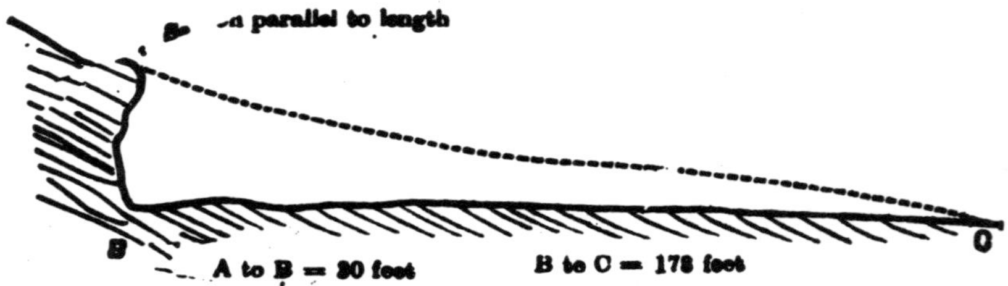

Fig 4. Section parallel to length

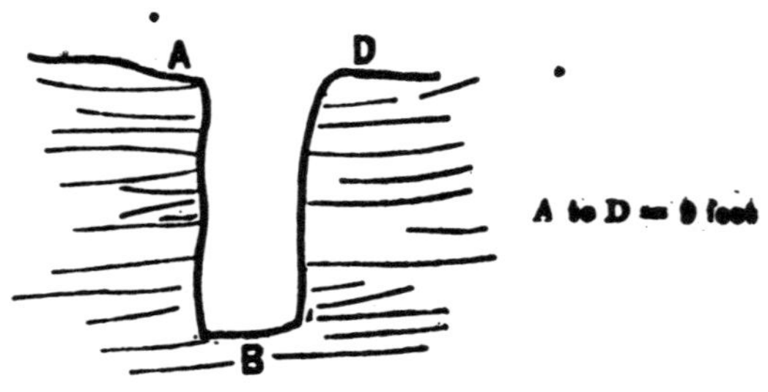

Fig 5. Section transverse to length at upper end

Another old working is in two "stopes" on the hillside, the nature of which will be gathered from figs 6.7 and 8.

The largest crystals from these pits measured 7 in. by 3 1/3 in. by 3 in.

On Cloyne estate and adjoining Crown land a considerable quantity of mica is exposed in the subsoil. Investigation showed that it occurs probably as a vein parallel to the foliation and forming the surface of a steep dip slope, and that the mica itself is for the most part flawed and valueless although the quantity obtainable is considerable.

At Budalawe mica is found in small branching veins in the band of crystalline limestone. The crystals are not large enough to be of commercial value, but the locality affords a valuable illustration of the mode of occurrence.

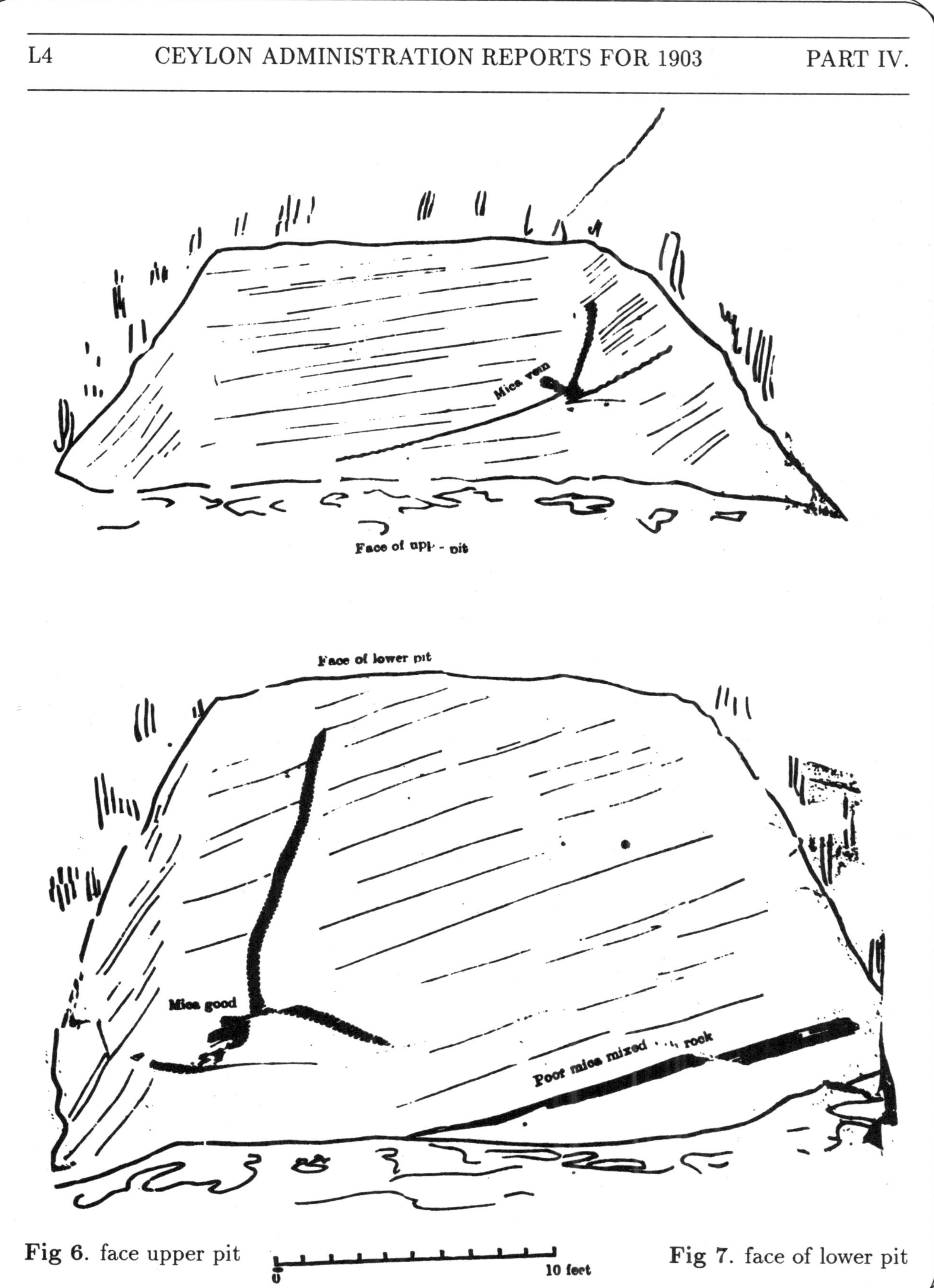
L4 CEYLON ADMINISTRATION REPORTS FOR 1903 PART IV.

Fig 6. face upper pit

Fig 7. face of lower pit

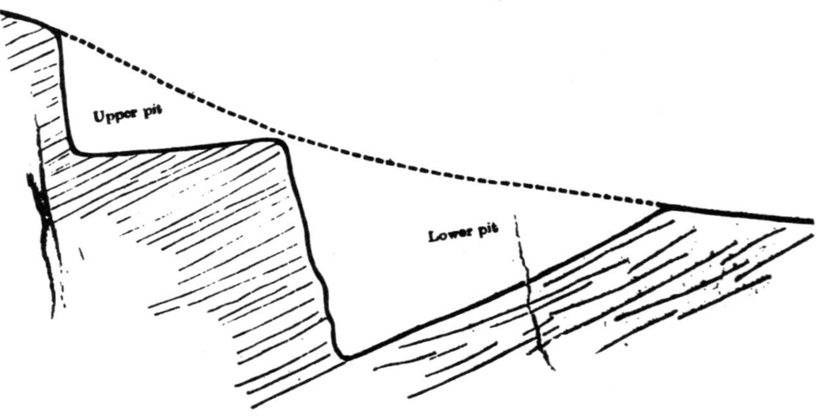

Fig 8. - Section of two pits

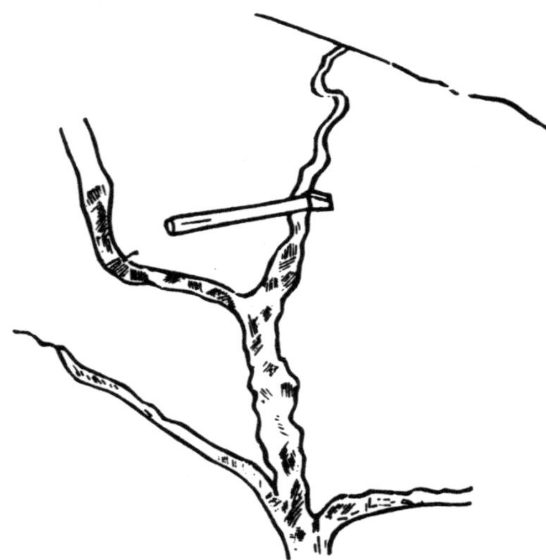

Fig.9 - Veins of mica in crystalline limestone (from photo) Budalawe near Badulla. (A geological hammer serves as scale)

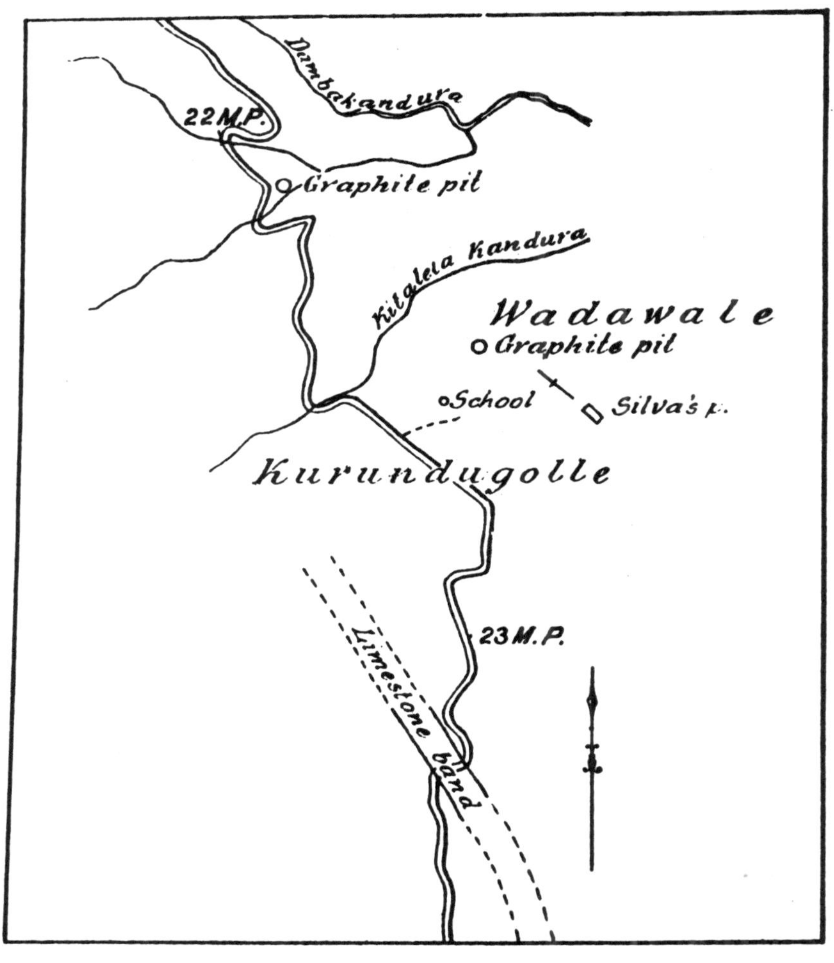
Dambakandura
22 M.P.
Graphite pit
Kitaleta Kandura
Wadawale
Graphite pit
School
Silva's p.
Kurundugolle
23 M.P.
Limestone band

A number of localities north of Badulla have been examined. Of these, the most important is Egodawela, half mile from the 5th milepost on the Taldena road and quarter mile south of Welihenakanda trigonometrical station. Another locality is at Migahakiwula, 12 1/4 miles from Badulla on the Taldena road.

In the Bogoda district mica was found at Mahakumbura about half a mile east of the junction of the Bogoda-oya with the Uma-oya, also near Landewela and near Lunugalla.

In the neighbourhood of Bandarawela no mica of value has been met with. The most promising locality is *Mirahawatta* (for Mineranwatta), six miles from Bandarawela on the track to Welimada. Mica is known to the Sinhalese as "potu-mineran," or "talatu-mineran," plumbago as "mineran ;" as distinction between mica and graphite is not however always made, and, as in this case, the term "mineran" is applied to the shining flakes which may belong to mica or graphite. At several places in this village quantities of mica are exposed in the soil (*e.g.,* in the path-cutting near the 6th milepost), and it is probable that if search were made, some of the exposures would be found to yield mica of commercial value. The largest (but much-flawed) crystal observed measured 8 inches in diameter.

In the neighbourhood of *Wilson's Bungalow* mica is being worked at Alutuwatugurugama, about 2 miles from the main road at Palugama. Bottle green mica giving trimmed pieces up to 4 inches in diameter is obtained and sold in Colombo for 14 cents a lb. This is the only locality investigated where mica is being worked at the present time, with the exception of the work done at Talatu-oya under my direction. The Alutuwatu locality is however of small importance.

Muscovite:- This mica has only twice been met with : on Shrubs Hill estate near Peradeniya it is found in flawed crystals not exceeding 3 or 4 inches in diameter in quartz rock ; at Ambalawa, Gampola, it has also been found, and occurs in a band of quartz rock, in which orthoclase is locally abundant, having a general north and south trend similar to that of the other rocks of the district ; the quartzose mica bearing band can be traced from a considerable distance. Its thickness could not be determined. The mica occurs in crystals scattered through the quartz and in veins and stringers running through the quartz and felspar, opening out here and there into pockets containing much mica, mostly decomposed, although sometimes of large size. Trial excavations have been made. The largest cut samples averaged 4 by 3 in. in size. Further development would be of a speculative character, but might possibly prove remunerative. The quartz rocks (sometimes felspathic) in which muscovite is found are to be distinguished from the typical quartz-felspar pegmatite veins in which biotite is the typical mica.

GRAPHITE.

Graphite (or plumbago) is, of course, by far the most important of the mineral resources of Ceylon. Only a few important graphite mines, however, are situated in the area investigated during the past year.

The mode of occurence and origin and method of working of the graphite will first be briefly dealt with as far as observations have been made, after which the more important mines examined are described. The price of graphite has varied much during the last few years. The high prices of 1899 led to correspondingly large output and export, which fell however suddenly in 1900, but has increased steadily in each succeeding year. There seems no reason to fear an exhaustion of the amount of graphite available in Ceylon. Graphite occurs in minute scattered crystals in many of the granulites and crystalline limestones of Ceylon ; but it is only where it is found in veins or beds, respectively crossing or parallel to the foliation planes of the crystalline rocks, that it has proved of commercial value. Occurring thus, it is found in beds, veins, or nests varying from the smallest size to a yard or more in width. Where the graphite occurs in small flakes as an accessory mineral in the granulites and limestones, it is unnecessary to ; suppose that it is not an original mineral.

It is clear, however, that the graphite occurring in veins has been deposited at a time posterion to the consolidation of the granulites. The veins are often of the most typical character. Usually they consist of pure graphite (sometimes there is evidence of more than one period of deposition, is a zoned structure of the vein) ; sometimes the vein shows a central zone of quartz or pyrite with graphite on either side. Sometimes the graphite is more irregularly associated with minerals such as felspar quartz, and mica, and with fragments of the surrounding rock. Metamorphism of the matrix in contact with the veins is found only on a very small scale; the rock surfaces in immediate contact with the veins are not impregnated with scales and flakes of graphite to a greater depth than half an inch.

As regards the source of the graphite, we see at once that its presence cannot be ascribed to the metamorphism of beds of coal or other carbonaceous depsoits, and indeed that the graphite can have had no direct organic source and is most probably of entirely inorganic origin. Like other minerals found in veins, it must have been deposited from vapours or liquids saturated with mineral (in this case carbonaceous) matter.

Methods of Working : - In the area examined the greater number of the pits are of quite small dimensions, and have been sunk by villagers or others in their own land. Most of these consist of rectangular shafts, some two yards square and 50 to 100 feet deep. Many of these appear to have been quite unproductive ; others were only worked during the "boom" in 1899, when the price of graphite was so high that every pound had an appreciable value. On frequent cause of failure, however, is owner neglects to observe the lie of the rocks and direction of the vein, and merely sinks a shaft can wherever signs of graphite are found on the surface, supposing that it is "bound to" a supposition on which it is most unwise to rely. In this way a shaft may be vein, which might have been productively worked if carefully followed nearer the direction required. It is always more profitable to follow the vein by sink a vertical shaft in the uncertain hope of meeting a good vein.

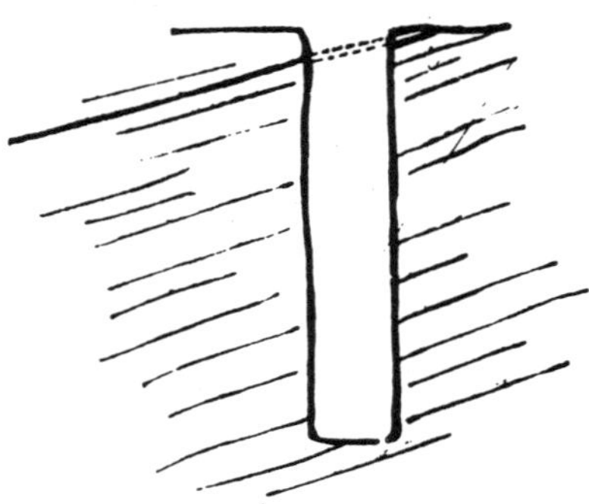

Fig 10. - Section of shaft sunk right through a graphite vein; useless expense.

A vertical shaft is of course to be recommended where the bed or vein of graphite is found to have a vertical disposition.

Whenever the graphite vein is parallel to the foliation planes of the granulites it is important to determine the dip and strike of the latter and to make use of the data so obtained in carrying on the work. From such observations the position of the vein underground should be calculated and due allowances made.

The foregoing observations apply for the most part to the small pits and shafts, which, with few exceptions, are the most frequent type of graphite "mine" in the areas visited. Two or three of the more important occurrences may be referred to in greater detail. The most important workings seen are Mr. Rettie's on *Glen Alpin* estate, Ballagalla, near Badulla. Extensive excavations were made two or three years ago in this place, and large quantities of graphite raised, the work however being stopped by water. The principal works are represented diagrammatically in the figure.

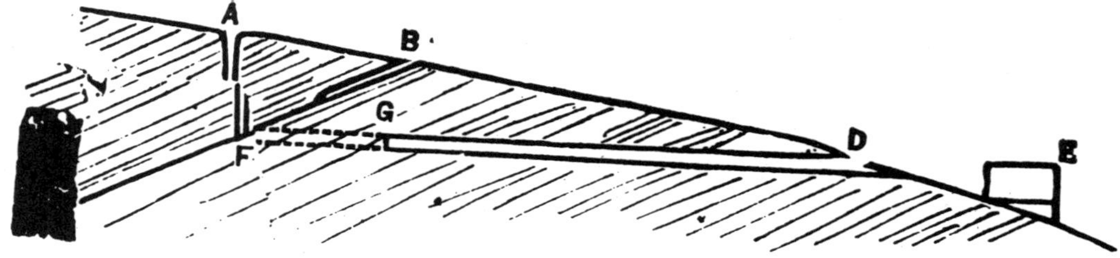

Fig.11: Diagram of graphite mine (vertical section) on Glen Alpine

A = Vertical shaft 80 feet deep
(work stopped by waterin both).

B = Adit along vein D = Sloping adit in progress

D F = 400 feet E = Engine house

B C = Graphite vein parallel to
foliation of rocks which dip into hill at 20.

The vein of graphite runs parallel to the foliation at BG, the beds dipping into the hill, and digging was carried down it in an inclined shaft until the accumulation of water in the shaft prevented further work. Work in a vertical shaft was also stopped by water. An inclined adit sloping at angle of 3 or 4 is now being driven into the hill to meet the vein ; this adit is large enough to allow of the use of mineral trucks running on rails.

A ventilating fan, worked by an oil engine, situated near the mouth of the adit drives in air through a pipe. It will be seen that this deposit of graphite is being worked on scientific lines, and that it will not be the fault of the method if the results are unsuccessful.

The same or parallel vein of graphite is exposed on the opposite side of the hill, where however the "beds" also dip into the hill (thus forming a syncline running through the hill), and work was here also stopped by water.

Mulhalkele (near 37th Milepost on Kandy-Nildandahena Road) - An important series of workings belonging to the Ceylon Tea Estates Co. has been made here. Four distinct diggings have been made ; of these, one is a vertical shaft some 22 feet deep, which appears to have been sunk right through a graphite vein which runs parallel to the foliation planes of the rock near the surface of the ground. The other three diggings, being by adits and " stopes" following the vein, yield much more satisfactory results. Fig. 12 gives a plan of one of the working, of which the main adit runs into the hill in a direction S. 75 W.

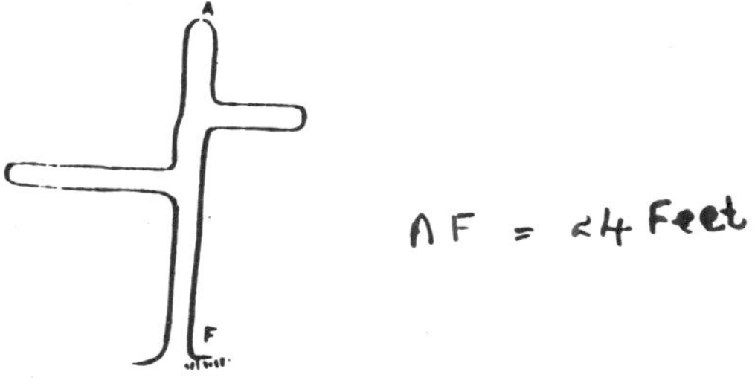

Fig.12

Kurundugolla (Kohoke Korale, Central Province) - An important mine (now now worked on account of the low price of graphite) has been worked at Wadawale, near Kurundugolla (22 1/2 miles from Kandy). The pit belongs to K.P. de Silva, and is said to have yielded 1,000 tons of graphite. The main vein runs parallel to the foliation of the granulites which have a vertical disposition, and strike in a direction 50 W. of N. A large open working has been made along the strike, producing a trough-like excavation. Besides the main vein there are lateral veins of graphite, one of which formed an excellent example of a typical vein crossing the foliation of the granulites, but dying out in the course of a few yards. A shaft said to be 40 fathoms in depth had been sunk at the bottom of the open working, right on the main vein, but it is now quite full of water. The graphite is associated with quantities of quartz, in which it forms veins and strings and occasionally rosette-like crystallizations. Two vertical shafts sunk very near the best pit have been unsuccessful, one wholly so, the other yielding only a small quantity of graphite. Perhaps a quarter of a mile north-west of the main pit and not far from the school house, two pits sunk by Mr. Gunaratne yielded few tons of graphite, while another near the 22nd milepost on the main road yielded 7 tons. It was not possible to locate these pits exactly on the map available but when put down in their approximate positions it was found that the three main pits were in a straight line, whose direction coincided with that of the strike observed in the large open working. This fact seems to indicate that the best results would be obtained by following up the deposit along the said strike.

The accompanying map may make the matter clearer.

IRON ORES

A number of occurrences of iron ore have been examined. It is well known that iron of excellent quality was formerly manufactured by the Sinhalese. The following extracts and accompanying remarks will explain the processes employed by them. The additional information given was chiefly obtained from a blacksmith of Kabilwela, near Bandarawela, whose account, moreover, confirmed in detail that of Sir Samuel Baker quoted immediately below:-

The bellows are formed of two inflated skins, like a double "bagpipe". Each foot of the "bellows-blower" is strapped to one skin, the pipes of the bellows being fixed in the air hole of the blast. He then works the skins alternately by moving his feet up and down, being assisted in this treadmill king of labour by the elasticity of two bamboos, of eight or ten feet in length, the butts of which, being firmly fixed in the ground, enable him to retain his balance by grasping one with either hand. From the yielding top of each bamboo a string descends attached to either big toe; thus the downward pressure of each foot upon the bellows strains upon the bamboo top as a fish bears upon fishing-rod, and the spring of the cane assists him in lifting up his let. Without this assistance it would be impossible to continue the exertion for the time required.

While the "Bellows-blower" is thus getting up a blaze another man attends upon the well, which he continues to feed alternately with fresh ore and a corresponding amount of charcoal, every now and then throwing in a handful of fine sand as a flux. The return for a whole day's puffing and blowing will be about twenty pounds weight of fine badly smelted iron. This is subsequently re-melted and is eventually worked up into hatchets, hoes, betel-crackers, & c.; being of superior quality to the best Swedish iron - From Sir S. Baker's *"Eight years in Ceylon" London 1855, pp 56-59.*

This account was confirmed in most of its details by the description of the process given me by a blacksmith of Kabilwela, near Bandarawela, from whom I was also able to obtain specimens of the iron and steel manufactured. The method of preparing steel practised in Uva was as follows:-

The method consists in introducing a small bar of good iron into a clay mould of a tubular form, which they call "covey" with pieces of the dried wood of the Cassia auriculata ("ranawara" of the Sinhalese). The open end of the tube is afterwards closed with clay, and is placed in a charcoal fire for two hours, by which process carbon is supplied to the iron, which is thus converted into steel. The proportions for making steel of the best qualities are as follows: - Seven parts of iron to three of dried wood.

They also use the wood of the *Toddalia aculeata*, the "kudu-miris" of the Sinhalese, in which case the proportions are three of iron to one of wood. This wood, however, produces an inferior steel, but by increasing the iron to five parts a better kind may be obtained. This kind of steel is not generally manufactured, as it is brittle and immalleable. The Hindus (*i.e.* Tamils) also have recourse to a similar method, using the wood of the cassia with a few leaves of the *Calotropis gigantea* or *Convolvulus laurifolia* for making their steel. I have been informed by an old Kandyan blacksmith that the manufacture of steel was one of the services rendered to the Sinhalese kings by the people, under the term "rajakaria" for the performance of which they wered allowed grants of land. The bars of steel generally manufactured and sold by the Sinhalese are small weighing 15 ounces. [*Note*:- One which I obtained from the blacksmith at Kabilwela weighed 1 lb., and measured 6 inches in length, and was of a tapering form, being 1 1/2 inch wide at the broad end. - A.K.C.] In the extensive district of Cuvah two blacksmiths of Dehigolla supplied annually the king's stores with twenty-four small bars of steel, "wana karal;" and twelve persons belonging to the same place furnished a sufficient quantity of charcoal for preparing the steel. In other places the same system was carried on. The following are the names of the places where steel was manufactured, viz., Dehigolla and Iwalla in Wellassa, Irewandumpalla in Kandapally. At present (1854) it is made at Horaguna, Hanahappawela Kammala, and Kosgama Kammala, belonging to Kandapally, also at Mahawalgaha in Saffragam district: *W.C. Ondaatje, "Government Calendar," 1854.*

To these localities I may add Walantharakumbura, Kabilwela, near Bandarawela. Localities where iron smelting was carried on include, besides the above. Medaperuwa, near Bandarawela; Yahalawelakanda, near Ella; on Naseby and "The Scrubs," Nuwara Eliya, as well as on the site of "Grimsthorpe," Nuwara Eliya; also Kendagolla, three miles east of Badulla. The ore was not smelted where found, but carried to the homes of the iron workers and there smelted; heaps of refuse slag with traces of charcoal are often met with on the old sites. Such deposits of slag are indeed of common occurrence, and have been mistaken for actual deposits of ore. The iron smelters were low-caste people, known as "Yamannu," who since the decay of the iron industry have turned their attention to lime burning. It is probably twenty-five years since any iron was manufactured in Uva. Specimens of the ores, the iron, and the steel can be seen in the Museum.*[2]

[2]* I found that a small quantity of iron and steel is still manufactured in the Balangoda district, and hope to give an out of the process (which differs only in detail from that described above) in

The iron was ols for 50 cents per "pig," the steel at 75 cents per bar. The iron was made up in large rounded "pigs" with scoriaceous surface, weighing nearly 6 lb.; these "pigs" were cut up as occasion required. The native iron has a high reputation for quality ; tools made of it are still preferred to English ones. The decay of the industry may be put down to the introduction of cheaper English iron, and the increasing scarcity of wood fuel as the chief causes ; and considering the limited quantity of ore usually available, these factors will probably prevent any revival of the manufacture of iron and steel.

Mode of Occurrence and nature of the Ores:- The ore consists usually of limonite; less frequently hematite is found. A greater or less quantity of psilomelane is almost always associated with the iron ores. Limonite is almost everywhere present in small pellets in the soil, and occurs locally in larger concretionary masses of value as a source or iron. The iron ore and psilomelane are alike decomposition products of the country rocks and occur with kaolin, & c., in beds replacing these. It is in particular the garnets which in their decomposition have given rise to limonite and in less quantity to psilomelane. Garnets are often extremely abundant in the granulite rocks, and all stages in their decomposition and conversion into pseudomorphous pellets of limonite can be readily observed.

The larger masses of iron ore are of a concretionary character, and result from the deposition of limonite in successive layers, often with a mammillated surface; concretionary psilomelane may be deposited on the surface of the limonite; the formation of limonite as a decomposition product in connection with the normal processes of weathering is probably still continued; it may often be observed as a thin deposit on the surfaces of well- weathered rock masses, having been leached out and deposited on the surface; but it occurs especially as a pseudomorphous replacement of garnet, and apparently of whole rock masses.

The quantity of iron ore available is on the whole small, notwithstanding its constant presence in small quantities. The downward extent of the ore cannot exceed the depth to which decomposition has proceeded in the country rock. The local occurrence of fair quantities of ore probably results partly from the presence of an abundance of ferriferous minerals in the original rock, partly on the drainage, for the ore is sometimes deposited in a way recalling iron-pan. The concretionary masses are sometimes exceedingly pure, but may be much mixed with decomposed rock material. Most of the localities examined are some distance from a railway station; whist it is possible that a supply of ore near the cost might be of value for export, there is little doubt that the additional expense of road and rail transport would be prohibitive.

The principal localities for iron ore are tabulated below:-

Ampitiya, 3 miles East of Kandy:- The ore consists of hematite and limonite ; the main bed is 14 inches thick ; the best exposure is on the western end of the Kelawala hill on a landslip scarp ; from here the most ferruginous beds can be traced eastwards at just somewhat below and south of the ridge, but afterwards along the ridge, for a

next year's report. - A.K.C.26-2-04.

distance of nearly one-third of a mile. Impure beds of kaolin are associated with the iron ore, the whole representing the results of decomposition *in situ* of the garnetiferous leptynites of the district. These rocks are altered to a soft crumbling mass in which the garnets are recognizable only as deep brown or purplish stains. Psilomelane associated with the iron ore is more than usually abundant. This locality is on the whole one of the best observed.

The other localities examined are in the Province of Uva. The best was on the Ganamatuwa estate, between Bandarawela and Haputale. Lumps of ore are especially abundant in an area one or two hundred yeards square near and just north of Alakolakanda trigonometrical station. The largest lumps are two or three times the size of a man's head. The locality was worked by the Sinhalese, the Ganamatuwa ore being smelted at Walantharakumbura (Kabilwela). Other localities near Bandarawela include:-

(1) Lebellewala, 1 mile from Bandarawela, a little north of the Badulla road.
(2) A place half mile 20 W. of N. of Halbaragala trigonometrical station
(3) A place about quarter mile on patanas from 6th milepost on Koslanda road

Near Ella

(1) Slope of Sripada Kota
(2) Neighbourhood of Yahalawelakanda (a little over a mile 10 N. of E. Ella trigonometrical station)

In the Badulla District

(1) Kendagolla 3 miles east of Badulla
(2) Mahakumbura (Bogoda korale)

Near Nuwara Eliya

About 2 1/2 miles along the path from Blackpool to Ambewela (hematite).

CORUNDUM.

The mineral (which includes ruby, sapphire, star sapphire, oriental amethyst, oriental topaz, white sapphire, and emery) is well known to occur abundantly in the gem-bearing gravels of Ceylon. The better stones are cut as gems ; those of no value are to some extent used as emery in cutting and polishing. In Southern India the mineral is found *in situ* and extensively quarried in the Erode district. It has been carefully searched for in Ceylon rocks; so far, however, only one locality has been discovered, viz., at Tennahena, about two miles from Talatu-oya near Kandy. Here a quantity of bright blue corundum is met with in one or two bands of felspathic granulite only a few inches wide. The corundum is not present in sufficient quantity to be of commercial importance, but is discovery is of interest as being that of the only known locality where it occurs *in situ* in Ceylon. The blue crystals (usually hexagonal prisms, with well-developed rhombohedral and basal cleavage, and often much hydrated) are scattered through the crumbling felspathic granulite. The mineral composition of the rock is orthoclase microperthite 64.2 per cent., oligoclase 23.5 per cent., biotite 4.7 per cent., corundum (with other heavy minerals) 7.6 per cent. The corundum appears to occur normally as a magmatic product, and its presence does not seem to be due to any sort of contact metamorphism. The corundiferous band occurs amongst normal varieties of granulite, there being nothing to lead the observer toe expect its presence. The experimental digging made three years ago by the Director is now filled up.

MOONSTONE.

Some of the most important localities for moonstone have been examined. The Dumbara district of the Central Province is the chief source of these stones.

The moonstone is a clear glassy variety of orthoclase (adularia), quite free from any microperthitic inclusions. The characteristic sheen is best seen when a plane parallel to the orthopinacoid, or nearly so, is turned towards the observer, and the gem should always be so cut as to allow of its flat underside being parallel to this plane, thus ensuring a central disposition of the moon-like light. The moonstone orthoclase occurs in large crystals in certain acid granulites (leptynites) composed of quartz and felspar (orthoclase, orthoclase microperthite, oligoclase). Very characteristic of these rocks is the great elongation of the individuals of quartz in the direction of foliation, and a tendency to a complicated graphic intergrowth of quartz and felspar. The large crystals of orthoclase may be several inches in diameter, and are usually to avoid form, not being idiomorphic. The crystals are well cleaved, and so usually extracted in fragments. A bright blue sheen is also occasionally seen in plagioclase felspars, *e.g.*., in labradorite form near Kurunegala, but no crystals large enough to cut were found. There is a local industry in cutting and polishing, six or eight men being employed in this way at Attaragala; the greater part of the stones found, however, are cut in Kandy. The moonstone bearing leptynites are usually associated with the crystalline limestones in which they occur as bands or sills of considerable width. The leptynite is sometimes intimately interbanded with the crystalline limestone with narrow intervening zones of diopside rock with mica, spinel, scapolite, sphene, serendibite, & c.

An inferior quality of moonstone can frequently be found in the coarse pegmatites, which are not uncommon in some parts of the Island. The inferior kinds are known to the natives as "beryle" moonstone.

Method of Working :- Shafts or trenches rarely more than 20 feet in depth are sunk on the moonstone-bearing granulite; the rock is, if necessarily blasted, then more completely broken up, and the fragments of clear moonstone picked out. In some cases the working has been carried on by means of adits driven along the best moonstone-bearing band, and not carried for any great distance. Water is sometimes removed from the pit by a bucket attached to a see-saw swinging pole.

The localities examined are tabulated below:-

Gangapitiya, Udagampaha Korale (in Uda Dumbara) :- Between Lower Rajawela estate and the Mahaweliganga at Horagama. Several moonstone pits; have been dug in this locality in bands of leptynite in the crystalline limestone. The only one worked in 1903 yielded white moonstone of fair quality. Another pit not far away from this one consists of a small shaft and adit driven along one of the wider banks of granulite, which alternates in bands rarely as much as one or two feet wide with bands of crystalline limestone. This pit is the original locality for the new mineral, *serendibite.* The adit measures some 10 feet in width by 5 feet in height, and follows the dip of the beds for some 30 feet. Work at this pit has been abandoned for some years.

Several remunerative pits in the same or a similar band of granulite are found on Upper Rajawela estate, but are now worked.

Very important diggings can be seen at Attaragalla, near the 11th milestone on the Katugastota road ; the band of acid leptynite has been followed along the strike for a considerable distance (the vertical ; beds striking in a direction almost parallel to that of the road) until the road was reached ; but although a shallow excavation was made on the further side of the road, no work is being carried on at present. The Attaragalla pits are the only locality other than Gangapitiya where *Serendibite* has yet been found.

Gangewatte Korale (Yatinuwara). A number of small pits for moonstone (some of which is of a particularly fine blue colour) have been worked at Kosinne, near Talatu-oya. The moonstone is obtained from bands of acid granulite occuring in the crystalline limestone. Work has been done at wo pits during the present year. Neither is very extensive, the larger not exceeding 0 or 12 feet in depth.

It may be added to the above account that the supply of moonstone available is more or less unlimited, and will continue for a long time merely to vary with the demand.

GARNETS.

Garnet is to a small extent collect *in situ* from the decomposed granulites in which it occurs so abundantly. Profitable pits were formerly worked on Victoria estate, near Teldeniya (Uda Dumbara) and also on Upper Rajawela estate, north of the cart road, a little before the junction near the 12th milestone is reached.

URANINITE (PITCHBLENDE).[*3]

This mineral, now of interests as the chief source of radium, has been found in small cubic crystals in gem washings in the Bambarabotuwa district, and also near Gampola,[+4] where Mr. Gaddum has detected it in a vein of pegmatite (consisting mainly of quartz, orthoclase, and biotite) in which apatite and tourmaline as well as limenite were also found. The quantity here met with is extremely small, but my examination will be of use in searching for it in the Bambarabotuwa district in 1904.

[3*] It now appears that mineral from the Bambarabotuwa district (and probably also that from Guranninite, but a new mineral, whose detailed composition is not yet certain - April, 1904.

[4+] See Spolia Zeylanica, Vol. I. Pat IV, page 112, 1904

DETERMINATION OF SPECIMENS SUBMITTED

A part of the work undertaken consists in the determination of specimens submitted by Government or by members of the public. A few specimens have been submitted by Government and duly reported on. Of those sent in by members of the general public, a number consisted of clays containing scales of finely divided mica, erroneously supposed to be or to contain gold. In the course of the field work, advice as to the best method of working and disposition of the veins of mica or graphite has frequently been given. Such advice is not infrequently sought and always freely given. No charge is made for the determination of specimens, but their examination is not allowed to interfere with the more systematic work of the Survey. A circular letter to the various Planters' Associations especially invited the forwarding of specimens for examination; this work is useful as leading to a more extended acquaintance on the part of the Survey with the mineral resources of the Island. Thanks are specially dur to Mr. W.D. Holland for the interesting minerals (some of which may be commercial value) which he has forwarded; several were identifies by himself, and are new records for Ceylon.

OCCURRENCE OF CINNABAR

Persistent rumours of the occurrence of cinnabar near Kadugannawa are afloat. Mr. Parsons has accordingly investigated the principal supposed locality, Sindurankanda, a hill about two miles south-east of Kadugannawa. The locality was visited by some Boers and by the Inspector of Mines. Samples were obtained by the Boers and several were assayed, the best yielding of 9 lb. mercury to the ton. Mr. Parsons remained a fortnight at Sindurankanda and made extensive excavations, finally examining the most promising material by retorting. No trace of mercury was found, nor does any of the ore resemble cinnabar, nor are there any signs of mineral veins or any probability of the occurrence of such. I have little doubt that the specimens obtained by the Boers were salted. This is not the first occasion on which I have found that their statements have shown entire ignorance of the geology of the country, or that they have "discovered" minerals which do not actually occur. Various specimens have been forwarded to London for further examination.

At Kotte, near Colombo, there seems to be no doubt that metallic mercury has been met with in some quantity in the ground, and has been obtained by washing and sifting the soil. I have examined the locality carefully and found that the soil consisted of ordinary laterite (cabook) resting no doubt on the ordinary undecomposed granulites. No mercury was obtainable at the time of my visit. Native mercury is very rarely found, and no signs of cinnabar were noticed, such as it might have been associated with. The explanation that if was stored by the Sinhalese kings and used for medicinal purposes seems most probable, though *a priori* rather a strained theory. There is no geological probability of there being ore of mercury at Kotte.

MUSEUM WORK

A period of ten days in July and again of rather over three weeks in December (recess) was spent at the Museum. A new room has been set apart as a mineral gallery. Plans were made and orders given for the necessary cases; these, with the exception of some glass, were completed before Christmas. During the recess visit the arrangements and cataloguing of specimens was proceeded with; labels have been printed for a few, but this work has been so slowly sent in by the Government Printer that the majority of specimens are still unlabeled. A total of between 400 and 500 specimens has been numbered and catalogued; this includes a part of the original collection, which was in many respects poor and limited. A very large number of specimens have been added to the collection by the Mineral Survey during the year.

A plan of the cases as no arranged is appended. Wall cases Nos.1 and 2 contain the systematic collection of rock specimens ; wall case No.3 is devoted to graphite; No.4 to mica, iron ores, kaolin, and other economic minerals. A fine series of commercial graphites has been presented by Mr. Peter d' Abrew; the series illustration the mode of occurrence is poor as yet, but it is hoped that it will be improved during the ensuing year. An interesting series of specimens illustrating the manufacture of iron and steel is, however, shown.

The table cases Nos.1 - 5 arranged around the walls will contain the collection of minerals (including gems) systematically arranged. The central table cases will contain (1) fossils; (2) specimens illustrating the physical characters of Ceylonese rocks and minerals, with descriptive labels.

A series of geological photographs will be hung upon the walls, as well as diagrams and descriptive ace of the rocks and minerals.

The collection of gems is at present poor; with the exception of an alexandrite and a cat's-eye, there are non of much value ; it is most desirable that a good collection should be shown, as visitors take an especial interest in these characteristic Ceylonese products, and should be able to see *good cut specimens* of all the kinds of gems,

---

L12 CEYLON ADMINISTRATION REPORTS FOR 1903. PART IV.

---

associated with the rough specimens of the same minerals. The gems specially needed include a ruby, sapphire (a really good one of each should be exhibited), also a series of garnets, moonstones, and the other less precious gems. A sum of not less than Rs.1,000/- is needed for this purpose, and I beg to suggest that a special grant be made. I regard the specimens as urgently needed.

Further work at the Museum will be done next year during recess; it may be very desirable also to spend two or three days at this work also during the middle of the year.

January, 1904

A.K. Coomaraswamy
Director, Mineralogical Survey

Plan of New Mineral Gallery

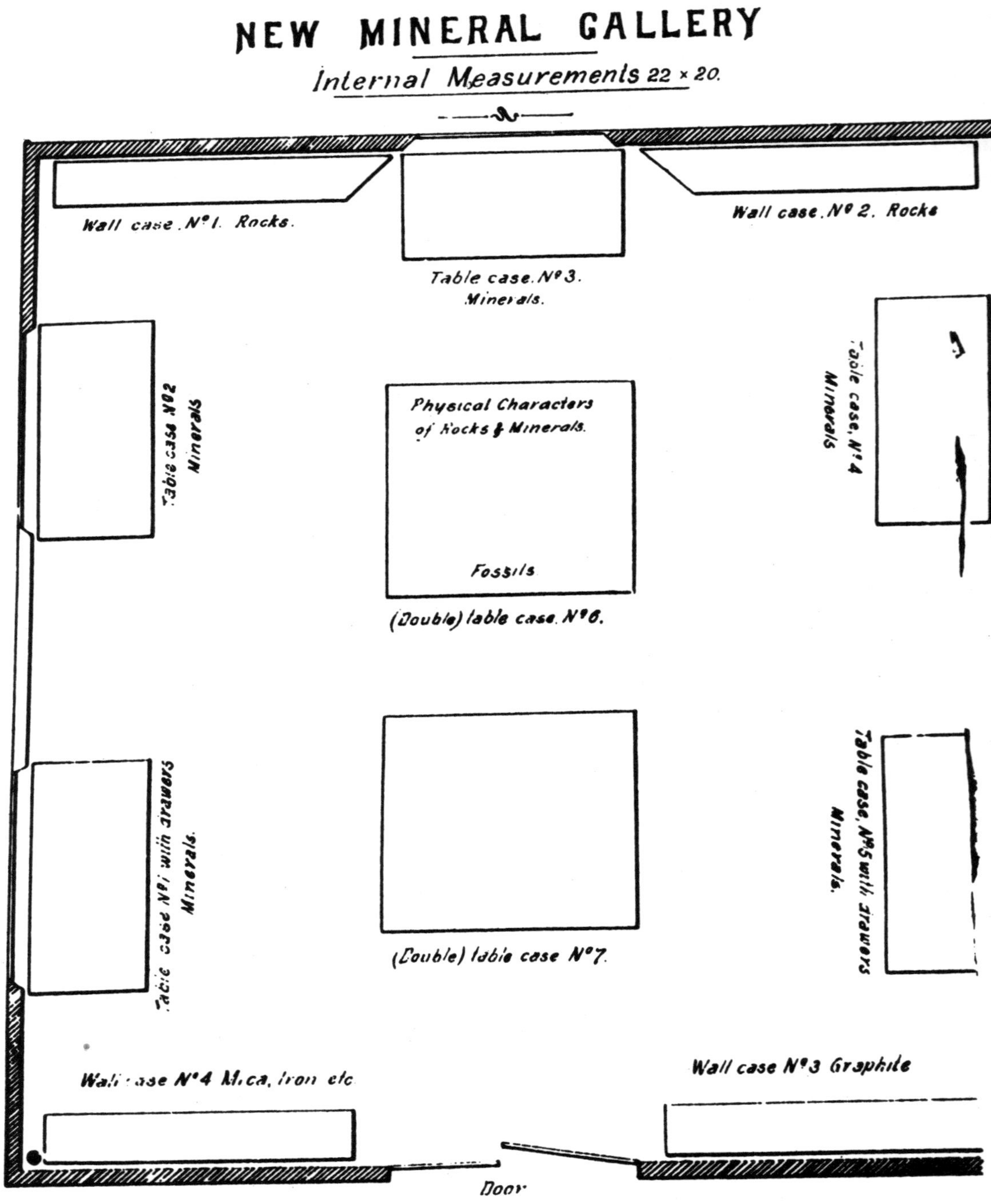

NEW MINERAL GALLERY

Internal Measurements 22 × 20.

Wall case. No. 1. Rocks.

Table case. No. 3. Minerals.

Wall case. No. 2. Rocks

Table case No. 2 Minerals

Physical Characters of Rocks & Minerals.

Fossils

(Double) table case. No. 6.

Table case, No. 4 Minerals

Table case No. 1 with drawers Minerals.

(Double) table case No. 7.

Table case, No. 5 with drawers Minerals.

Wall case No. 4 Mica, iron etc

Wall case No. 3 Graphite

Door

GEOLOGICAL SURVEY & MINES BUREAU

MINERALOGICAL SURVEY

CEYLON ADMINISTRATION REPORT
FOR YEAR 1904

BY
ANANDA COOMARASWAMY
1995

MINERALOGICAL SURVEY

Director : ANANDA K. COOMARASWAMY
B.Sc., F.L.S., F.G.S.

Assistant Director : **JAMES PARSONS, B.Sc., F.G.S.**

REPORT OF THE DIRECTOR OF THE MINERALOGICAL SURVEY FOR 1904

DISTRICTS EXAMINED I 1904

(a) By the Director

I began work in January at Haputale and continued thence to Balangoda, stopping at Haldummulla and Bilihuloya, after which a stay of four weeks was made at Bambarabotuwa in March to investigate the occurrence of thorianite there. In May I returned to Bambarabotuwa, but the break of the monsoon rendered work impossible there, so that I returned to Balangoda and went across country to Kolonna, stopping for a few days at Weligepola, pallebedda, Godakawela, and Ratkarawa. After twelve days spent at Kolonna I proceeded to Madampe, leaving the neighbourhood of Rakwana to be examined by Mr. Parsons, and thence to Pelmadulla ; but my work during June and July was much hindered by heavy rains. At the end of July I visited Maturata to inspect the workings for gold being carried on there. In August tend days were spent at Museum, and I then returned to Sabaragamuwa and worked in the Kuruwita Korale, making stays of week or a fortnight at Gilimale, Pallabadaela, Ratnapura, Kuruwita, Eratna, and Pussella and returned to Kandy by road through Kegalla in time for recess, the months of November and December being occupied with the preparation of the tenth interim report and of the annual administration report, the determination of minerals, and Museum work. The health of the staff has been excellent, with the exception of some days in October and November, when the preparation of reports was somewhat delayed by fever, & etc.

(b) By the Assistant Director

Work was commenced in January at Akuressa, Southern Province, attention being chiefly given to the plumbago of the district. The next district examined was Maliduwa and the Dedivagala - mukalana gemming lands from Kohugoda as centre. Kottawa was next visited, thence to Nakiyadeniya and Udugama. The Gin-ganga was ascended by boat as far as Hiniduma : a week was then spent at Habarakada, and the district up to the Pasdun korale boundary examined. An excursion was made to Panagoda to examined the plumbago pits of the place. From Habarakada the route was taken through Opata to the main road at Hulandawa, from which the gemming lands east of the Nilwala-ganga were examined. A stay of a week or so was made at Morawaka, and from there the route was to Urubokka *via* Kotapola. A month was spent at Urubokka examining the gemming lands. A few days were then spent at Deniyaya, after which Viharahena was visited "The gemming of the southern portion of the Sinha Raja forest was investigated and the country traversed up to the Province boundary.

After a week's leave and a week occupied in work at the Colombo Museum a month was spent in Rakwana. From here the route was taken to Bultota and Illebekanda. At these places mist and heavy rain much interfered with the work. Returning to Rakwana a traverse was then made into the Kukulu korale; some time was spent in the villages of Rambuke and Potupitiya, and the various gemming lands examined as far as Delgoda.

CORUNDUM

By the Director

Corundum was discovered in two localities in the Haputale district. It occurs on Ormiston estate, about one fifth of a mile S.S.W. of Dambaketiya trigonometrical station ; violet pink to purplish hexagonal crystals are here found on an in the surface soil, over ; quite a small area, evidently corresponding to the outcrop of the corundum-bearing rock; this matrix, however, could not be detected either *in situ* or in loose boulders, although very careful search was made, and the mineral traced to the point of greatest abundance on a slightly sloping ground ; and above this pint none could be discovered. The corundum has been sent to England, and a value of $ 12 to $15 a ton quoted. Unfortunately there is no prospect of commercial quantities being available here.

A much more important occurrence is that on Haldummulla estate, about five miles below Haldummulla. The corundum occurs here in violet pink and purplish, very rarely blue-gray, hexagonal crystals scattered in varying abundance in the surface soil, over a great part of Haldummulla estate and extending on to the western portion of Kalupahane estate, e.e., for a distance of about a mile horizontally and perhaps a thousand feet vertically. The area within which is abundant on Haldummulla estate is about 30 acres or more ; that on Kalupahane much less.

E2 *Ceylon Administration Report for 1904* *[IV-EDUCATION.*

The loose crystals are accompanied by blocks of rock containing corundum. The minerals sillimanite, corundum, orthoclase-microperthite, garnet, rutile, and ilmenite enter into the composition of these rocks. The principal types met with are : sillimanite-corundum rock, sillimaniter-garnet rock, and sillimanite rock ;

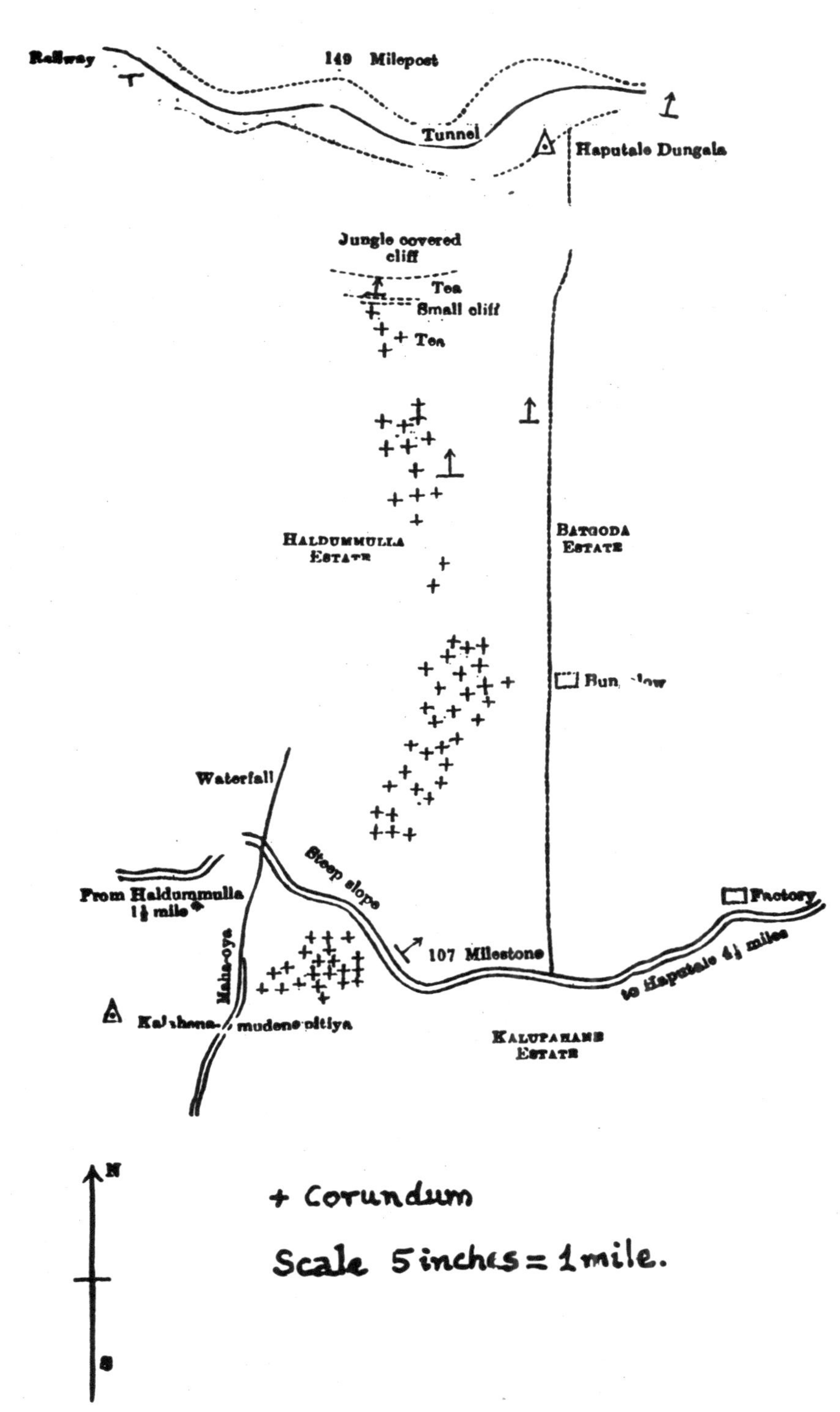

Fig 1 - Sketch-map showing distribution of Corundum near Haldummulla by A.K. Coomaraswamy, 1904

orthoclase-microperthite is present in subordinate and varying amount ; ilmenite and rutile are accessory and do not occur together, but mutually replace each other ; the corundum and garnet also do not occur together in the same rock. The corundum forms violet-coloured hexagonal crystals, usually less than three-fourths inch in diameter ; the crystals have often a tabular development, the forms c (0001), a (1120), and r (1011) are characteristic ; phombohedral cleavage is well developed. The sillimanite occurs in parallel and radiating groups and single individuals, the latter

Science and Art] *Mineralogical Survey.* *E3*

generally idiomorphic (prismatic, giving rectangular cross sections). The crystallization is much coarser than is usual for sillimanite ; the largest crystals may reach a length of two inches or more and diameter one-tenth inch. The sillimanite has in the rock a pale gray colour, but is colourless in thin flakes.

The corundum was traced up to the foot of the jungle covered and inaccessible cliff which forms the highest part of Haldummulla estate ; there can be little doubt that the corundum and blocks of corundum-bearing rock are derived from exposures within that area.

The crystals occurring scattered in the soil are quite freed from the matrix, which readily dis-integrates, and they can be very easily collected, especially after rain, and it is possible that it would be worth while to do so for purposes of export, but it is not likely that more than a few tons at the most would be obtainable. It is just possible that the matrix could be found in the cliffs referred to, but the search would be costly and difficult ; if found, it would probably be worth working. The sample sent to England was pronounced not sufficiently uniform for valuation, but it is very like the Ormiston material and probably has a similar value. It is not unlikely that corundum might be met with elsewhere in the same district.

Sillimanite has been found to occur in some abundance in the garnetiferous leptynites (which may also be called quartz-felspar-garnet sillimanite schists, or khondalites*)[1] over a great part of the areas examined in 1904. The sillimanite-bearing rocks are found at many places between Bandarawela (where they were first detected) and Gilimale; they are found also much more sparingly in other districts, *e.g.*: at Kandy (Central Province) and near Kolonna (Sabaragamuwa); while a very coarsely garnetiferous sillimanite-bearing rock has been collected by Mr. Parsons near Matara (Southern Province). Very interesting varieties of similar rocks have been described by Lacroix;[+] [2] some of these were collected by Leschenault de la Tour in 1819 and others formed a part of various old collections in Paris. "A biotite-sillimanite-gneiss" is stated to occur near Colombo, while a remarkable andalusite-sillimanite rock came from Fort King, *i.e.* Attapitiya in Galboda korale, Kegalla District ; a corundum-sillimanite rock is also described, but no exact locality quoted.The occurrence of an isolated crystal of sillimanite in gem refuse or

[1]* Walker, Mem.Geol, Sur. India, vol.XXXIII., part 3, 1903, p.8

[2]+ Lacroix, Bull.Soc.France.Min.,vol. XII., pp 203-211

nambu has been recorded by Grunling++ 3 A specimen of sillimanite-bearing rock was also sent in to the mineral survey from the Matale District during the present year.

The sillimanite-bearing rocks have a strong resemblance to the khondalites of Southern India; they do not, however, occur in the same way above the charnockite series, but completely incorporated with it, forming probably bands and lenticular masses ; the rocks have however, been rarely seen *in situ,* and it has not been possible to study closely their relation to the charnockites proper. It must be pointed out that there is complete transition to the ordinary granulite type. On the one hand we have a rock composed wholly of sillimanite, and all varieties from this through quartz-felspar-sillimanite garnet schist, to garnetiferous leptynite with very little sillimanite, and ordinary garnetiferous and non-garnetiferous leptynites, can be collected. The coarsely crystalline character of many varieties is also very noticeable. It is probable, if it be considered that these aluminous rocks represent the remnants of some ancient sedimentary series (perhaps to be associated with the crystalline limestones in this respect), that the original rocks have been melted and recrystallized, a conclusion to which a study of the crystalline limestones also leads. The abundance of felspar and transition to ordinary igneous types shows that the aluminous rocks have been more completely incorporated with the latter, and that they have been more mixed with igneous material than would appear to have been the case in Southern India, where they form more individual masses and occur bordering or overlying the charnockite series proper.

The presence of sillimanite can hardly be taken as any general indication of the occurrence of corundum as it is only at Haldummulla that the corundum sillimanite type rock; has been found. It is rather courious that the corundum found in these rocks, as well as that from Talatu-oya, near Kandy (described in my last report), occurs in columnar hexagonal crystals, while most of that to be got from the gem-bearing gravels occurs in doubly pyramidal crystals. It is evident that corundum occus in Ceylon in several types of rock, and of these only two have been found as yet. It may be mentioned here that the said blue there is not the least prospect of commercial quantities being obtainable. Samples of corundum have been received from Hatton and Dikoya and forwarded to the Imperial Institute; but the exact localities were not stated, and valuation reports are not yet to hand. Beside the actual and possible occurrences above referred to, corundum is to be found in most of the gem-bearing gravels of the gemming districts ; it includes of course the gems ruby, sapphire, star ruby, and star sapphire, white sapphire and yellow sapphire, oriental amethyst and king topaz, as well as the poorer material called *dalam*, and that known as *kurundugala* and useful only for abrasive purposes. The better sorts of corundum, coloured and transparent, but not so valuable as to be worth selling separately, are called *dalam*, and fetch locally from Rs.5 to Rs.70 for a pound. The opaque sorts, useful only for cutting gems and known as *kurundugala*, fetch from 10 cents to Rs.5 per lb. Some at least of the *dalam* is exported to Europe, the "corundum" only very occasionally. Gems apart, corundum probably never occurs in sufficient quantity for the gravels to be worth working for it along ; but it sometimes happens that the dalam is relied on to pay working expenses, while there is always the chance of getting something really good as well.

3++ Grunling, F. Zeit.fur.Kryst,Bd.XXXIII., Heft., 3-5p.236

Pattalewaladeniya, opposite Paradise estate, Kuruwita (Sabaragamuwa), is a noted locality for corundum and may be taken as a place where it is rather more than usually abundant. Two men digging a hole 3 feet square and 4 feet deep and working half an hour got about a quarter of a pound of corundum; and it was stated that four men in four days could get ton. Again, at Kinagahadeniya, Kuruwita, a pit 17 feet by 9 feet in area and 8 feet or 9 feet deep sunk; by four men, working three days, yielded 13 oz. of corundum besides other minerals. These figures do not suggest that it would be very profitable to collect corundum for export from the localities where it occurs in gem-bearing gravels. Mr. Parsons states that it is generally absent altogether in the Galle and Matara Districts of the Southern Province.

E4 *Ceylon Administration Report for 1904* *[IV-EDUCATION.*

MANUFACTURE OF IRON AND STEEL

By the Director

All over Ceylon, or at the least in Kandyan districts, the manufacture of iron and steel was once and important industry ; the heaps of slag (now occasionally mistaken for actual iron ore) marking the sites of ancient furnaces are everywhere abundant. It will not therefore be worth while to attempt any enumeration of the sites of ancient furnaces actually observed during the present year. The account given last year of the distribution and character of the iron ores needs but little modification. The essential facts were observed as early as 1681 by Robert Knox (An Historical Relation of the Island of Ceylon), who states that "the country affords plenty of iron, which they make of stones, that are in several places of the land ; they lay not very deep in the ground, it may be about four or five or six feet deep;" he gives also an excellent account of the process of smelting, identical with that still followed.

One sample of ore (hematite) has been analysed at the Imperial Institute. It was collected at a point on the path from Nuwara Eliya to Ambewela, about 2 1/2 miles from Blackpool.

Analysis

SiO2	9.14	S	Nil
A12O3	9.85	P2O5	.05
Fe2O3	72.30	P	.022
H2O	8.40	... ...	—
			99.96

Of some other ores from Uva, it was stated that they appeared to be of good quality, but somewhat more hydrated.

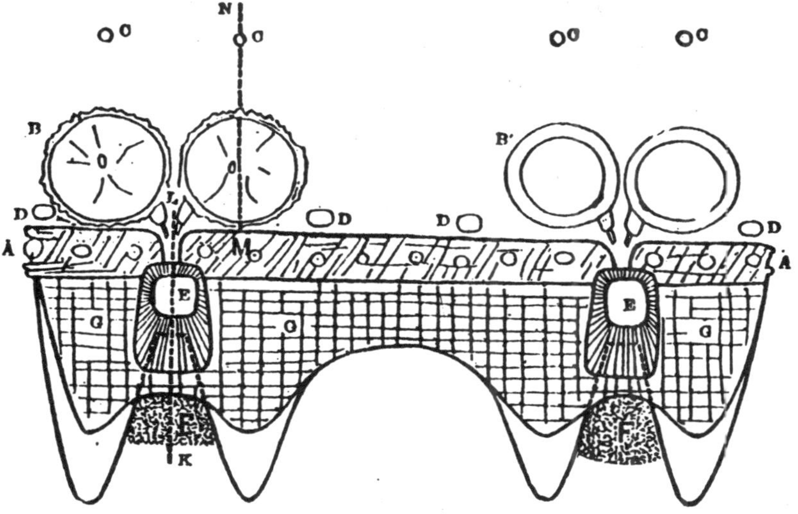

Fig 2. Combined plan of double furnace at Hatarbage at the three levels. I., II., III. of fig.3

AA	=	Wall if mud and wattle protecting bellows blower from heat of fire
B	=	Bellows, B'do without deerskin B" do in section
C	=	Sticks acting as springs to pull up the bellows
D	=	Stakes to which the strap supporting the bellows blower's back is attached
E	=	Furnaces. In fig.2 filled with ore and charcoal.
F	=	Slope of sand, over which the slag runs out when a stick is inserted
G	=	Flat surface of furnace about 2 ft. 9 in. above ground at k (the squared shading marks the limits of the level area in the plane II. in fig 3).
KLMN	=	Line of section of fig.3.

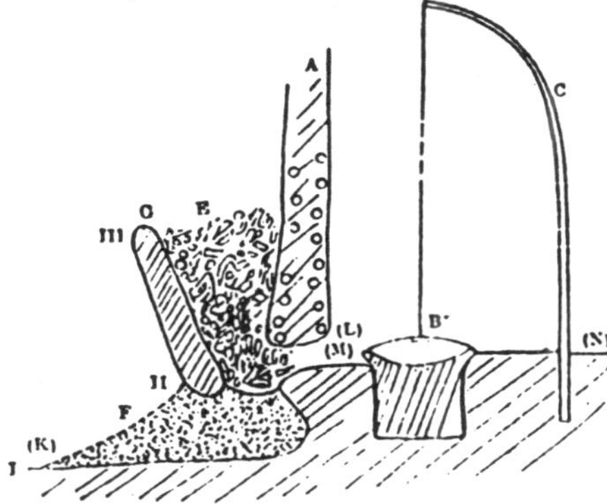

Fig.3 - Section of furnace, along K L and M N of fig.2
Scale of figs 2,3,4 – 1/2 inch = 1 foot

The manufacture of iron is still carried on to a small extent near Balangoda, and steel can still be made, though it is not now done as a regular trade. This appears to be the only part of Ceylon where these crafts survive, and even here only three or four families are employed in this way. I have seen furnaces that must have been quite recently abandoned. The industry has largely suffered from the competition of cheaper European iron and steel; but at the present time
it seems to be mainly the difficulty of obtaining charcoal due to the strict preservation of Crown jungles that is giving the finishing touch to this dying industry. The Government Agent of Sabaragamuwa informs me that no arrangement can be made by which the small amount of wood required by the remaining smelters can be provided, although the smelters were anxious to obtain and pay for it. It appears to me that this is much to be regretted.

Process of iron smelting - The furnace inspected was at Hatarabage, about a mile from Balangoda hospital. The furnace ( ) is worked be an old man, Kiri Ukkuwa, and his sons. The iron smelters are called yamannu ( ), and belong to a low caste. The furnace is sheltered beneath smelter and his family; as was always the case it being more convenient to collect and bring the ore to the furnace than to build a furnace near the ore. Near the furnace is a large heap of slag, the accumulation of very many years. The furnace is a double one, one part only being used for my information ; its essential parts are the well or actual furnace, a wall of sticks and mud protecting the bellows blower from the heat of the fire, and the bellows behind this wall. The well has two openings, one in front, from which the slag runs out, and through which the bloom of iron is ultimately extracted and a smaller one behind, through which a blast of air from the bellows is forced into the well, and through which a rod is inserted from time to time to pike and test the iron to see if it is ready. The ore, mainly limonite, is broken up into pieces a little larger than a walnut, and roasted previous to the charging of the furnace, a large part of the water being thus driven off.

The bottom of the furnace is covered thickly with sand, which also fills up the front opening of the well. A charcoal fire is started at the bottom of the well and covered with layers of ore and charcoal until the well is filled, a continous blast of air from the bellows keeping up the fire meanwhile. Additional ore and charcoal are added as the work proceeds, the stoking with charcoal, together with the occasional letting out of slag ( ), being sufficient occupation for one man.

The bellows ( ), from which a continuous blast is kept for three hours or more, are worked by two other men, who share the work by turns; they are of a primitive bu ingenious type. Behind the furnace are embedded vertically in the ground two logs of wood, about 17 inches in diameter, and hollowed out above to a depth of 3 or 4 inches ; over each of these a piece of deer skin is tied having a small hole near its centre, and at the centre a piece of cord ( ) attached, which latter is fastened 6 or 7 feet above to a springy stick, the lower end of which is embedded in the ground. From each of these two sections of the bellows a small pipe conducts the blast into the hole ( ) leading to the well of the furnace. The bellows blower stands with his feet, one on each of the deer skins, facing the wall and lifting and putting down his feet in turn, drives a continous blast of air into the furance; for as he puts down his foot to press out the air from one skin, he covers the hole in it with the sole of his foot and meanwhile the other skin is filled with air through the hole, which is uncovered when he lifts his foot, the skin being pulled up by the tension of the bent stick upon the string. On either side of blower is a stake which he grips hold of, and to which a strap is fastened that passes round behind his and supports his back.

After three hours or more, when the iron is found to be ready, the sand is cleared away and the bloom pushed out through the opening, and then follows the process of "cutting the iron".

The bloom of iron is taken up in long tongs made of greenwood sticks tied together near one end, and it is then beaten a little into shape with thick sticks; then while it is held down on a log with two of the wooden tongs a third man takes a katty and cuts the lump of iron nearly half through; the gash so made is widened by the insertion of a green log which is beaten in so as to force the edges further apart; this is done that the quality of the iron may be examined, for the outside is spongy, and the whole texture far from homogenous. The bloom is then thrown into water and afterwards taken out and left to finish cooling. The bloom thus made is soft and malleable, and weighs about 6 lb.; it is sold to the blacksmiths for 50 cents. Some was formerly sold to the steel makers ( ), a higher caste than the iron smelters. At the present time about 200 pieces per annum are made by three families.

Steel making: - I inspected the manufacture of steel at Alutnuwara, five or six miles from Balangoda. It is not regularly made, but the process is not yet forgotten. The process is somewhat more delicate than that of smelting iron. The furnace ( ) is much smaller than the iron smelting one. The wall and bellows are identical, but the hearth is merely a semicircular depression filled with charcoal, and into which the blast of air is conducted from the bellows. Around this actual hearth is a low clay wall rising five or six inches from the gound. The steel is made in clay tubes. Each tube ( ) is about 8 in. long, nearly 2 in. in diameter, and the clay is about 14 in. thick. Into it is put a piece of iron and some chips of *ranawara* wood (*Cassia* auriculata). In the present case the proportion was 12 1/2 oz., iron to 5 oz., wood. (See W.C. Ondaatje, "Government Calendar", 1854, quoted in Mineralogical Survey Administration Report for 1903 p.8, for further details as to the woods employed and amounts required.) The tube is then closed with a lid of the same material, but with several small holes pierced for the escape of gas. The tubes are embedded in the charcoal and a fire started. Very soon the gases from the wood burn off, and while this goes on the blowing is stopped. Then the blast is kept up continuously, while the tubes are turned about and more charcoal added, the great object being to keep up a constant and even distribution of heat. When the steel is likely to be ready, a hole is opened in the front part of the hearth, so that the blast goes right through the furnace, and the tubes are lifted up one by one in long iron tongs and shaken to see if the steel is quite liquid. Any which are not quite ready are returned to the furnace for a time. The others are laid down to cool and subsequently broken upon and the bar of steel removed, the steel is highly crystalline, and is used for the best cutting tools. The bars are worth from 75 cents to a rupee each.

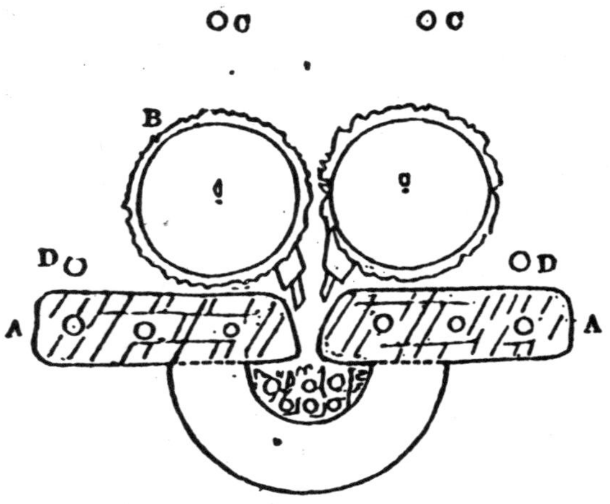

Fig.4:- Plan of steel-making furnace. Lettering as in figs. 2 & 3

An almost identical method of manufacture is practised in India ("Iron Smelting in Mysore," a lecture by V.S. Sambasiva Iyer, Madras, 1902), where both iron and steel have been manufactured from time immemorial.*. [4] The methods employed in Ceylon probably date from the earliest aryan times, and were doubtless brought by the Sinhalese themselves. Indian (and I am told also Ceylon) steel has a great reputation in medieval times, and was much sought after by the makers of swords, being even exported to Damascus for the manufacture of the famous carpons there produced.

Specimens of modern Sinhalese steel have been sent of the Imperial Institute for a technical report. It will be much to be regretted if the arts of smelting iron and making steel should become quite extinct in Ceylon, as they are on the verge of doing, both on account of their historical interest and because they do still affort a partial livelihood to a limited number of persons.

The amount of charcoal required for the very few furnaces remaining could not possibly have any serious affect on the Crown forests, if some arrangement were made by which certain quantities were allowed to be taken by the smelters, combined, if necessary, with some amount of replanting, it seems that in India it has even been considered that iron smelting "will facilitate the formation, protection, and improvement of forests" (Sir D. Brandis, quoted in the lecture already referred to.)

MINERALS CONTAINING RARE ELEMENTS AND ASSOCIATED MINERALS

By the Director

[4*] Cf.H. Parker. Report on Archaeological Discoveries at Tissamaharama, Journal C.B.R.A.S., vol VIII., No.27, 1884. p.153(61)

A group of minerals occurs in Ceylon, a number of which are distinguished by their content of rare elements. Those which can be associated together as certainly or probably derived from the intrusive granite rocks to which the name "Balangoda group"+[5] has been given, include the following:- allanite, baddeleyite, beryl, cassiterite, chrysoberyl, fergusonite, geikielite, ilmenite, magnetite, rutile, sphene, some spinels, thorianite, thorite, topaz, tourmaline, and zircon, as well as some undetermined minerals apparently belonging to the samarskite and aeschynite groups. Of these minerals a short account will be given in turn, commencing with thorianite, as being of the greatest commercial importance at the present time : a few of those mentioned of course are by no means peculiar to the Balangoda group.

Thorianite++[6] This newly discovered mineral, at first supposed to be uraninite, but now found to contain over 70 per sent, of thoria and only 12 per cent to 15 per cent of uranium oxide, occurs in moderate quantities near Kodurugala in Bambarabotuwa, Sabaragamuwa. The principal deposit occurs in and near the bed of the upper part of the Kuda Pandi-oya, a small stream which at first occupying a small north-west and south-east strike-valley north of the Hopewell-Hapugastenna bridle path, turns through nearly a right angle and joins the Maha Pandi-oya a little below the same path. For the localities referred to the accompanying map should be consulted. In the bed of the Kuda Pandi-oya and in a small "deniya" (swamp) just below the camp, the thorianite is to be obtained in considerable abundance. It could not be discovered *in situ,* but it can hardly be doubted that it is derived from some rock outcropping not far distant from the highest part of the little stream; the stream is so small that after a few days of drought (not very usual in this wet district) no running water is met with above the camp. The matrix, is no doubt a rock of granitic type similar to those containing zicon, allanite, & c., which have been referred to as belonging to the Balangoda group, and in which thorianite was actually met with near Gampola.

The thorianite occurs in the form of very heavy black cubic crystals more or less waterworn ; associated with it is a large quantity of zercon in good crystals not much waterworn, and rolled pebbles of ilmenite ; there is also a much smaller quantity of a brownish-yellow to organe-coloured mineral identified as thorite, and occurring in crystals usually very much worn, this mineral (at first identified as monazite has a chemical composition and specific gravity corresponding to those of thorite ; but the shape of the comparatively few crystals whose form is well preserved is quite unlike that of a tetragonal mineral, and I do not at present feel satisfied that the mineral is really thorite. All these minerals are also found in broken and rolled graments of a sub-recent ferruginous conglomerate which has formed in the bed of the stream and been subsequently broken up. This conglomerate contains rounded pebbles of rock, heavy minerals, & c., held together by a ferruginous cement. Not very much of this material is found, the greater part of the heavy mineral occurring in loose crystals.

---

[5]See contributions to the Geology of Ceylon, III., The Balangoda Group by A.K. Coomaraswamy, - Geological Magazine, 1904, pp. 418-422

[6]++ W.R. Dunstan, Nature, 1904 p.510

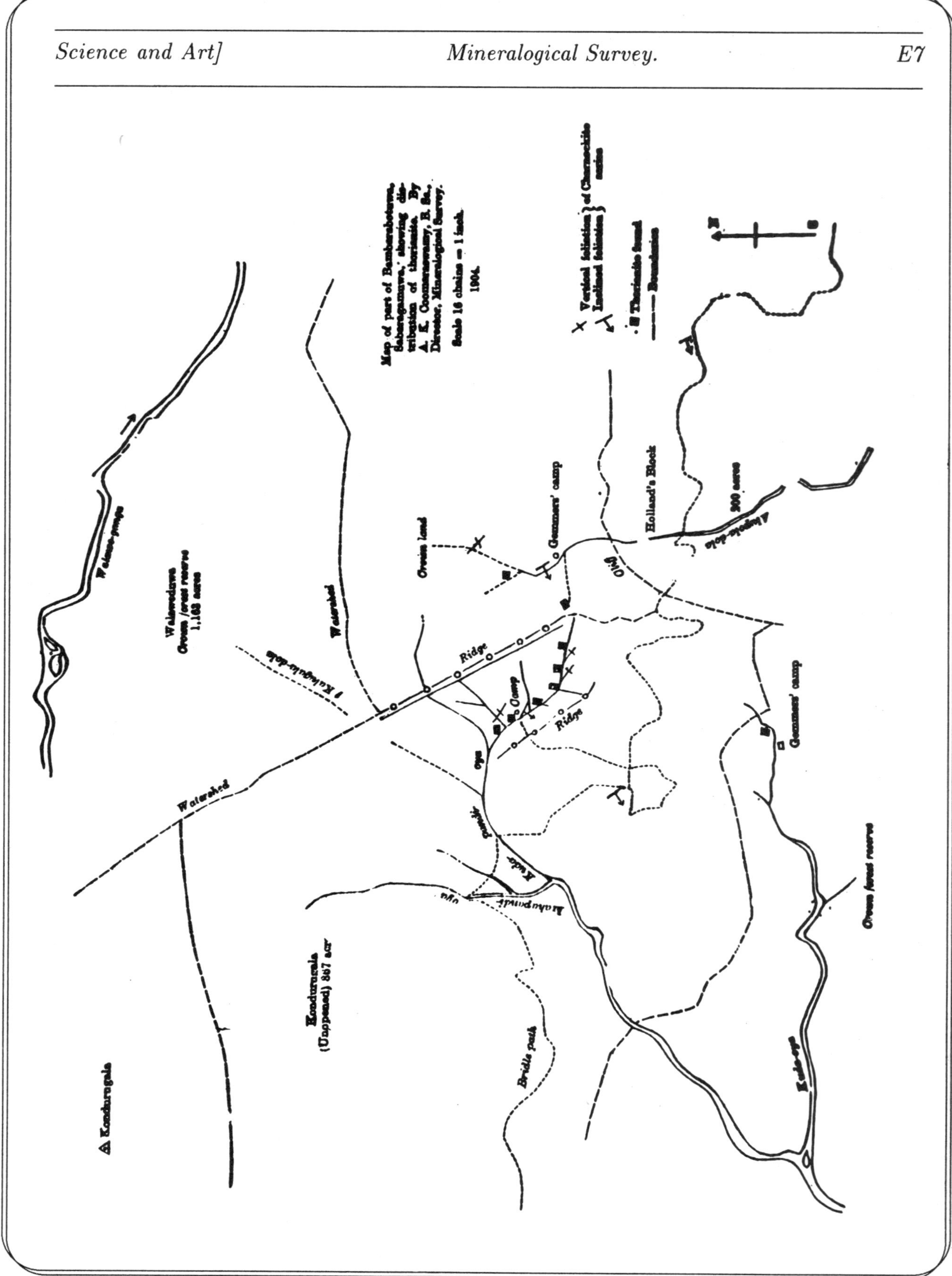
Map of part of Bambarabotuwa, Sabaragamuwa, showing distribution of thorianite. By A. K. Coomaraswamy, B. Sc., Director, Mineralogical Survey.
Scale 16 chains = 1 inch.
1904.
Vertical foliation } of Charnockite series
Inclined foliation }
Thorianite found
Boundaries
N
S
Kondurugala
Kondurugala (Unppened) 867 acr
Walawedawa Crown forest reserve 1,163 acres
Walawe-ganga
Watershed
Watershed
Crown land
Gemmers' camp
Gemmers' camp
Camp
Ridge
Ridge
Holland's Block
200 acres
Bridle path
Kudo-pandi-oya
Kudo-oya
Crown forest reserve

The thorianite is probably derived from a granitic rock with no very large outcrop, but in which it occurs in considerable abundance; perhaps together with the zircon, thorite and l-ilmenite, which quite possibly, however, occur separately in other rocks of a similar character. The outcrop of this rock could not be discovered owing to the dense jungle and thick soil and landslips; rock (mainly decomposed granulite) is indeed exposed at several points in the beds of the Kuda Pandi-oya and of the small streams joining it ; but any search for the outcrop of a particular rock in the adjoining jungle could only be expected to succeed at the cost of a considerable expenditure of time and money, and might even then result in failure ; if found it would probably well repay working.

The physical properties of thorianite are briefly as follows :- Specific gravity ; 8.98 - 9.32 (Mr. Holland says he obtained one crystal of specific gravity approaching 10) ; hardness 5.5-6; colour black, streak greenish gray; in this sections translucent, with strong red-brown colour; crystallized in the cubic system ; isotropic ; usual form, a simple cube, but penetration twins (tw. pl. o) (like those of fluorite) common; fracture uneven to conchoidal ; brittle; lustre resinous to submetallic ; soluble in dilute nitric and sulphuric acid ; with borax or salt of phosphorus gives reactions for uranium ; radio-activity about twice that of pitchblende. The chemical composition if given below ; the analysis was made by Mr. G.S. Blake at the Imperial Institute : -

				Percent		Percent
Thorium oxide*	...	ThO_2	...	76.22	...	72.24
Cerium oxide	...	CeO_2	...		...	6.39
Lanthanum & Didymium oxide	...	La_2O_3 Di_2O_3	...	8.04	...	0.51
Zirconium oxide	...	ZrO_2	...	trace	...	3.68
Uranium oxide	...	UO_3	...	12.33	...	11.19
Ferric oxide	...	Fe_2O_3	...	0.35	...	1.92
Lead oxide	...	PbO	...	2.87	...	2.25
Silica	...	SiO_2	...	0.12	...	0.41+
				99.93		99.93

* Specially separated fragments, subsequently analyzed, contained nearly 80 per cent of thoria (W.R. Dunstan, Reports on the Results of the Mineral Survey in 1903-04. London 1905)
\+ Insoluble residue.

A partial analysis made in Mr. Curie's laboratory in Paris gave 79 per cent oxides of rare earths (principally thoria) and 14 per cent of uranium oxide. A rough preliminary analysis made by Professor Sir W. Ramsay is as follows : - Oxides like thorium oxide 76.4; oranium oxide 14.9; iron oxide 6.1; lead, arsenic & c., 2.0; insoluble (not SiO2)

0.7. More complete investigations are being carried on at the Imperial Institute and by Professor Sir. W. Ramsay.

The specimen of thorite analysed was one of two which were brought to Mr. Holland before he had discovered the source of the supposed uranite ; they were stated to come from Bambarabotuwa, and appear to be similar to the material actually obtained from the Kuda Pandi-oya ; of this actual material, however, no analysis is yet available.

It is not likely that any very extensive deposit of thorianite occurs. In the Kuda Pandi-oya valley a total of five tons or somewhat more or less might be obtained ; from the Alupola - dola or the Kuda-oya, where traces were also found, it is doubtful if half a ton could be profitably extracted. The whole amount removed from Bambarabotuwa hitherto is less than 30 cwt.

In the immediate neighbourhood of Kondrugala, the Walaweduwa jungle seems the most favourable district for further investigation. An examination of the Rat-ganga valley for the west showed that thorianite is absent there, and that even zircon is very scarce. Beside the localities referred to, thorianite has been actually observed, but in the smallest possible amounts (1) in washings from a gem pit beside the Pita-ela between Walawe and Morahela estates near Balangoda; and (2) in a pegmatite vein at Gampola, Central Province ; and specimens have been inspected, but not actually obtained by the survey; (3) from Massena estate near Balangoda ; (4) from Ellagekumbura, Muwagama, near Ratnapura; and (5) from Iriyagahadeṇiya in Bodhimaluwa, near pussella ; while a massive specimen (6) was presented to the Colombo Museum by Mr. Jacob de Mel, who tells me it was found in a mass of graphite from his mine at Ragedara, near Kurunegala. In all these localities it is evidently present only in very small amounts. The thorianite is actually obtained by washing as if for gems, and at the Kondrugala locality is so abundant as to form with a less amount of zircon, thorite, and ilmenite, practically the whole of the *nambu* or heavy residue from gemming. It has been found useful however, to complete its separation from other mineral by further washing in a "jigger".

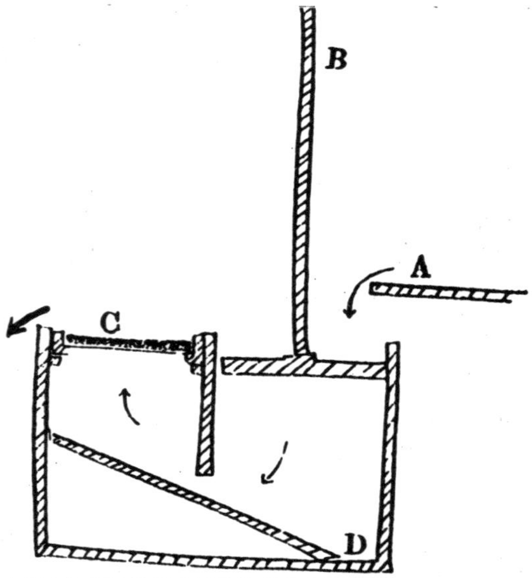

Fig: - Section of "Jigger".
A - Spout with small stream of water.
B - Piston working up and down.
C - Layer of thorianite on wire gauze tray.
D - Small stuff slipped through the meshes is collected here.

The jigger consists of a wooden box partially divided into two compartments, one carrying a wire tray, the other containing a wooden piston worked up and down by hand. A small trickle of water is required to replace the overflow. The pistons is worked up and down and a current of water is forced through the meshes of the wire tray which has been covered with a layer of thorianite mixed with zircon, & c. A thin layer of pure and large material should be laid first on the tray and the material to be cleaned laid over this ; the small crystals are thus prevented from sinking through the meshes. After a short time the lighter minerals accumulate on the surface of the layer, and may be strained off with a spoon. For an illustration of the jigger in use see plate 2, fig. 3 and for and explanatory section see text, fig.6 above.

Thorite:- A mineral consisting mainly of thorium silicate occurs in association with the thorianite at Kondrugala. The minerals has a somewhat resinous lustre and yellow or brownish to orange colour. Its specific gravity is 4.98. The analysis made by Mr. G.S. Blake at the Imperial Institute is as follows :-

				Per Cent
Thorium oxide	.. ..	ThO_2	..	66.25
Cerium oxide (and Ce earths)	.. ..	CeO_2	..	7.18
Zirconium oxide	.. ..	ZrO_2	..	2.23
Uranium oxide	.. ..	UO_3	..	0.46
Ferric oxide	.. ..	$Fe_2\,O_3$	..	1.71
Calcium oxide	.. ..	CaO	..	0.35
Phosphoric oxide	.. ..	$P2O_2$	..	1.20
Silica	.. ..	SiO_2	..	14.10
Water	.. ..	H_2O	..	6.40

Most of the crystals are very much waterworn; but some in which the crystalline form is still fairly clear do not at all resemble a tetragonal mineral, but rather suggest an orthorhombic or monoclinic one, which renders it uncertain whether it be actually thorite.

The same mineral was met with at Bulatdiwanella, Durayakanda, in Gilimale South, at the lower end of the Bambarabotuwa valley. Here it formed nearly one-tenth by weight of the nambu obtained by gemming in a small rice field. Perhaps a hundred weight of the material could be got here. The mineral is similar to that from Kondrugala, but many crystals are better preserved and are clearly unlike a tetragonal mineral. One specimen has a specific gravity of 5.06. The remainder of the nambu consists namely of zircon (including good specimens of the irregular crystals described by Mr. Spencer) with a little corundum, a mineral allied to samarskite, & c.

Allanite:- This mineral has been met with *in situ* in three places near Balangoda, [*7] in coarse pegmatites composed of felspars, quartz, biotite, and horblende, with accessory pyrite, and allanite in crystals up to 3 inches in diamter. The content of allanite varies much even in the same exposure. The localities are (1) in the Weweldola on Dikmukalana estate (discovered by Mr. Holland); (2) a dyke, 3 or 4 feet thick, on lower Denegama

[7*] Coomaraswamy, A.K., Geol. Mag., loc. cit

estate ; (3) a dyke at Weligepola, nine miles from Balangoda. No analysis of the mineral is available, but it may be expected to contain a trace of thorium besides other rare elements.

E10 *Ceylon Administration Report for 1904* *[IV-EDUCATION.*

Samarskite group :- Minerals belonging to this group and giving good reactions for niobium and uranium have been found in several localities, of which details are given below:-

Bulatdiwanella, Gilimale:- The nambu contains a very small quantity of a very dark brown or black mineral, with conchoidal fracture and very brilliant resinous lustre on the fractured surface. Several crystals have a tetragonal form ; some are grouped together in slightly readiating, or nearly parallel groups. The specific gravity is 5.63 (Westphal's balance.) The mineral gives good reactions for uranium and niobium but not for manganese. It would seem to be possibly annerodite.

Kinagahadeniya, Kuruwita:- A very similar mineral, but in larger pieces and without crystalline form occurs here somewhat more abundantly. A gem pit about 12 feet by 9 feet in area yielded about 1 1/2 lb. of this mineral. Similar or identical minerals were also met with in very small amount at Diyahonda-ela, Kuruwita, and near Pussella.

Walaweduwa:- A tributary of the Kalugaladola (see map, p.6), yields a considerable amount of an iron black mineral with specific gravity 4.9 to 5, and apparently hexagonal orthohombic. It may be allied to aeschynite or polycrase.

Fegusonite: - This mineral has been recorded by Mr. G.T. Prior.* [8] who identified it amongst some specimens brought from Rakawana by Mr. Joseph Baddeley. The locality given - "the gem mines at Rakawana" is of course excessively vague. The result of mr. Prior's analysis is given below:-

	I	II	
Nb_2O_5		44.65	
Ta_2O_5	49.82	4.98	
UO_3			
Y_2O_2		24.67	
Er_2o_3	39.84	13.24	Atomic weight 112
CaO_2		2.02	
Fe_2O_3	0.50	0.51	
H_2O		4.58	
F?			
		99.76	

Specific gravity, 5.023

[8]* Prior G.T. Fergusonite from Ceylon, Min. Mag., vol.X.No.47, pp.234-238, 1893

The mineral had a dark brown to black colour, hardness 5 to 6, vitreous to submetallic lustre and conchoidal fracture, in these particulars resembling most of the heavy minerals met with in Ceylon.

Baddeleyite:- This mineral was described by L. Fletcher,[+ 9] and was brought from Rakwana by Mr. Joseph Baddeley. It is an iron black mineral of hardness 6.5 and specific gravity 6.025. It consists of zirconia, ZrO2. Only a single crystal was found in Ceylon, but the mineral has since been discovered in Brazil.

Geikielite:- This mineral was described by Mr. Allan Dick[++ 10] in 1892, and is peculiar to Ceylon. It is a bluish or brownish black mineral with hardness 6 and specific gravity 4. It consists of magnesium titanate MgTiO3.

Casiterite:- This valuable ore of tin has not previously been definitely recorded frm Ceylon. Some black nambu (specific gravity 6.91) brought to me from Niriella, near Ratnapura, proved to be cassiterite, and the district will be investigated during the early part of next year, as the occurrence of any quantity of this mineral would be of great economic importance.

Rutile is very often met with amongst the residue of heavy minerals from gemming operations, but only in small amount. A very fine crystal was presented to the Museum by Mr. D.I. Jayatileke a (100), e (101),s (111), and subordinate in (110) and twinned on ell1 (011) and e (101) of the principal individual. Mr. H.T. Armitage has forwarded some prismatic crystals from Hatton; these are rather broken and worn, but several crystals show the same forms a, e, s, m. A crystal from near Guruluwawa, Gilimale, was vertically straited and otherwise of similar character. Rolled fragments occur in localities too numerous to be specified. All the specimens are iron black and have a conspicuous adamantine lustre.

Beryl, Chrysoberyl, Topaz, Tourmaline, some Spinel:- These gems are probably derived from rocks of the Balangoda group, and but tourmaline is the only one that has been actually met with in a granitic rock in Ceylon. As to spinel it is not improbable that some of the dark opaque spinels approaching magnetite in chracter are derived from granitic rocks; especially as spinel was once observed as an inclusion in a mineral apparently allied to aeschynite found in Walaweduwa, and also the very darkest spinels have not been seen *in situ* in the crystalline limestones, where the spinels are pink, green, or blue. Topaz is in some places quite abundant ; *e.g.,* in the Haldola on Dikmukalana estate. it is usually colourless and clear and transparent, serving for "water sapphires" so called (properly cordierite) and white sapphires ; but a green colour is not usual, and the stones are then cut as aquamarines (peroperly beryl). Brownish yellow and yellowish pink topaz, indistinguishable from Brazilian specimens, are occasionally found, but the pure yellow topaz does not occur. The stones sold as topaz are of course yellow sapphire. The pinkish yellow topaz differs from that of Brazil in that the colouring matter is driven off when the stone is heated while the Brazilian specimens actually turn pink when heated [information kindly supplied by N.W.C.Wild].

[9+] Fletcher, L. "On Baddeleyite (Native Zirconia), a new mineral from Rakwana, Ceylon," Min. Mag., vol.X., pp.148-160, 1893

[10++] Dick, A. "Opceikielite, new mineral from Ceylon," Min.Mag.,vol X., pp.145-147.

General ACCOUNTS OF Gemming IN Ceylong

By the Director and Assistant Director

With the exception of moonstone, which is almost invariably obtained by quarrying a peculiar adularia-bearing leptynite in the Dumbara district of the Central Province (Administration Report, Mineralogical Survey, 1903, p.10), and garnet, which is also occasionally obtained *in situ* from disintegrated granulites, the Ceylon gems are obtained by washing gem-bearing gravels. This is done in parts of Sabaragamuwa (the Ratnapura District) and the Southern Province ; in the Western Province only one gemming license is in force at present, while there are two in force for pits near Hatton in the Central Province; but no work is now done near Nuwara Eliya or in the Horton Plains, in the Central Province, though these places were once regarded as good gemming grounds.

The gems of Ceylon are transparent and well-coloured varieties of the minerals corundum, spinel, zircon, tourmaline, topaz, garnet, chrysoberyl, cordierite, amethyst, felspar, and beryl. Non of these (except felspar and granet) have hitherto been found in the parent rock in a sufficiently perfect state to be useful as gems, and only corundum spinel, zircon, garnet, felspar, and tourmaline have been found in any condition. All gems must, however, have been derived from a parent rock by disintegration and deposition in the gravel beds. It would hardly be necessary to point this out but for the very prevalent opinion among all classes that gems "grow" *in site* in the grvels, presumably by some sort of crystallization; the perfection of a stone being regarded as a sign of its age, and small and opaque stones being considered immature ! Owing to the very few exposures of rock which can be found in many districts (it is sometimes impossible to record a dozen observations of dip and strike in a week or fortnight) it is perhaps not altogether surprising that the sems have never been traced to their origin in the parent rock ; and, indeed, it is not likely that they will be found *in situ* either by the Mineralogical Survey or by any one else, save by good fortune in a few exceptional cases; and it is probably the existence of this difficulty which has led to the idea of the gems having been formed where they are now found.

The gem-bearing gravel or *illam* occurs in beds, patches, or pockets deposited by streams and rivers, and may be found at any depth, upto 120 feet, the greatest depth observed, viz., at Botiyatenna, Rakwana. Where the *illam* is found in its typical form it consists largely of white quartz pebbles, ranging in size from small short to a football, indeed in all gradations of size up to that of the boulders associated with the *illam.*

The empirical observation that gems and other heavy minerals are associated with such gravels (the founded quartz pebbles being regarded as an infallible sign of gems) is not altogether easy to explain ; it is probably the case, however, that floods that would round and convey the pebbles to a distance would usually be able to carry the heavier minerals also, which would otherwise remain near their place of origin; and a process of natural selection based on the superior hardness and toughness of the gem minerals would somewhat reduce the relative proportion of other minerals present.

It is obvious, however, that the typical rounded *illam* stones do not of necessity always accompany the gem minerals, but the character of the *illam* must, as is actually the case, vary from place to place according to its distance from the place of origin and according to the nature of the ground in which it rests. In the course of a hill stream it may consist of irregular and small quartz grains mixed with abundant little ferruginous concretions and with organic matter. In small swamps in the hills it is often sandy or mixed with disintegrated material and with organic matter.

In paddy fields or similar alluvial flats the *illam* or *illams* lie beneath a considerable depth of muddy alluvium; and somewhat resemble it in being mixed with kaolin or laterite. In the latter case, where the founded quartz, pebbles and other material are present, the whole may be cemented into a hard rock, difficult to excavate, forming in fact a regular "puddingstone". Iron pan, too, is sometimes met with at various depths, either just above, or in or below the *illam.*

The gravels forming the present beds of may rivers in gem-bearing districts, especially of those which are shallow and swift flowing, are often rich in gem minerals, and in this case the actual reiver gravel or that just below the surface is known as the *illam.*

In places where the alluvium is of any considerable depth (say over 6 feet) there are often two (or even more) layers of *illam*, upper and lower, the intervening beds of sand, kaolin, or *Makul*, being sometimes termed the *malawa* of the *illam* above it, by and extension of the meaning of the term, which, as used above, is usually applied to the decomposed rock *in situ* on which the gravel rests, and below which it is useless to dig. When there are two or more *illams*, the lower one resting on the actual *malawa* consisting or rock *in situ* is always considered the richest in gem minerals, and to some extent this seems to be actually the case, though it would be difficult to assign a reason for the act. Where there is a considerable thickness of *illam* in a bed, the heavy minerals have undoubtedly a tendency to accummulate near the bottom, and especially is this the case where the *illam* lies between boulders, between which the heavy minerals may be supposed to slip and lodge.

When the deposits lie in a valley with concealed rocky sides having a rapid slope, the gem minerals tend to occur more abundantly in the *illam* in the central line of the valley.

The absence of gem minerals in a particular part of a valley, otherwise having a good reputation for gems, is thus often due to the pit having been sunk on a concealed slope of the underlying rock.

In level or only slightly undulating country, such as large flat deniyas and rice fields, the *illam* is rarely thick and is evenly distributed over the area, and is usually covered with five or six feet of sandy alluvium. In a well-gemmed district such swamps are riddled with old gem pits, now filled with water and partly overgrown with water plants. In such cases a long thin iron rod, called a *diga illam kuri*, is used to discover where any unworked *illam* still remains.

Where a water-course is silting up in excess of the erosion of its bed, the *illam* lies, as a rule, below the present surface level of the water (*e.g.* at Alpaladeniya, Morawak korale, Southern Province); where erosion is in excess of silting, it may be several feet above (*e.g.* The maha-oya at Dehipala, Kuruwita, Sabaragamuwa). Most of the small rivers in the terai have attained their regimen, and the top of the *illam* is at the same level as the furface of the nearest water-course. In estimating the time and expense of sinking a pit it is often fairly safe to suppose the *illam* will be at this depth.

E12 *Ceylon Administration Report for 1904* *[IV-EDUCATION.*

The following are sections of typical gem pits:-

(1) Rambuke, Kukul Korale.

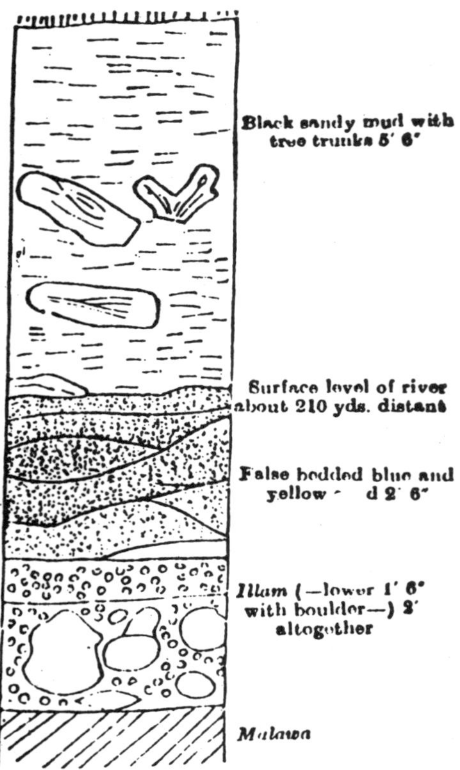

Fig.7

The section in fig.7 is one in a gem pit in a deniya, near the Koswatta-ganga, and through which flows a small gem-bearing stream, which doubtless contributed the gem minerals and part of the pebbles and boulders of the *illam* and san overlying it. As the top of the sand is at the same height as the surface level of the river, it must have been laid down in a backwater of the river which was slowly silting up. The stratum of false bedded sand represents the last stage of the combined silting of stream and river when the water was shallow, it appears probable that the blue sand was derived from the stream, and that the yellow (*i.e.* with the ferruginous matter oxidized) was derived from the main river, with the present deposits of which it corresponds in appearance. When the silting was complete, the muddy soil with vegetable matter formed- [J.P.]

Science and Art] *Mineralogical Survey.* *E13*

(2) Ukuwatta, Rakwana.

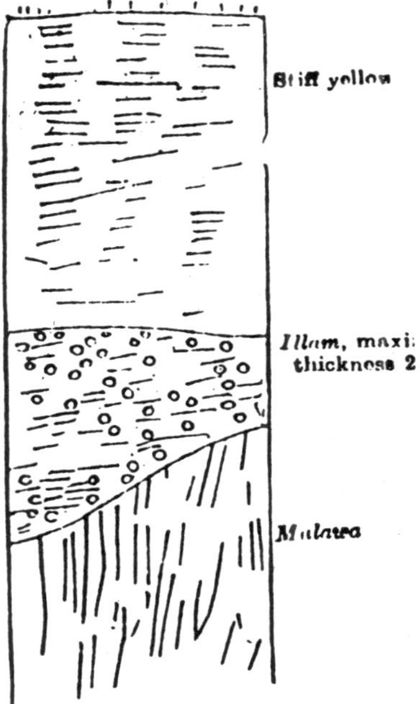

Fig.8

The section in fig.8 at Ukuwatta, near the Rakwana-ganga, Rakwana, illustrates the lenticular form which a bed of *illam* may assume. In the same pit is varied from 2'6" to nil. The top of the *illam* was marked by typical rounded quartz pebbles, but lower down it was more and more clayey and passed almost imperceptibly into the clayey laterite beneath, being only separated by an indefinite and irregular line of limonite. The vertical ferruginous lines of the laterite were continued almost imperceptibly into the vertical or almost vertical folia of the underlying granulite - [J.P.]

(3) Hapugatenne estate

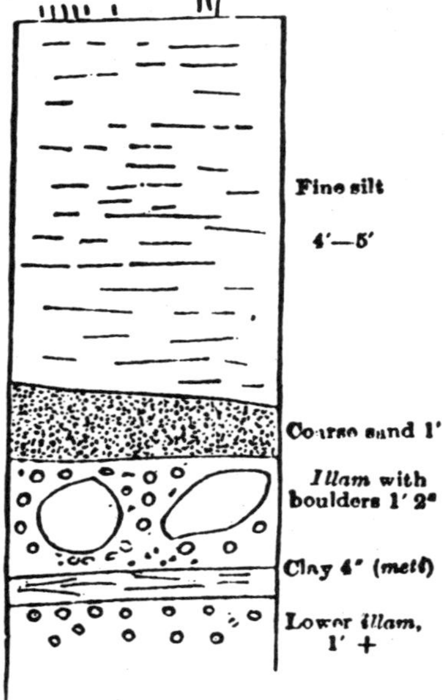

Fig.9

The section in fig 9, a pit sunk in the large swamp on Hapugastenna, Sabaragamuwa, is a characteristic section showing a double *illam*; the upper and lower *illam* are separated by a four-inch layer of clay. Only the upper illam countains large boulders of rounded quartz and other rocks. The heavy mineral present are spinel, garnet, and rarely zircon ilmenite, and tutile (A.K.C.).

(4) Rico field at Karawa, Kalaella, Madampe

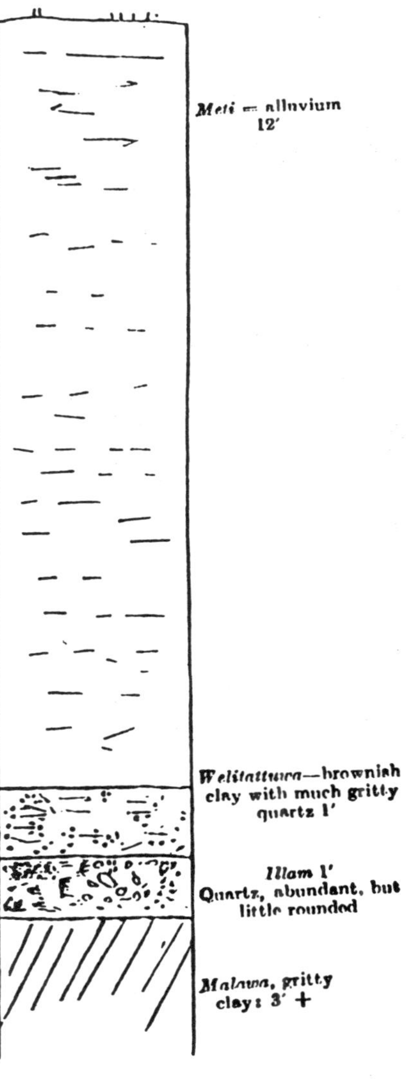

Fig.10

The section in fig.10 shows the deposits in a long straight valley at Karawa, Kalaella, near Madampe, Atakalm korale, Sabaragamuwa. The valley is properly a rice field, but has been gemmed from end to end in a most systematic manner. The alluvium is unusually deep. The pits sunk have to be carefully timbered; they measure 2 or 3 by 4 yeards in area. Water is removed by an *andiya* (see below). Gems of all sorts are said to occur. The nambu contains a large proportion of pyrite (little cubes), brown zircon, and spinel. [A.K.C.]

Some other pits may also be referred to. Very important pits are those at Ukkatadeniya in Weralupe, near Ratnapura. The pits are on the edge of a wide sketch of rice field. The section was not visible, owing to timbering of the shaft in use and to the other shafts being filled with water. I was informed, however, that in the shaft now worked, and which showd about five fathoms alluvial sand and three fathoms water at the time of my visit, the *illam* occurred beneath 48 feet of alluvial sand; while in an abandoned shaft a short distance away the section had been somewhat as follows:-

Sandy alluvium	12.00	
Good *illam*	1.00	to 3.0
Weli malawa (sandy clay)	6.00	
illam (not very good)	3.00	
Malawa		

but the details are probably far from accurate. These extensive deposits are actually those of the Kalu-ganga; it will be seen that their thickness and character differ a good deal from those of the deposits of much smaller streams, such as those already referred to.]

The deepest pits known appear to be those at Botiyatenna, about 4 miles from Rakwana and on the watershed between the Koswatta-ganga valleys. The actual section of the pits could not be examined, but reliable information obtained indicated that they had reached a depth of 120 feet.

Methods of obtaining the illam:- It will be clear from the above account of the various modes of occurrence of the *illam* that the methods of obtaining it must be equally varied. When the gemming land is in a narrow valley or in an alluvial flat or swamp (the word deniya applies to either case) a pit is sunk. In the case of large excavatins, it is very usual for the gemmers to erect a little altar (*malpela*) and make offerings, usually to Saman Dewiyo. In any case a little prayer is generally made previous to the actual cutting of the *illam*. The spot where the pit is to be sunk is first examined with a long iron rod, 8 feet or more in length (*diga illam kuri*). With a little practice the nature of the underlying deposits may be distinguished by the sound and resistence as the rod is thrust into the ground, and beds of sand, *makul, kirimetiya, illam*, if present, and *malawa*, made out and their depths determined. If there is but a small thickness of alluvium overlying the *illam* the shorter and stouter iron rod (*illam kuri*), which serves the general purposes of crowbar and pickaxe, suffices for these preliminary investigations.

Often, if there is much surface water, a small trench is dug round the proposed site of the pit and connected by a drain with the nearest water-course. The pits may be any size or shape, but are usually square in plan, and are usually less than three or four yards square in area. In digging a trial pits it is good plan to adopt on oblong shape, which, while involving a minimum amount of material to be excavated, allows two men to work in it easily, an important matter when one has to be constantly baling out water, while the other has to dig with mamoty and *illam kuri.*

Where the top of the *illam* is on the same level with or below the nearest water-course, or in rainy weather, (when the swamps are often under water), the inflow of water is the main difficulty to be contended with. In small pits it can be removed by coolies with buckets as fast as it accumulates, but in the case of large pits it is advisable to erect an *andiya,* which serves equally for removing the *illam.* The *endiya* consists essentially of a long pole swinging some distance above the ground, on a scaffolding; one end of the pole overhanging the pit has attached to it a long stick, bearing a bucket which passes down into the pit; at the other end is a counterpoising weight; the bucket is alternately raised from the pit and emptied at its edge and again lowered and filled.

The use of a tub and windlass is less usual. At Weralupe, near Ratnapura (see above), there are two shafts, one now abandoned and formerly worked with an *andiya*; the other now being worked with tub and windlass. This present shaft, 12 feet by 6 feet in section, is timbered in two square divisions each 6 feet square, and divided by the central timber partition; over each section is arranged a tub and windlass, which serve to bring to the surface the water and sand, and ultimately the gem-bearing *illam.*

Mr. Jayatileke's pit at Potgulkanda is of quite and exceptional character. The pit is a large quarry, cut on the side of the hill, and the material exploited the *illam* is not a river deposit, but consists of the "run of the hill", talus material, and huge boulders. The water supply is very limited. The *illam* is not washed in the usual way, but passed into a sluice having three sloping grids, which serve to keep back all the large stones and bits or rock. A stream of water is conducted in to and through the sluice, and the *illam* with mamoties as the water passes over it. The material in the third and last division is of such a size as the gems may reasonably be expected to attain; the whole of it is taken out and spread on a board and examined by hand by a responsible person. The disadvantage of this process seems to be that so much material has to be gone through; while, by the ordinary process of gem washing described below, the greater part and sometimes all of the lighter minerals are mechanically removed, and only the gems and *nambu* remain to be examined.

When a pit is sunk through loose and muddy soil, it is necessary to timber the sides, as in many of the cases referred to above. This is done by fixing stout beams horizontally against the side of the pit, two sets being at first required at a suitable vertical distance. Behind these and between them and the soil are thrust long thin sticks, closely places, and these prevent the sides from falling in ; they are pressed down as the excavation deepens. At a greater depth another set of horizontal beams can be arranged of needed. Large pits are often divided into two compartments, as in several of the cases already referred to, by transverse beams arranged across the pit from side to side so as to form a partititon wall, which also provides a convenient means of ascent and descent, the successive beams acting as the rungs of a ladder ; a long creeper is tied vertically on to the wall so as to serve as a hand rail or holdfast.

When *illam* is being looked for beneath or between boulders by the side of a small stream or river, the course of the stream is directed from the spot by a dam of stones, sticks, and mud and the *illam* extracted as can best be managed under the circumstances.

It often happens, when the *illam* is found above the present stream level, that pits are sunk some distance apart and tunnels driven through the *illam* from one shaft to another, *e.g.,* at Ratkarawa near Timbolketiya, Atakalan korale, Sabaragamuwa, where the section is as follows:-

Soil and alluvium	4.6
Illam with well-rounded quartz pebbles rarely as big as a man's fist	4.0
Pargugala (decomposed granitic rock)	4.0

The gravels are some 15 or 20 feet above the present river level, and are almost always dry. A good deal of illicit gemming has been done; the method of tunnelling from shaft to shaft has been largely employed. The same thing was observed in the illicit gem pits beside the Alupoladola at Bambaragbotuwa, Sabaragamuwa.

In small hill streams the *illam* is often obtained practically on the surface with mamoty and *illam Kuri.* The smaller streams, if gem-bearing, have often been quite gemmed out in this way, the gemmers having proceeded systematically up stream, washing all the loose material obtainable in the bed of the steam, and sinking pits in any small deniyas which may interrupt the continous slope of the stream; this holds for many streams in the Gilimale and Bambarabotuwa districts. If *illam* is obtainable directly in the bed of a more considerable river, as in the Rakwana-ganga at Pallebedde, a special form of mamoty, the *poru udella,* is often used, having a broad, slightly concavo-convex blade, and a handle 6 feet or more in length. This is dragged through the river gravel at a sufficient depth, the curved form of the blade allowing the *illam* to be retained in some quantity.

Washing:- However obtained, the usual practice is to pile up the *illam* until the pit is exhausted, when it is washed in the presence of the owner of the pit or other responsible person. When it has to be left some time, the pile is covered with leaves and branches, the disturbance of which would indicate and attempt at theft. Washing is carried out in the nearest water of a suitable depth, preferably in a (*wewel*), and are about 2 feet in diameter and I foot in depth. Inside the edge extgends a rim which should not be less than 2 1/2, inches wide, the purpose of which is to prevent the heavier contents from being too readily washed out of the basket; for the whole process of washing is based on the mechanical rejection of the lighter stones and retention of the heavier, including the gem stones. The basket is held in both hands by this rim.

The *illam* is carried from the pit to the water in small baskets, and in the best practice thrown into a small basket (*keta kude*) with a bottom of coarse mesh resting in the gemming basket. The washer then shakes the *Islam* through the *keta kude*, the large stones of *keta* remaining in it and are thrown away. Otherwise, after a preliminary wash, the large stones are removed by hand.

The process of washing consists of a peculiar turning movement given to the basket in the water, the rim being depressed below the surface of the water once in each turn, allowing the lighter sones to be swept out over the edge of the basket by centrifugal force. Washing is continued, while baskets of Islam are brought until the contents of a sufficient or lucky number, say fifteen of twenty-one, have been washed. The whole is given a final hard washing, and the resulting gems and heavy minerals, mixed usually with some gravel and ferruginous concretions, remain behind. If there is an abundance of heavy minerals, corundum, ilmenite, zircon, & c., a good washer will succeed in mechanically removing the whole of the lighter minerals, such a quartz and felspar, and nothing remains but the gems, if any, and other heavy minerals. The *nambu* or residue of heavy minerals remaining after any gems present have been picked out is usually thrown away; this practice is to be deplored, as it is in the *nambu* of certain districts that the rare thorium-bearing minerals recently discovered in Ceylon occur, beside other minerals of scientific interest.

some men are unable even with practice to acquire the knack of washing gems, so that either all the heavy minerals are washed out of the basket, or the washing is so incomplete that the *nambu* is half gravel.

The following general conclusions as to the distribution of gem minerals in the gemming districts appear to be well founded:-

(1) When the streams on one side of a watershed are gem-bearing, those on the other side are usually also gem-bering and yield the same sorts of gems.

(2) While in some districts a certain uniformity in the gem and other heavy minerals found is noticeable, in others the variety is very great, each stream or deniya yielding *nambu* of a different character and the gems also are very locally distributed.

(3) The heaviest minerals are rarely transported far from their source, tending rather to accumulate in small deniyas near their place of rogin. They occur but sparingly in thick or very extensive deposits of gravel; this applied especially to minerals with specific gravity over 5; the gems, with specific gravity from 2.63 (cordierite) to 4.7 (zircon) are more readily transported; they may be found, for example, in otherwise barren gravels at points where streams from gem-yielding areas enter the main valley. At the same time it is very usually the case that the gems and heavy minerals are of exceedingly local distribution, showing that they have been derived from numerous separate exposures of the rocks in which they were originally formed.

GLOSSARY OF TERMS USED IN GEMMING

By the Director and Assistant Director

Andiya	Well-sweep; a long pole swinging between two uprights fixed in the ground; bearing at one end a bucket and at the other a stone for counterpoise. Used for removing water and *illam* from gem pits.
Deniya between	The word is used in two senses: (1) A swamp on level ground; (2) A narrow valley running the spurs of a range of hills.
Diyaratran	"Water gold," viz., iren pyrites.
Dona	A narrow tunnel driven along the *illam* from a gem pit Dola.
Dola	A small streem
Ela	A drain
Garanawa	Literally, to sift, hence to gem; the word entering into many combinations, *e.g.* garan idama ( ), a gemming place.

E16 *Ceylon Administration Report for 1904* *[IV-EDUCATION.*

Illamkura prospecting	Light crowbar used for loosening the *illam*, and also for detecting *illam* when not too far below the surface; diga *illam kura* ( ), long thin iron rod used in for *illam* a greater depth.
Illama	The gem-bearing gravel
Katta	Opaque and worthless gem stone; *e.g., toramalli katta* worthless zircon.
Keta	Pieces of rock in the *illam*
Kude carrying from	Basket in various combinations, as *punchi kude* ( ) small basket used for *illam ; garan kude* ( ) small basket for sifting out larger worthless stones the *illam*
Mada	mud

Makulu	Mixture of decomposed felspar and quartz grains resulting from dis-integration acid granulite.
Malawa	The decomposed rock *in situ* on which the *illam* rests.
Malpela	Small shrine of sticks and cadjan built near a gem (or plumbago) pit, in which are placed offerings of betel, flowers & c., to the presiding dewiyo.
Menika	Gem
Metta	Clay, *kirimetta* ( ) kaolin.
Nambuwa	The residue of heavy minerals after washing the *illam* and from which all valuable gems have been removed ; *wediber* ( ) *nam* consists of angular grains of black ilmenitbesides rutile, & c.
Patale	Pit
Parugala	Name sometimes applied to *malawa* when in the form of harder, little decomposed rock; solid rock underlying malawa ; subsoil.
Tattuwa	The different layers or beds exposed in digging a get pit, *e.g., welitattuwa ( ) madatattuwa ( ), Makultattuwa ( ), illamtattuwa.*
Tiruwanagala	Milky white quartz.
Udella	Mamoty; *poru udella* ( ), and *garan udella* ( ), mamoty with broad and slightly curved blade and long handle used for dredging *illam* from rivers.
Wala	Pit
Wattiya	Gemming basket
Weli	Sand.

NAMES OF GEMS

Corundum (name derived from the Tamil Kurundam, Sanskrit Kuruvinda).

Kurundugal	Opaque corundum used in cutting gems.
Dalam	Semi-transparent, pale-coloured varieties of corundum, sold only by the pound.

Nila	Sapphire
Rata, Ratugal, or Tatuketa	Ruby
Tarukagaloa or Arunul	Star stone
Ratnapushparagaya	"king topaz" (the "king topaz" of Ceylon is usually pink of flesh coloured corundum, and not true topaz.
Padmaragaya	So-called topaz, actually yellow sapphire; pure yellow topaz is not found in Ceylon.

Topaz

Padiyan	White topaz, so-called water sapphire; the true water sapphire of Ceylon, however, is cordierite (iolite), and seems to be very rare ; it is blue
Pachchapadiyan	Green topaz as well as true beryl are included here; both are sold as aquamarine.
Ratnapushparagaya	"Kind topaz", see above, also p.10

Chrysoberyl

Wayirodiya	Cat's - eye
Pasingal	Alesandrite

Beryl

The quamarine of Ceylon includes both green topaz (see above) and pale beryl, both called *pacchapadiyan*; a true emerald colour is extremely rare.

Spinel

Kirinchi	Spinel

Garnet

Rawwa or rabhaha	Garnet, Ratnagomeda, Cinnamon stone.

Zircon

Zircon from Arabic zaqun whence Tamil jargoon.

The majority of the stones described as *toramalli* ( ) are zircons, the name being applied indifferently to zircon and tourmaline. The varieties are described according to appearance, as pachchatoramalli ( ), usually green zircon, rarely green tourmaline or chrysoberyl; peni toramalli ( ), pale brownish zircons usually, sometimes tourmaline. This variety is usually burnt to form the colourless "Matara Damond." Many greenish zircons acquire a good yellow colour when "burnt." Golden yellow zircons are called *kanaka toramalli* ( ), and good fiery red stones *gomeda* ( ).

Tourmaline

See above. The English word is derived from the Sinhalese through the Dutch. Ceylon tourmaline is usually brown. Green varieties are scarce ; most of the stones sold as green tourmaline are zircon. The name *Mangu toramalli* ( ) is applied to brown (real) tourmaline.

Quartz

Palinguwa	Rock crystal
Sewandiya	Amethyst

Felspar

Punakkan	Moonstone

MORE DETAILED NOTES ON SOME OF THE GEMMING DISTRICTS VISITED

(a) By the Director

Balangoda:- Very little or no gemming is done in the immediate neighbourhood of Balangoda now. Gemming has been done from time to time on Massena estate, six miles from Balangoda. Some old pits are to be found in the neighbourhood of Walawe estate also ; a single crystal of thorianite was obtaned from a trial pit near the Pita-ela, near Walawe estate. Good gemming is reported from place called Etbedde, said to be eight or nine miles below Balangoda on the banks of the Walawe-ganga. A large number of cinnamon garnets were obtained recently in a land not far from Mahawalatenna, but hough large they were too flawed to have much value.

Bambarabotuwa:- The gemming here is mostly ilicit. The Walaweduwa and Kondrugala jungles have been thoroughly worked in this way. The gems obtained belong chiefly to zircon, but others, such as sapphire and spinel, are also got. Thorianite was first brought to light by illicit gemmers; these men take provisions for a week or two and build a booth in the jungle or live in a cave till they have exhausted the neighbouring streams. I have actually found them at work myself.

Ratgama, Pallabadaela, Eratna:- A belt of barren country (so far as gems are concerned) is met with here. In Ratgama there are some old illicit pits in the jungle; and a very little gemming has been done on what is now Dehenakanda estate, but there seem to be few gem minerals obtainable ; and zircon and heavy minerals of the rarer sorts are almost totally absent. All of these areas are situated along the main mountain range, which rises in a series of steep scarps and precipices, alternating with platforms of flatter rock covered with dense jungle. The streams are steep and violent. No proper deposits of *illam* occur. Gem washing yields only an abundance of worthless garnet. Magnetite, sometimes in good octahedra, is more abundant than usual.

Pallebedde, Timbolketiya, Kolonna:- At the two first-named places a little illicit gemming is done in the Rakwana-ganga, or its gravels. Practically no gems are obtained in the Kolonna area, though a little gemming is carried on occasionally.

Madampe:- The only pits of importance are at Karawa, where gemming is carried on very extensively. See 13. Gems of all kinds are stated to occur.

Ratnapura:- This is the centre of the gemming districts, and here also a great part of the trade in gems is located. Cutting and polishing is largely carried on. The district has only been briefly examined by the Survey during this year. The most important pits bear Ratnapura are those at Weralupe (p.14) and Mr. Jayatileke's pit at Potgulkanda (p.14). A specimen of *nambu* stated to be from Niriella, five miles from Ratnapura, was found to consist of cassiterite (tin ore). This important discovery will be looked into early next year, when the Ratnapura, when the Ratnapura District will be further investigated.

Gilimale: A limited amount of gemming is done I Gilimale South, a few good sapphires being sometimes obtained. Thorite was met with at Durayakanda.

Kuruwita is a very important centre, both for gemming and for cutting and polishing and for trade in stones. The gemming is done chiefly by villagers in lands belonging to the Maha Saman Dewala; and it forms their regular work for a great part of the year. The gemming is in various swamps and deniyas, the material composing the gravels being of local origin, as is indicated by their diversity in different localities; the gravels of the Kuru-gange are barren.

Talawitiya, another important gemming village, forms with the exception of Getaheta, the northern limit of the gemming district.

(b) By the Assistant Director

The most southerly gemming district was that of the Dediyagalamukalana in the Weligam Korale, Southern province. This consists of low tableland covered with jundgle, in which rise tributaries on the Gin-ganga, Polwatta-ganga, and Nilwala-ganga. The gemming has been carried put in small pits dug at the side of the dolas. Gemming has been and stil is very occasionally carried on by illicit gemmers. Good zircons and cat's-eyes are reported to have been found, but no varieties of corundum.

I have seen a very small quantity of gold obtained from the southern side of the forest, but was unsuccessful in finding any traces myself.

E18 Ceylon Administration Report for 1904 [IV-EDUCATION.

More gemming has been done in the neighbourhood of Kottawa, where gemming from Galle have quite exhausted several dolas. I have myself found abundant zircon near

Kottawa. The Galbandudola flowing from Dediyagala to the north-east yields zircon at Nakiyadeniya, and a considerable amount of gemming has been done at Udugama.

North of the Dediyagalamukalana is important gemming land in the villages of Alapaladeniya, Derangala, and Rambukkana. The best land is in the long strike valley or deniya of tghe Naya-dola. Part is private land and part Crown land, a portion of which was claimed and gemmed by Mr.Le Mesurier, late C.C.S. It has been largely gemmed, but a considerable amount of *illam* is still left, and is rather rich in good zircons. Many cat's-eyes and alexandrites are reported to have been found. Parallel to the valley of the Nayadola are two other strike valley, those of the Etakala-ela and Atu-ela, both gem-bearing streams. The Atu-ela and Naya-dola meet at the boundary of the villages of Derangala and Rambukkana. Here extensive gemming has been carried on in the past, and the whole of the allivium has been gemmed out.

An isolated stretch of gemming land occurs on the east side of the Nilwala-ganga in the villages of Dankaluwa and Dehigaspe. There is still some gemming carried on in the paddy fields by the side of the river, and good zircons and cat's eyes occasionally obtained. Two beds of *illam* are usually found , the bottom at a depth of about two fathosm. As no gems are gound in the gravels of the Nilwala-ganga for four or five miles north or south of the villages of Dankaluwa and Dehigaspe, it appears probable that in this district they have been brought down by gem-bearing streams from the hills in the neighbourhood of Siyambalagoda. Most extensive gemming has in the past been carried on at Urubokka.

A number of streams converge from the hills which form the bounbdary of the Southern and Sabargamuwa Provinces and from Aninkanda into marshy land, from which rises the Urubokke-ganga. To prevent floods which constantly occurred in Nilwala-ganga and to provide water to the south-east, a dam has been built across the end of the marsh, transforming it into a large branching tank, from which a channel has been cut, carrying the water to the south-east into the valley of the Walawe-ganga. The marshy land round the tank is riddled with gem pits, some of them three fathoms deep. The course of the Urubokka-ganga below the dam is left as a deep channel with rocky sides 15 or 20 feet high, in which now runs a comparatively small stream. *illam* is abundant in the bed of this stream and in patches up the sides of the channel, and extends up the deniyas opening into it. As the lateral stream flowing into the river at this point are not gem-bearing, the beds of *illam* were probably deposited by old floods before the dam was built.

The new Urubokka channel, which is deply cut in *kirimetiya*, has abundant zircon in its gravels, but I do not know if this is derived from the surrounding alluvium or is directly brought down from the hills. Zircon, toumaline, and cat's-eye have been found at Urubokka, but I have not heard of any variety of corundum being obtained. At only indication of the mineral I have seen in the Southern Province, though I understand it is found near Galle.

A dola at Naminkanda, near Deniyaya, which in part of its course flows through a deniya and forest in Crown land and in part through jungle belonging to Campden Hill

(Kandapana) estate. It yielded abundant and valuable stones, but has the Crown land portion. Tourmaline, cat's-eye, alexandrite, and topaz were abundant.

In old *nambu* brought me from this place I found a fragment of what appears to be thorite, but I was unable, as the place is completely worked out, to obtain any *illam* there. The *nambu* had been thrown by the washers into the stream and I failed to recover any.

The range of hills dividing the Southern from the Sabaragamuwa Province and covered by the Sinha Raja forest yields gem-bearing streams throughout its length. It is worked both from the southern or Morawak korale side and from the northern or Kukulu korale side.

All accessible places appear to be worked out if the *illam* is good, but doubtless in the heart of the forest are good places still left. There are numerous local rumours of such places. Unfortunately, the weather was too bad during my stay near the southern portion to investigate these fully in person. Cat's-eyes and good zircons are the chief gems found. I have not seen or heard a reliable report of any variety of corundum being found in the Sinha Raja range or the streams flowing from it. The northern side of the Sinha Raja may be considered as belonging to the Rakwana district, the steams at Illebekanda being mainly derived from the range.

It is the Kabaragala range, however, which is parallel to the Sinha Raja and stretches from Suriyakanda to Beregalakanda. This range, which is parallel to the strike of the rock foliation, forms the north side of the valleys of the Rakwana-ganga and the Koswatta-ganga from Botiyatenna to Rakwana and from Botiyatenna to Rambuke respectively, the watershed of the wo rivers being a transverse ridge across the valley at Botiyatenna. At Botiyatenna and Golden Grove the alluvium, mixed probably with talus, attains the depth of at least 120 feet, at which depth *illam* was excavated. The whole place is riddled with deep pits and tunnels (*donas)* going in all directions.

It is difficult to account for such an accumulation on a ridge, and I can only suppose that the boulders and the talus generally from the hills on each side sufficiently hindered the flow of the streams to allow the formation of this deep deposit.

The streams flowing from Kabaragala are all gem-bearing. About two-thirds of the way up Kabaragala is a flat in which several streams arise. This place, which is in Crown jungle, has been most extensively gemmed. A similar flat occurs at the end of the range at Galkanda. Some spots have *illam* still left here, where there is an abandoned tea estate, but such places are being gradually worked out by illicit gemmers. The Rakwana-ganga has gem-bearing gravels throughout its length from Botiyatenna to Rakwana village, where it is joined by the Rakwana-oya. All the paddy fields in Rakwana have gem-bearing gravel, though good stones are rarely found. The *illam* lies at a depth of 10 to 15 feet. North-west of the watershed at Botiyatenna is gemming land as far as Dependence estate and Rambuke.

The Kabarangala range and the streams flowing from it yield all varieties of corundum, especially sapphires, for which the Rakwana district was framed. Alexandrites and chrysoberyls have been occasionally found, but I have not heard of or seen any trace of cat's-eye. Spinel is abundant in all varieties, the pale pink obviously derived from the crystalline limestone of the district, dark green and

purple, and the black opaque variety with semi-metallic lustre. Tourmaline is usually abundant, but zircons suitable to be cut for gems are rare. A word should be said as regards the extent to which gemming has been carried on in the districts I have described. It may be said that in one of these places is any good gemming left.

A gemmed out deniya should be seen to be appreciated; the place is turned into a rabbit warren of pits and tunnels, and if the *illam* is good every pound has been excavated and washed. In marshy land, where it has been difficult to drive tunnels, small quantities may be left between the pits, but speaking generally, only those places are left where the *illam* not rich.

This applies even more to the Crown forests than to private land, where unless a certain profit is assured, the villagers are unwilling to disturb their paddy fields while having no scruples concerning poaching on Crown land, where risk of detection is slight; and even if the offenders are known the headmen, as a rule, for reasons which are not far to seek, display little energy in bringing them to book. The extent to which the country has been gemmed has been a great hindrance to prospecting for thorium-bearing minerals, the great difficulty in a district being not to find gemming places, but places that have not been gemmed out. Though there is doubtless a tendency to conceal the existence of gemming pl,aces, I am convinced that very few gems are now obtained or could be obtained from the districts I have described. Gems must still be derived from the disintegration of their parent rock and *illam* slowly being deposited; but as regards the beds of *illam*, the accumulation of ages, there can be no doubt that it is only a question of time when they shall be exhausted in all parts of the Island as thoroughly as they have been in the Kukulu korale and in the central and western parts of the Southern Province.

CRYSTALLINE LIMESTONE

By the Director

These rocks, which in spite of their very impure (siliceous) character are used for burning for lime, were sparingly met with in Sabaragamuwa, except in the Kolonna and Palabedda districts.

Hardummulla (Uva) - A small exposure on Kalupahana estate has been burnt for lime.

Bilhul-oya (Sabaragamuwa):- A considerable band of crystalline limestone is exposed below the bridge over the Kadawat-oya, on the main road bear the 99 1/2 milestone.

Balangoda- Crystalline limestone is exposed in some quantity beside the cart road near Mahawalatenna, six miles from Balangoda. A small quantity is to be seen near the Herimitigala on Herimitigala estate, about seven miles from Balangoda in the opposite direction. There are more considerable exposures on Detengala estate (near the

Sabaragamuwa-Central Province boundary on the path to Kotiyagala). A small exposure near Morahela estate has been burnt for lime; one variety is noteworthy as containing platy crystals of ilmenite, a mineral not previously recorded from Ceylon limestones.

Bambarabotuwa:- Limestone occurs in huge boulders not far from *in situ* on Hapugastenna estate, in the bed of the Rat-ganga near the infall of the Hatdara-ganga.

Gilimale North :- Three is an exposure of crystalline limestone at Kudawe, with a considerable cave, in which are quantities of very fair stalactites, superior to anything hitherto me with in Ceylon.

Eratne:- Crystalline limestone expopsed in a small stram called the Hunugala-ela (which joins the left band of the Kuru-ganga, a little below the Bimwatura-ela) has been burnt for lime.

Kuruwita:- Limestone is exposed in a deniya behind Paradise estate, but was not actually seen by me.

Kolonna:- There are several exposures of crystalline limestone near Kolonna; one is a short distance from Kolanna on the Rakwana path; another and rather remarkable on is on the Hayes-Rakwana path, nar the Nedullakanda corner; here the inclusions of mineral aggregates and seemingly rock fragments are so abundant as to give the limestone exactly the appearance of a coarse agglomerate

Pallebedda:- There are many exposures of crystalline limestone on both sides of the village, but especially a couple of miles south; and on the north exposures are frequent on the Gansabhawa path to Balangoda, and also on the track to Kolonna and on the main road to Madampe.

Rakwana:- Thin beds of limestone occur on Rangweltenna estate, chiefly exposed in the bed of the Rakwana-ganga. One part consisting of almost pure calcite has been burnt for lime. Dark brown tourmaline occurred in one variety of limestone, together with the more usual accessory minerals. Limestone is also met with at Pannila, about six miles from Rakwana:-[J.P.]

MICA

By the Director

Mica:- Only two localities for mica of commercial value have been met with during the present year.

Haldummulla (Uva):- An old pit below Kalupahana estate, near Haldummulla, appears to have been the most important mine in Ceylon, and the only one regularly worked by means of shafts and galleries. The mine is now abandoned, save for a little illicit work which is still carried on. The mica is a golden brown or amber-coloured phlogopite, very pale in thin flakes. It used to be obtained in large and flawlesssheets, and a good deal of money was, made out of the mine before its partial exhaustion and abandonment; it was worked by the late Mr. Edgar Heanly and subsequently by a Mr. De Winton. The best samples obtaibale now are not more than four inches square. The spoil heaps consist of decomposed and kaolinized rock, and as the shafts are now inaccessible it is impossible to say exactly how the mica occurred. A few fragments of orthoclase with mica seem to suggest that it may have occurred in a sort of orthoclase-mica pegmatite, a different mode of occurrence from that met with elsewhere in Ceylon.

Balangoda:- There is an exposure of muscovite-quartz-orthoclase rock near the Kanmudu-oya in Piddeligannawela, some nine or ten miles from Balangoda; but although the muscovite is in fairly

large individuals, it is so irregularly grown that it is impossible to obtain clear unflawed pieces exceeding 2 inches in diameter; and there would of course be no prospect of an improvement in this respect by going further down. A large vein of brown phlogopite is exposed in the Mineran-oya, just within the boundary of Sherwood estate (about nine miles north of Balangoda) in the south-west corner. The vein is about 5 feet wide, and strikes 20 N. of E. and hades 70 S.S.E., and crosses the foliation of the granulites which are inclined at low angles. The oya has for some distance cut its way along the mica vein which is exposed in the bed of stream. The vein has been superficially worked, and a small profit obtained. The largest crystals are 18 inches in diameter, but somewhat flawed. Better material could probably be got by digging lower of further into the hill.

GRAPHITE

(a) By the Director

There is not much to be said on the subject of graphite in the districts examined by the Director; in the greater part of the area none is obtained. The district between Haputale and Balangoda is barren, with the exceptions of small pits not now worked near Denegama and other small pits on Kinagaha-ela estate, which have yielded a few tons of graphite and are now being worked. No graphite is obtained in Bambaraboutuwa or on the north side of Ratnapura in the Gilimale districts, nor at Kuruwita nor Eratna.

In the Kolonna korale there are important mines at Kumburugamuwa (about three miles from Kolonna); a number of small pits are worked at Kolonna itself, and others slightly more important at Erep oruwa. An open working at Nedullakanda, Kolonna, was of some interest, as the graphite occurs in this fibrous films on the surfaces of the lenticular fragments into which a zone of the country rocks has been broken up; possibly some shattering of the rock was due to earth movement and the graphite deposited in the crevices and on the surfaces thus exposed to penetrating vapurs; no graphite occurs in the actual rock, which is a white leptynite. This is in agreement with the usual observation that the graphite of commercial importance in Ceylon has been deposited in veins and crevices in the country rock, and must be regarded as of secondary origin, deposited subsequent to their consolidation.

In the Atakalan korale there are a number of important mines with many shafts at Werahera, near Godakawella, and near Rakwana at Masimbula.

(b) By the Assistant Director

A number of plumbago workings and veins were examined, including all of any importance in the districts traverse. Details of these are given in the interim reports. Advice was given as to the best methods of working and the directions in which veins might be expected to continue. It is pointed out in the interim reports, how often, labour and money is wasted in sinking deep shafts in a purely speculative way in the hope of reaching plumbago, of which the surface indication are most feeble or altogether, absent, the idea ingrained in the Southern Province villager being that to dig deep enough is to ensure success.

While it is impossible to foretell with any certainty the existence or continuance of a vein in any direction, yet it would save much wasteful working if it were realized that the main veins always follow one or more of the three determinable direction in relation to the strike of the rock foliation, viz., parallel to the foliation dip joints (vertical planes at right angles to the strike of the foliation), and strike joints (planes at right angles to the plane of foliation), also instead of indiscriminate sinking of deep shafts, it is more profitable to follow up surface indications by tunnels, meeting well-placed vertical shafts for supply of air, and easy removal of water, country rock excavated, and plumbago.

The following is a list of the principal plumbago mining places vistited:-

In the Southern Province

Henagama and Maragoda:- Extensive works, in the first place by shafts and in the second by shafts and open working; in this the veins are difficult to trace, and the plumbago in places injured by a well-marked fault.

Neraluwa:- In Crown jungle. A vein following strike for nearly a mile and exposed practically at the surface in many places.

Naramba and Lenama:- Considerable workings now abandoned.

Bopitiya:- Two pits working; chiefly "yabura" plumbago being obtained.

Tellambure and Nakiyadeniya:- Extensive series of pits, abandoned in the first place owing to exhaustion of the veins, in the second temporarily abandoned (February, 1904).

Koralagama:- Two pits, from which a relatively small quantity but excellent quality of plumbago has been raised.

Habarakada:- Two series of workings, distant three miles from each other, but both yielding similar plumbago in well-marked veins between bands of "iron stone" (an intimate mixture of limonite and quartz). This appears to be due to decomposition of the country rock on each side of the veins.

Panagoda and Dewalagama:- Extensive series of small workings over a large district. The veins are small, but the plumbago of good quality.

Gurudiyapota:- Plumbago chiefly in granular quartz rock. Work has been carried on a large scale, but the great distance the mineral has to be conveyed by cooly and boat to Galle render profitable working difficult.

Gorakawela:- Extensive excavatins were made for plumbago here by mr. Le Mesurier, late C.C.S.,It is probable that the veins are not exhausted.

Kotoapola:- A strike vein, practically exposed at the surface for about half a mile. It is on Crown land, but much plumbago was removed buy villagers before the work was stopped.

Pasgoda:- A number of pits sunk on two or three veins following strike joint, which may be traced over half a mile.

Kondapana (Campden Hill) estate:- Extensive wroking by shafts and adits in the hill-side south of the Mahadola. Absence of ventilation has now practically caused stoppage of work in the more important series of pits.

Pannila (Kukulu Korale). Sabaragamuwa:- Work now in progress (October 1904) in two series of pits and adits.

MUSEUM WORK

About ten days were spent by the Director and Assistant Director at the Museum in August, and a little more than two weeks in December. The arrangement of the mineral gallery is now fairly complete. Over 600 specimens have been catalogued, and printed labels have been set out for over 500 of these. The gems have been carefully gone through and accurately determined; they were formerly named in a very haphazard manner. The best are a ruby (value about Rs.200) and a cat's-eye and alexandrite of somewhat less value. A number of the cheaper sorts have been purchased out of sums available for contingencies in the Mineral Survey vote, and these have filled up a few of the gaps in the series. A large number more is required however to give a fair idea of the colour variations; and beside this some specially good specimen of some of the most conspicuous gems are much needed. The absence of a good sapphire is especially noticeable; really good spinels, garnets, and zircons are also needed; the varieties of chrysoberyl, on the other hand, are for the present sufficiently represented, so far as cut stones are concerned. The collection is especially lacking in uncut specimens however. If a special sum of Rs.1,000 or Rs.1,500 could be provided to purchase a few more expensive stones, and a further sum is spent in 1905 from the Mineral Survey vote as in the present year (and I recommend that all balances available at the end of the year and not merely the avilable balance on contingencies be allowed to be used for this purpose), so as to obtain a more complete series of the less expensive sorts, the collection will be able to progress as fast as can be expected. It is very desirable, however, that a moderate sum of money should be spent yearly for many years in this way, because the visitors to the mineral gallery take a far greater interest in the gems than in any other part of the collection, and the existence of a representative collection which can always be consulted is sure to lead to the expenditure of money in the Island.

Of other additions to the collection, the following donations are worthy of special mention:- Photograph of graphite mine by Messrs. Skeen & Co., an exceptionally large mass (400 lb.) of ilmenite from Mr. W.D. Holland; and a number of cut gems from Mr. W.C. wild. Amongst the more important additions for which the Mineralogical Survey Staff is directly responsible may be mentioned the large masses if sillimanite-corundum rock and sillimanite-bearing garnetiferous-leptynite or quartz-felspar-garnet-sillimanite schist from Hadummulla. The latter specimen very closely resembles the khondalites of Southern India. The Director of the Geological Survey of India has kindly sent specimens of these rocks for comparison.

Sets of photographs illustrating the processes of gemming and of iron-smelting and steel-making by the Sinhalese from negatives taken by the Mineral Survey Staff and Mrs. Coomaraswamy have been placed in the gallery. The recent fossils have been as far as possible identifies (by Dr. Willey) and arranged. The minerals containing rare elements. & c., are temporarily grouped together pending their further examination and identification. Anumber of gems have been purchased at the end of the year out of the balance available for contingencies on the Mineral Survey vote. A list6 of these is appended below. Most are of the cheaper sorts, but were much needed too fill up gaps; one of the most remarkable specimens is one of white topaz, and uncut waterworn pebble, perfectly clear and transparent and weighing about 3lb. Of the other stones, the two moonstones may be noticed as specially good of their kind:-

List of Gems cheaper sorts purchased out of Balance available on the Mineralogical Survey Vote

Catalogue No.	Descriptin	Rs.P. (Cost)
629	5 moonstones	12.00
584	1 moonstone	17.50
583	1 moonstone	60.00
625	1 rought topaz (3 lb. weight)	50.00
620	1 green topaz	36.00
628	1 black cat's-eye	3.00
623	3 corundum (kind topaz)	
621 & 639	2 zircon	27.00
639	4 zircons (Matara diamond)	4.00
632	1 oriental amethyst	1.50
633	1 oriental amethyst	
634	1 king topaz	26.00
635	1 tourmaline	
636	1 spinel	
522, 523	2 sapphires	20.00

A large number of minerals have been submitted by members of the public and have been identified; no charge is made for work of this sort. The minerals submitted have in most cases been iron pyrites and mica, which are not infrequently mistaken for gold.

A.K. COOMARASWAMY
Director

December 31, 1904

CEYLON.

ADMINISTRATION REPORTS, 1905.

PART IV. — EDUCATION, SCIENCE, AND ART.

MINERAL SURVEY.

REPORT of Dr. A.K. COOMARASWAMY, F.L.S., F.G.S., Director.

Geological Survey & Mines Bureau
1995

MINERALOGICAL SURVEY

Director: ANANDA K. COOMARASWAMY, D.Sc., F.L.S., F.G.S.
Assistant Director : JAMES PARSONS, B.Sc., F.G.S.

REPORT OF THE DIRECTOR OF THE MINERALOGICAL SURVEY FOR 1905

(1) DISTRICTS EXAMINED IN 1905

(a) By the Director

DURING January the Director was on leave. February, March and part of April were spent at Niriella and Ratnapura, mainly in the examination of river sands, many of which were found to contain thorianite in very small crystals. A short examination of part of the Kegalla District was made in May. The Director was laid up with fever during the first part of June, and then started for a tour in the North-Central and Eastern Provinces, in order to make a reconnaissance in the drier districts during the south-west monsoon. A further attack of fever at Alut-oya delayed matters again ; but the tour to Trincomalee and up the coast to Kuchavelli, and finally back to Anuradhapura, was completed before August. A week was then spent in museum work, and three weeks' leave was taken to recruit. In September and October the Director carried on work in the Ruanwella and Avisawella districts, where it was found that monazite is generally distributed in the river sands, and galena was also detected. The months of November and December were spent in the preparation of reports, museum work, and a further period of leave.

In October it was decided that the Mineralogical Survey, originally appointed for three years ending March 7, 1906, should be continued to the end of 1906. With respect to the future there appear to be three courses open to Government:-

(i) To establish a Geological Survey
(ii) To appoint a single permanent Geologist and Mineralogist
(iii) To discontinue all mineralogical work

The work of the present Mineralogical Survey, regarded merely as a reconnaissance or flying survey, will be more or less completed by the end of 1906, while, on the other hand, no really detailed and systematic examination of the Island has been possible or even expected. The establishment of a regular Geological Survey is scarcely recommended, if only because suitable large-scale maps are not available for the very detailed mapping which along would be satisfactory. I have accordingly prepared and submitted a scheme for, and recommend the establishment of, a single permanent office as Government Geologist and Mineralogist, and have recommended Mr. Parsons for the post.

(b) by the Assistant Director

Work was commenced in January at Bentota (Southern Province), where the rock exposures on the coast were examined. From this place I proceeded into the Pasdun korale, spending some time examining the graphite of the district with Matugama as centre. A week was then spent in camp at Pelawatta, and the gold and gemming of the district investigated. An expedition was made to Boralugoda, with the object of examining the country on the borders of the Western and Southern Provinces, and so connecting with my work in the Hinidum pattuwa early in 1904. Proceeding thence through Badureliya, I crossed the boundary of the Western and Sabaragamuwa Provinces into the Kukulu korale. I then proceeded to Ratnapura, stopping at Kukulugama, Kalawana, Karawita, and Noragalla. While here I first observed the thorianite deposits of Erabadda which necessitated a return to the district later in the year. After a week spent in office at kandy I went to Nuwara Eliya in order to examine the gem-bearing deposits of the Moon Plains and surrounding district. During the south-west monsoon it was decided that I should make a rapid traverse through parts of the north-west and north of the Island. I proceeded from Kurunegala to Puttalam, where I examined the salt pans and process of salt manufacture, thence to Anuradhapura, from there north to Vavuniya. From Vavuniya I made a short traverse to the west to Paraiyanalankulam, and returned to Anuradhapura *via* Madawachchiya. About a week was then spent in office at Knady, after which I returned to the Ratnapura District, stopping for about two months in the village of Dela, making a thorough examination of the important mineral deposits in Dela and the neighbouring villages. I then returned to Kandy for recess, stopping a short time at Pelmadulla and Pussella. November was spent in Kandy preparing reports, and December in Colombo for the purpose of museum work.

(2) GRAPHITE

By the Director and Assistant Director

Considerable attention has been paid to the occurrence of graphite by the Mineral Survey during the year. The well-known graphite producing districts of Kegalla, Ruanwella, and the Pasdun korale have been thoroughly examined as well as other localities. It is now possible to make the following general statements as to the mode of occurrence of graphite.

E2 *Ceylon Administration. Reports for 1905.* *[IV.-Education.*

(1) The veins of graphite occur in the natural planes of division in the country rock, *i.e.*, they are parallel to the foliation planes of the granulites or other country rock, or they fill, strike, or dip-joint cracks, [For a definition of these terms see earlier reports.] Occasionally a small vein may be seen following an irregular crack for a short distance. Usually in any single pit or series of adjacent pits there is a single main vein, or several parallel veins following one of these directions, sending off minor veins or stringers along the other planes of division. Fig.2 shows graphite occuring along contorted foliation plans.

Even where the graphite occurs in pockets and not in regular veins, the disposition of the pockets is usually parallel to one of the main directions. This is well shown in the once extensively worked pocket deposits along the Agalawatta-Badureliya road (Pasdum korale). Here the general direction of the pocket is at right angles to the general rock strike of the district, and hence probably follows dip-joints. As a general rule, however, the trend of deposits in a district is parallel to the strike of the rock foliation. Graphite in the Island appears to occur in a series of "belts", but further work is required to determine their extension exactly. On a small scale the arrangement of veins that may be found in pit of even in a district is often shown even in a hand specimen. Such a specimen from the Ruanwella district here figured (Fig.3) illustrates this. Here "main veins" are shown parallel to the rock foliation, and sending off "stringers" along joint cracks. In one place of veins coalesce to form a "pocket". There is also a vein, not shown in the figure, in a plane parallel to the face of the specimen (strike joint).

(2) *Mode of occurrence of graphite in the veins*:- In a small vein the graphite usually occurs as a parallel aggregate of small platy needles set at right angles to the wall of the vein; in larger veins it tends to occur in thicker flakes, which from their resemblance to jak seeds are known as *maduluminiran.*

In a large vein, zones having a different habit of crystallization may often be observed, the zones corresponding on the two sides of the vein.

In pocket deposits the mineral tends to occur in irregular plates.

An spherulitic structure is sometimes observed throughout the mass of good-sized pieces of graphite, but is not particularly common. Occasionally the structure is more perfectly exhibited when the graphite occurs in association with quartz, as shown in Gif.1, where it is developed in beautiful spherulitic rosettes on the surface of a piece of quartz. The specimen figured is from Wadawala, between Hanguranketa and Padiyapelella (see Mineralogical Survey Administration Report, 1903).

(3) *Associated minerals*:- Very often the outer walls of a vein are composed of more or less hydrated ferric oxide, and sometimes this is intimately mixed with the graphite. This sometimes seems due to the decomposition of the country rock by percolation of water along the joint cracks, and sometimes to have been introduced as a deposit from percolating waters. Occasionally it seems to owe its origin to the decompostition of iron pyrites.

Iron pyrites is an almost constant associate of graphite in veins. When it occupies a definite position in the vein it fills the central portion. The pyrites is often found disseminated through the graphite, between the individual flakes and needles, usually in a decomposed state. This gives a dull earthy appearance to the graphite, which is then called *yabara-miniran.* Occasionally a curious intergrowth of selenite and pyrites is found in a graphite vein, as at Ratmale, Pasdun korale, of which a specimen is gifured (Fig.4). In this case the selenite has crystallized first and the pyrites is moulded on it; the different portions of selenite are similarly oriented. Graphite veins are very commonly associated with quartz veins, and quartz often forms partof the vein material, filling the central portion. Felspar is sometimes found, when it seems derived from the pegmatite material of an adjacent intrusion.

Quartz is also very frequently associated with graphite; it is usually milky (tiruwanagala), and tends to occur in the centre of veins. Quartz and iron pyrites are the only two accessory minerals which are almost always present ; the other minerals are rarer. Biotite is not uncommon in the veins, and pyroxene is sometimes found. Apatite is occasionally met with. A peculiar amorphous black substance, perhaps carbonaceous, and known as "burnt plumbago", is often met with in pits. A sample has been forwarded to the Imperial Institute for analysis.

(4) *Associated rocks*:- Graphite is most commonly met with in the acid granulites of the charnockite series; it is often found in granular quartz rock, not so often in typical charnockite. We have only observed one occurrence actually *in* pegmatite (at Talatuoya, Central Province), and have never observed it actually in limestone except in small scattered flakes; though some pits at Udawela (Central Province) have been sunk through limestone.

A very common country rock, and one almost always associated with graphite in the Southern Provice, is an acid granulite containing pink garnets and rich in iron pyrites, usually more or less decomposed.

These observations apply only to vein graphite. Flakes of graphite are very common in all rocks, but such flakes form most probably an original constituent of the rocks in which they occur, and have no direct relation with the graphite veins. There is no evidence that the graphite of the veins penetrates the surrounding rock, except possibly locally on a very small scale. (A.K. Coomaraswamy, Quart. Jour. Geol. Soc., vol. LVI., 1900, pp.610-612).

Graphite veins are very commonly associated with quartz veins, which in their turn may often be traced as apophyses of pegmatite veins. The connection with pegamatite is often more direct, and in the Pasdun korale, graphite almost invariably follows pegmatite intrusions. The boundary of a pegmatite and the rock into which it is intruded is often very undefined, the pegmatite re-melging the original rock, in some cases almost completely destroying or distorting its original foliation and the pegmatite material becomes mingled with that of the re-melted granulite. Near the boundary the felspar and quartz individuals of the pegmatite may be seen isolated in the rock so formed, but further from the boundary the acid constitutents of the pegmatite and those of the granulite are indistinguishable.

Such intrusions may be well studied on the foreshore opposite Bentota resthouse, which may be looked on as a small centre of intrusion, as proceeding north and south of this point the boundaries of the pegmatites are found to be increasingly better defined.

It is alsong such ill-defined boundaries of pegmatites and granulite that graphite very commonly occurs in the Pasdum korale.

(5) *Origin of graphite in Ceylon*:- The data of the preceding four paragraphs are practically all that are available on which to form a theory of the origin of graphite, and it will be seen that they are scanty. It appears certain that the graphite was not aggregated in veins and pockets by a process of "excretion"

*

Modes of Occurrence of Graphite
(Figures of Naturalsize)
In Figure 4, G. represents Graphite, P. Pyrite and S. Selenite.

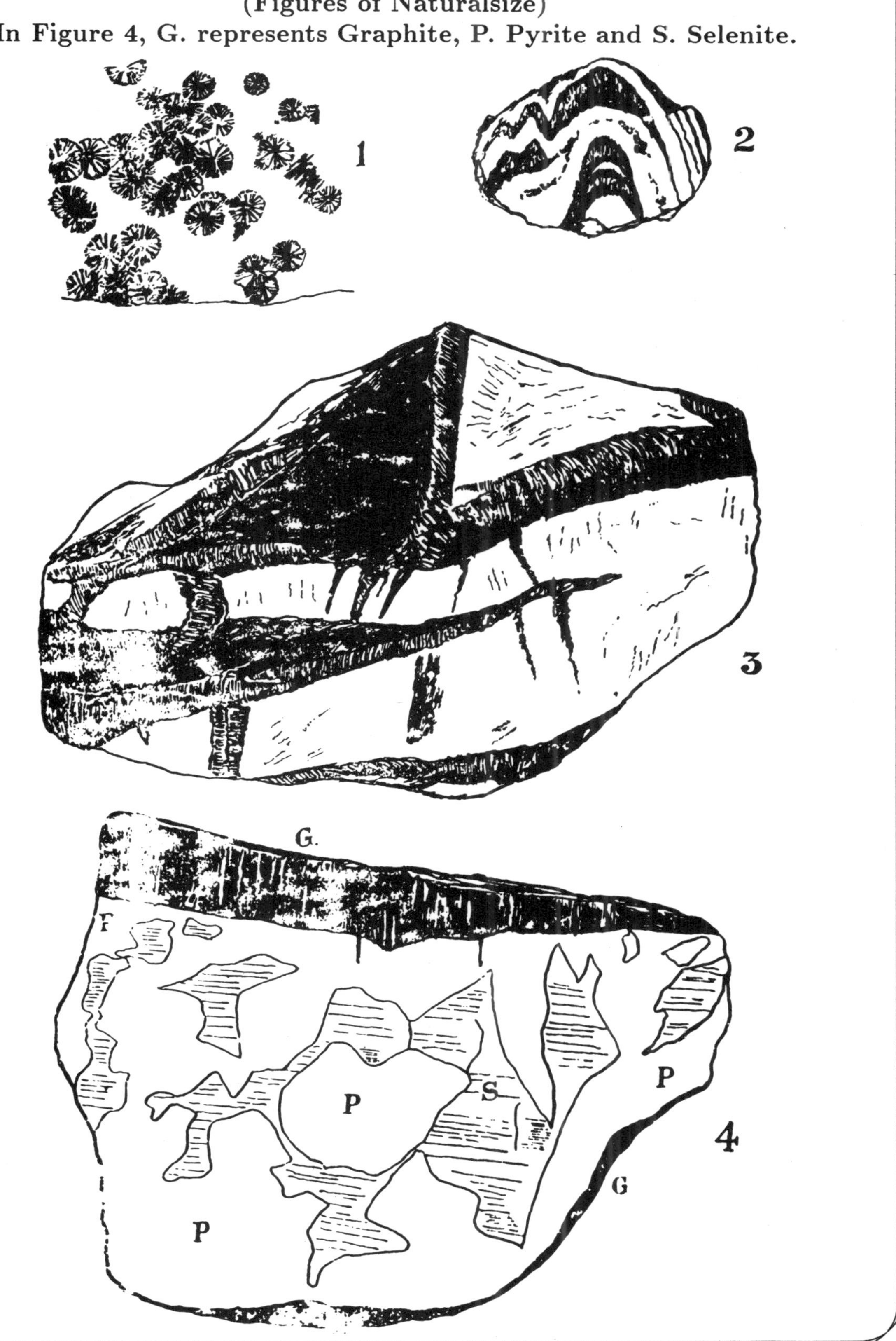

from surrounding rocks, but on the contrary it seems certain that it was introduced under high pressur into pre-existing cracks in the containing rock, or planes of weakness in the same, in either a liquid or gaseous form. That it should occur in rock, which at the time of introduction was sufficiently cool to have developed cracks or lines of weakness, precludes the hypothesis that the carbon uncombined with other elements was introduced in the volatile condition.

It therefore appear probable that it was introduced in the form of an unsaturated carbon compound, or a saturated compound which dissociated at a high temperature on the diminution of pressure due to the shrinkage and formation of cracks in the containing rocks, but what this compound was, in the present state of our knowledge, it seems rather idle to speculate.

Professor Weinschenck considers that emanations of carbon monoxide, with or without cyanogen bearing compounds, may have given rise to the graphite veins, and further suggests that it may in part be due to the decomposition of metallic carbonyls. The general absence, however, of metallic oxides, which can be definitely stated not to be due to the percolatio of water decomposing the surrounding rock, would not support this view. (See E. Weinschenck, Zur Kenntniss der Graphitlagerstatten. III. Die Graphitlagerstatten der Insel Ceylon Abh. Bayr. Akd. Wiss. 21, 1900). In general Professor Weinschenck advocates the view here outlined. Thus Dr. A.K. Coomaraswamy, in an abstract of the above-mentioned work, writes (Geol. Mag. Decade IV. vol.III., No.442, p.175):- "Professor Weinschenck would suppose the following to have been the sequence of events in Ceylon. a Fluid magma intruded into beds of unknown age consolidated as a peculiar *schlierig* rock while contact metamorphic structures were developed in the surrounding beds. Contraction joints developed on cooling, allowing the formation of pegmatites including pure quartz veins to some extent. But contemporaneously with the formation of the pegmatites there were emanations of carbon monoxide and eyanogen-bearing compounds, which followed the same paths as pegmetites and then gave rise to the graphite veins.

In view of the almost constant occurrence of pyrite with the graphite it is tempting to suggest that the carbon may have been introduced in the form of sulphide, and this in contact with oxides of iron to have given rise to the pyrites. however, the quantity of the pyrites is much too small to account for all the sulphur which would been combined with the carbon, and if the greater part of the sulphur had not been so absorbed one would expect that deposits of sulphur would be found in the neighbourhood of the graphite. Sulphur does indeed occasionally occur, but in such cases appears to be due to a subsequent decomposition of the pyrites. However, a sulphide of carbon may have been introduced in relatively small quantities. If so, it might account for the curious selenite-pyrite intergrowth mentioned above, by the oxidation of the sulphide in contact with corbonate of lime. In unweathered specimens the pyrites itself shows no signs of a decomposition which would give rise to a sulphate.

Though there is certainly no direct connection between the graphite disseminated through the rocks and that in the vein, that the former should occur in such abundance in a country rich in graphite veins cannot surely be due to a remarkable coincidence, but indicates some genetic connection. If so, following the nalogy of the quartz veins, we may regard the graphite-producing substance as corresponding to an ultimate residual intruded magma consolidating later than the quartz, this view being supported by the connection between graphite veins, quartz veins, and pegmatites noted above. Should this be the case we naturally seek among the rocks an analogue of the silicates and find this in the crystalline limestones. Whatever view may be held as to the ultimate origin of the crystalline limestones, weather we regard them as original magma streaks or as the metamorphosed products of ancient sediments, that they in common with the whole charnockite series were once in a state of fusion, may be regarded as certain, and hence may be treated as forming part of one viscous magma. Such being the case it is not unreasonable to regard the carbon of the limestone and the carbon in the graphite veins as having a common origin. should the limestone ultimately appear to be organic in origin, then we may also attribute an indierect organic origin to the graphite.

Without going to far as to say that graphite and calcite are mutually exclusive, we may note the very rare occurrence of calcite in graphite veins (none has been observed by the Mineral Survey) and the general absence of graphite veins in proximity to limestone. Professor Osann described occurrences of graphite in Quebec ("Notes on certain Archaean Rocks of the Ottawa Valley", Geol. Surv. of Canada, Part O, Annual Report, Vol.XII., 1902, pp.66-84), which seem to bear a general resemblance to those of Ceylon, though the resemblance is rather less than is there indicated. In Quebec the graphite is intimately associated with limestone in which the veins occur, and calcite is common in the veins.

Professor Osann's statement that "coarsely crystalline calcite also plays an important role in the Ceylon veins" cannot be substantiated. Minerals other that graphite seem to occur more frequently in Canada and extensively than in Ceylon veins, especially apatite, which as stated above, is very rare and not, as Professor Osann states, a common occurrence in Ceylon veins. In Quebec, as in Ceylon, the graphite is supposed to be deposited by fumerole action after the cooling or solidifying of the eruptive rock.

Though the graphite from the veins interpenetrates the surrounding rocks much more than in Ceylon (so much so that in one place the limestone is mined for the sake of its contained graphite), nevertheless graphite occurs independently of that disseminated through the limestone which professor Osann considers to be entirely organic in origin and to have no genetic connection with that in the veins. That similarly to the Ceylon graphite it has no direct connection appears probable, but no reasons are given why an ultimate genetic relation should not be maintained.

(3) Short Account of Pits Visited

(a) By the Director

Nambapana:- An important mine was being exploited.

Pelpitigoda, near Horana, Western Province - Here are a number of good veins, ramifying in all directions. The curious carbonaceous material mentioned in section 2 is well seen here.

Kegalla District:-

Batalegala:- Small joint veins occur here associated with quartz.

Hatgampola:- An important series of working with properly arranged shafts and adits, including main shafts and drainage adit. The graphite occurs in joint veins partly associated with granulite, partly with pegmatite material. The locality has been much affected by earth movements.

Diganewata:- A number of pits, the veins being associated with quartz.

E4 *Ceylon Administration Reports for 1905.* *[IV.-Education.*

Walakadeya:- Small veins, apparently parallel to strike of rock foliation.

Siyambalapitiya:- Many shafts are in use in the workings here. The veins are ( ) The amorphous carbonaceous (?) material mentioned above is well seen here.

Ruwanawella District:-

Wahakula:- Nagahahena ; Kekirihena.,

Siddamulla:- Joint Veins, often with quartz occupying the central part of the vein.

Pushena, Ampe:- These pits were visited by Dr. Grunling, and his specimens described by Professor Weinschence (see above). Especially at Ampe the graphite is intinately associated with quartz and pyrite.

Mapitigama:- Two pits are worked at Megagoda. The veins strike north and south.

(b) By the Assistant Director

Moragala, Kiriwattahena:- A number of pits worked here. The veins run E.S.E., and appear to follow strike joints. The veins occur in garnetiferous leptynit, but are associated with pegmatite, which appears to contain diopside (pieces were found in the waste heaps). The leptynite contains strings of quartz felspar intergrowth from the pegmatite.

Two pits were being worked in other parts of Moragala. The veins strike east and west.

Yahalihena:- One pit is worked here with tunnels following the veins in southerly and easterly directions.

Liniyawa:- Old digging, but now occasionally worked illicitly. The veins are associated with much limonite, and have a approximate S.W. to N.E. strike.

Molahena- Graphite occasionally worked here illicitly. The hanging wall of an exhausted vein had a strike of N.20° E.

Ratmala (Kukuluwalakanda):- The vein strikes about east and west; it is associated with graphite granite and contains much pyrite, which is sometimes intergrown with selenite.

Botalawa-Manama:- Three pits having good veins with approximate east and west strike. Pegmatite (graphic granite) occurs in the country rock. One of the pits yields graphite in radiating plates covered with a grayish bloom of decomposed pyrite (*yabara-miniran*). In one of the pits the graphite is directly associated with a black sandy mud, containing graphite, mica, and decomposed pyrite.

Pelawatta:- Here are the most extensive workings in the Pasdun korale. The water in the pits is moved by steam-driven pumps, and the mineral is raised by steam power. There is one main vein striking north and south and at its best is one foot thick. The country rock is a coarse granulite with garnet, mica, and pyrite.

Ingahakotena:- Formerly extensive poaching was carried on here, the occurrence being on Crown land. As usual in such cases a number of shallow pits have been dug. The graphite appears to be irregularly distributed in "ironstone"

Baddawila:- Formerly the graphite was worked by license on the Crown land here. At the time the place was visited it was obvious that illicit digging was being extensively carried on at night ; between 40 and 50 recently sunk small pits were found. There are several small parallel veins striking north and south, which are probably interfoliar.

Along the Agalawatta-Badureliya road:- There are may pits along the road, the land near the 5th and 6th milestones being riddled with old workings, which are dug again and again, and graphite is even washed out of the old mud from the pits.

Between the 2nd and 3rd miles are also many old pits. The graphite occurs in a country rock of acid leptynite, with numerous injections of quartz-felspar intergrowth obviously derived from pegmatite. In one of the pits much hallosite or similar mineral is obtained, doubtless corresponding to the nontronite of Weeinschenck. It is not, however, in actual contact with the graphite. All the pits contain pocket deposits of graphite and not veins.

Yakkupitiya:- The vein which strikes north and south is in garnetiferous granulite and associated with an orthoclase-quarts (—) garnet pegmatite.

Kukulugama:-

Two important workings by the side of the Wewa-ganga, about half a mile from its junction with the Kukuluganga. The veins are apparently both interfoliar and filling dip joints. Near the pits the micaceous granulite exposed in the river is seen to contain numerous pegamtite veins, which do not however bear any definite relation in their direction with the strike of the granulite foliation.

Graphite has been found as far north as Vvuniya in the Northern Province, but he furthest north deposit which is profitably worked is at Wagakunwewa near Talawa, about twelve miles south of Anuradhapura on the Kurunegala road. The graphite was first found in the Wan-ela near its junction with the Hunu-oya, and a pit was sunk striking the vein, which appears to strike east and west, and to follow a dip joint. Some of the graphite was much mixed with clay and decomposition products of pyrite; some was associated with limonite. No graphite is to be seen in adjacent rock exposures, but intrusions of pegmatite are numerous.

(4) GOLD

By the Director and Assistant Director

Mr. G.G. Dixon, when acting as gold expert for Government in 1902-1903, found that alluvial gold very generally occurred in small quantities in the central, western, and southern portions of the Island. Its occurrence has also been noted by many observers in the last fifty years, but never in really paying quantities. These observations have been confirmed by the officers of the Mineral Survey in the Province of Sabaragamuwa and the Central and Western Provinces.

A number of such observations have been made by the Director, following the native method of dredging the rivers with special mamoties : removing the superficial sand and gravel from the bed of a river and scraping up the surface of the decomposed rocks, over which the river flows. For detailed account of this process reference should be made to the section on dredging which has been applied by the Survey, not only for the purpose of obtaining gold, but also to obtain sands containing thorianite and monazite.

Niriella:- Dredging was carried on for two weeks in the Niriella (Karawita) ganga, which flows into the Kalu-ganga. Small quantities of gold were obtained. It was in this dredging that thorianite and monazite sand were first discovered (see under).

Minirandeniya, Elapata:- Here a small stream yielded gold by dredging, which was peculiar as occurring in a dendritic form. Probably the whole deniya would yield gold, but it has only been found as far in the stream.

Weralupe (katugaha) ela (Ratnapura):- This is traditionally well-known place for gold and for gems, and it has been much worked for these in the past. At the time of investigation it was difficult to carry on work satisfactorily, on account of insufficient water due to drought, also the lowest deposits are hard to work, being protected by a layer of iron-pan, or in places the *illam* itself is cemented by feruginous material which is difficult to break up. However, some gold was obtained. The stream is remarkable for the extraordinary abundance of old shot which it contains.

The *We-ganga (at Marapona)*:- This is a river of considerable size and one of the chief tributaries of the Kalu-ganga. Dredging was carried on here with difficulty on account of the large number of boulders in the river bed. A little gold was found.

The *Walawe-ganga near Balangoda*:- Owing to the break of the little monsoon work could only be carried on here for four or five days. The quantity of gold obtained, however, was greater than usual, and one piece weighing 415 gram was larger than any hitherto obtained by the Survey.

Ruanwella District:- Heavy sands collected by dredging from the Gonagala-oya and the Lawpitiya-oya at Mapitigama were found to contain traces of gold.

Avisawella District:- Several dredgings were carried out in the Getaheta-oya and the Sitawaka-ganga and yielded gold. a little gold was also found in the sand underlying the *illam* in the alluvial deposits of the Getaheta-oya. The gold from the Sitawaka-ganga is remarkable in that sore of the grains; are wholly or partially covered with a film of silver like metal, which dissolves readily in nitric acid. The Sitawaka is reported to yield gold in paying quantities, *i.e.*, to repay dredgers at the rate of Rs.1 or so per day. The Karawita-ganga, about six miles further up the stream than Niriella, was dodged and yielded small quantities of gold. The work is difficult here on account of large boulders in the river bed.

We-ganga at Dombagammana:- A dredging was carried on here for upward of a month, but the work was stopped several times, and the excavation in the river bed filled up owing to floods due to heavy rain during the S.W. monsoon. Only a small proportion of gold was found in the heavy sand, but the place cannot be said to have been tried fairly.

This completed the account of gold found by the dredging process, and it cannot be said that in any place the quantity found would justify dredging *for the gold alone.* Further reference will be made to this under the head "Dredging".

Two other districts where gold was obtained by following the ordinary gemming methods will be described.

Pelawatta (Pasdun Korale East):- Gemming pits sunk in many places in the alluvium of the valley of the Pelawatta-ganga showed that gold is widely distributed in the top layers of the *malawa.* In the *malawa,* which may be found by digging small pits in the banks of the river itself, it is especially abundant. Dredging was not carried out, as, except in very dry weather, it is stated that the current is too great; however, it could no doubt be managed if the methods employed in Sabaragamuwa were followed.

Nuwara Eliya:- An account of former digging for gold, mostly on the Moon Plams, is to be found in the "Manual of the Nuwara Eliya District", by C.J.R.Le Mesurier, Esq., late C.C.S. (Ceylon, 1893), pp. 21-25. For this see also "Eight Years in Ceylon," by Sir S. Baker (pp.229-305, ed.1874). Search was made for gold under the direction of Sir S. Baker in 1854, and further search was made by mr. Le Mesurier about ten or fiteen years ago. Since that time there appear to have been a few sporadic attempts at gold washing and gemming in the district. Mr. G.G. Dixon found traces of gold in the streams following into the lake, but does not seem to have tested the alluvium on the Moon Plains. Before the British occupation the district was undoubtedly dug for gems by the Sinhalese and possibly also for gold.

As mentioned above, Sir S. Baker carried out prospecting for gold on the plains on a comparatively large scale, and there appears to have been at one time something approaching a "gold panic". A description of the section he exposed in sinking a shaft is given under the head of "Alluvial Deposits". Gold was found in the top gravel at a depth of 18 inches, but more abundantly in the "lower gravel" at a depth of 18 feet. The bottom of the alluvium does not appear to have been touched at this point, but the shaft was stopped on account of influx of water, and judging by the experience of the Survey it is very hard and slow work digging in the stiff clay. Another shaft was sunk to a depth of 40 feet and gold was found in increasing amounts from practically the surface. The only definite statement given as to the amount of gold found is that 11a good washer ... can earn from 7s. 6d. to 10s. a day". In the first mentioned pit the pieces of gold were the size of small rice grains. There are some rather vague statements to the effect that small nuggets were found.

The work was ultimately stopped owing to lack of funds, and "what really deep-digging might effect in the discovery of gold in appreciable quantities remains still to be discovered".

Sir S. Baker points out that considerable quantities and nuggets could only be expected to be found in the alluvium close to the bed rock, which never seems to have been reached. The experience of the Survey in the gold-bearing gravels of the Ratnapura District would entirely endoorse this view. We have only found gold in the top of the decomposed bed rock, or *malawa.*

Following the rapid prospecting methods of the Survey it was impossible to sink shafts even approaching the depth Sir S. Baker reached. In all seven small trial pits were sunk on the Moon Plains, with a maximum depth of 10 feet. Traces of gold were found in three of these in the clay associated with the top *illam* at a depth of form four to seven feet. Pits sunk on the rifle range in the course of the small stream there did not yield any gold, though more satisfactory as gem pits.

(5) PLATINUM

By the Assistant Director

A few specks of a hard silver-white mineral, insoluble in nitric acid, and probably platinum or some allied metal, were found by the Assistant Director associated with gold in the Karawita-ganga and in the Narunkandura dola, Ratnapura District.

This appears to be the first authentic observations of platinum in the Island.

E6 *Ceylon Administration Reports for 1905.* *[IV.-Education.*

(6) THORIANITE

(a) By the Director

The public interest in thorianite has rapidly increased during the present year; indeed there has been almost a "boom" in thorianite, and a good deal of prospecting has been carried on, with the result that a number of small occurrences have been met with and a total of 179 cwt. valued at Rs.74,315 exported during the present year. Not all of this, however, can be regarded as pure thorianite ; a great deal of money has been wasted on all sorts of other minerals, such as ilmenite, magnetite, & c., and a part of this material has even been exported. A great many requests for the identification of minerals supposed to be thorianite have been attended to. Of these some were thorianite, but the majority consisted of other black minerals such as ilmenite, zircon, magnetite, and even black tourmaline, a mineral with specific gravity about a third that of thorianite. Villagers, however, are now thoroughly acquainted with the appearance of the mineral and regularly collect and sell it to local kade-keepers or in Ratnapura at from Rs.2.50 to

Rs.3 per lb. The thorianite fetches from Rs.4 to Rs.9 per lb, in Colombo, according to percentage of thorium present in the sample and its freedom from admisture with foreign mineral. Several Colombo firsm have paid special attention to thorianite. Messrs. Finlay, Muir & Co., are carrying on regular work at the original locality in Bambarabottuwa, and find the mineral scattered in surface soil on the slope of the little Kuda Pandi-oya valley as well as in the bed of the stream itself. mr. W.D. Holland has leased considerable areas in Bambarabotuwa and has obtained thorianite from the Kuda-oya from the locality indicated in last year's report and elsewhere. A considerable amount of thorianite has been received by Government as a 10 per cent, royalty on thorianite obtained by Mr. W.D. Holland and others from crown lands. This material has been sold through the Crown Agents. It would perhaps be simpler to arrange for its sale by auction in Colombo.

It is reported that a discovery of thorianite has been made in the Hinidum pattu of the Galle District, and that the deposits are being worked there. This occurrence will be investigated early in 1906 and district prospected for the rare minerals more thoroughly than was póssible in the preliminary gramination carried out in 1904.

An important discovery made by the Director is that of the occurrence of thorianite in tiny, almost microscopic, crystal in sands of various rivers, particularly the Niriella-ganga at Niriella, the Denewaka-ganga above Malwela, and the Walawe-ganga near Balangoda, in the Bambarabotuwa-oya, and in the Weralupe-dola.

The Assistant Director has observed a number of occurrences of thorianite in the Ratnapura District, both in the form of fine sand in rivers and in stream gravels. These are described below. It is possible that it would pay to work these sand for thorium-bearing minerals (monazite and thorianite) especially as gold is also present in small quantities. There can be little doubt that all these minerals would be met with also in the Kalu-ganga itself. The concentrated sands examined have all been obtained by the difficult process of hand dredging described under the section "Gold", but some much simpler method would be necessary if work on a commercial scale were to be thought of. The deposits in an ordinary minor river, such as the Niriella-ganga, consist of 6 or 8 feet of sand and a foot of sandy *illam* (gravel) resting on *malawa*. The material resting immediately on the *malawa* and mixed with the upper part of its is richest in gold and gems, but it would probably be worth while to put the whole of the stuff through some mechanical separator, that is, if such operations paid at all. In the dry part of the year not more than one or two feet of water is found in such streams, and the sands could be worked without any floating dredger, the use of which would be, indeed, impossible. On the other hand, it would probably be necessary to work such deposits as those of the Kalu-ganga by means of some sort of floating dredger.

Some observations by professor Dunstan on the "Extraction of Heavy Minerals from Ceylon Sands" appear in the Bulletin of the Imperial Institute, Vol. III., No.3. It would generally be quite impossible to divert the river beds as there suggested, owing to the narrowness of the valleys and the value of the lands adjoining the rivers. Supposing, however, that the sands are dredged or dug in dry weather, it is pointed out that they can be washed free from quartz and felspar in suitable sluices, and that magnetite, limonite, fergusonite, and garnet can be removed by magnetic methods. The heavy sand remaining will consist mainly of zircon, spinel, rutile, and cassiterite, thorianite, thorite, monazite, gold, and other heavy rare minerals when these are present. It is suggested that the material thus obtained could be treated in a tubular separator, on the principle of the gemming machine, the establishment of which at Rakwana was so unsuccessful. The possibility of finding the thorianite *in situ* is also mentioned, though it is little more than a possibility in most cases, owing to the lack of exposures in the districts where it occurs in the alluvial deposits. It is, however, incorrect to say that "these minerals are usually found in ancient alluvium laid down by *former* streams", for there is no absolute proof of *any* such occurrence, and certainly in most cases the gems and other heavy minerals occur in the gravels deposited by existing streams, and often in their very beds.

It is obvious that the exploitation of river sands in Ceylon is not to be lightly undertaken, and would necessitate the outlay of some capital and the employment of experienced superintendents. As Professor Dunstan says the "complicated machinery would require constant adjustment according to the minerals present and their size, and would require skilled labour and highly paid superintendence .... Too much stress cannot be laid on the importance of systematic work, adequate machinery, and skilled superintendence in the exploitation of these heavy minerals, which vary so much in their characters, contents, and value".

(b) By the Assistant Director

During August, September, and part of October a thorough examination was made of about three miles of the valley of the We-ganga, south of waht may be termed the Kiribatgala hill range, in the Ratnapura District. A cursory examination of this district earlier in the year, which showed that a number of the streams flowing into the We-ganga both from the north and south yield thorianite and other rare minerals, was considered to justify a fuller investigation, though rather a departure from the usual practice of the Survey make so prolonged a stay in any one district. The result of the work was to show an exceptionally wide distribution of thorianite in the streams, and in two cases to discover deposits which, if not of any great commercial value, at least thoroughly repaid working and are now exhausted.

Science and Art.] *Mineralogical Survey.* *E7*

The district lies in the valley of the We-ganga, as stated, one of the chief factors of the Kalu-ganga. The general trend of the valley follows the strike of the rock foliation *i.e.* N.W. to S.E., the general direction of dip being to the N.E. To the north of the river (see map) lie the villages of Panagama, Dombagammana, and Erabadda, to the south Dela and Niwitigala. There are high hill ranges to the north and south of the river; that on the north stretches from Ma-udellakanda to the Kiribatgalakanda, which is over 4,000 fet high and forms a landmark for many miles ; intermediate parts of the range are named after the villages situated thereon, *e.g.,* Dombagammanakanda and

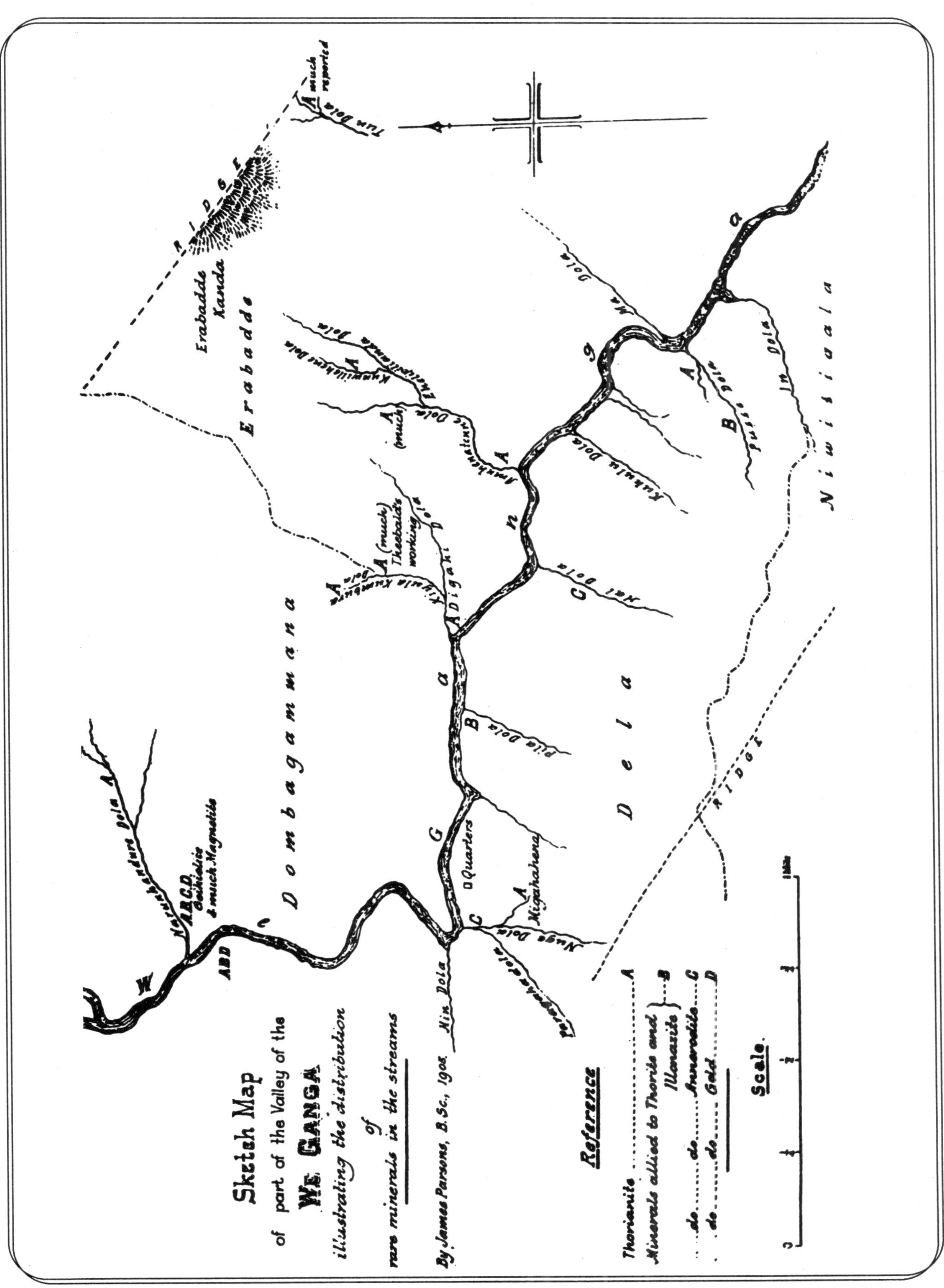
Sketch Map
of part of the Valley of the
WE GANGA
illustrating the distribution
of
rare minerals in the streams
By James Parsons, B.Sc., 1905.
Reference
Thorianite .... A
Minerals allied to Thorite and Illmenite .... B
do ... do .... Amerodite ... C
do ... do ... Gold .... D
Scale
W e G a n g a
Dombagammana
Erabadde
Erabadde Kanda
Dela
Niwitigala
RIDGE
Quarters
Migahahena
Min Dola
Nuge Dola
Pila Dola
Nai Dola
Kukulu Dola
Digahi Dola
Kiyula Kumbura Dola
Theebald's working
A (much)
Pusse Dola
In Dola
Ma Dola
Tun Dola
A much reported
ABD
ABCD
Geikielite
& much Magnetite

Erabaddakanda, which form precipitous scrap ridges, often bare of vegetation, and over which numerous small streams flow on their way to form the larger streams ultimately joining the We-ganga. The first part of the ascent is however gradual, and terminates in a narrow flat cultivated in paddy fields and gardens extending along the ridge, above which tower the bare rock scraps. The range to the south of the river being a dip slope is less precipitous and broken by numerous deniyas and flats.

The numerous streams on both sides of the valley join the main river at right angles, but the greater part of their course is along strike lines dividing the main ranges into small subsidiary ranges or foot hills.

The rocks of the district consist of ordinary acid leptynites with numerous pegmatite intrusions. It is in these pegmatites that one might expect to find the thorianite *in situ.* Careful and continued search, however, failed to reveal any rocks bearing rare minerals.

Taking first the streams flowing into the We-ganga from the north-

Narukandure-dola- A water-worn fragment of thorianite was found in *illam* collected between the boulders of this tream about a mile from its junction with the We-ganga. In consequence of this search was made up and down the stream wihbout success, the *illam* both between the boulders and on each bank being exhausted by previous gemming operations by villagers. Finally, the sandy bed of the stream near its junction with the main river was dredged, and the resulting "concentrate" was found to contain thorianite in minute crystals and in rounded fragments. As previously mentioned, this stream also yielded monazite sand, and possibly the sands might be worth working. There is, however, hardly half a mile of the stream that could be dredged for sand, as above this the bed is full of large boulders. It has also been mentioned that a little gold, a trace of platinum, as well as "thorite" were found here. A mineral, apparently geikielite, was found in the course of dredging, as well as a cerium-bearing mineral not yet determined. The *nambu* was curiously rich in pieces of magnetite, and contained a fair proportion of varieties of corundum and zircon.

Kiyulakumbura-dola:- this is a small stream, a tributary of the Digani-dola, flowing into the We-ganga. Rumours of my work in the district in the early part of the year having spread, I found returning later that native prospectors had discovered a considerable deposit of thorianite in the deniya through which the stream runs. It was worked by a European in partnership with the native owners of the land and yielded, I understand, about 5 cwt. of the mineral, which practically exhausted the deposit, but the small *maladolas* draining into the stream should be carefully examined, as in one of these I observed a small pocket of thorianite. I sank many trial pits in the alluvium of the Digani-dola without result, it being difficult to find a place which has not been gemmed out, though I found a little thorianite in some old *nambu* below the junction of the Kiyulakumbura-dola with this tream. It might possibly be worth while to dredge the sand of the stream near its junction with the main river.

Amuhenatenna-dola:- It was in the alluvium of this stream near its junction with the We-ganga that the first indications of thorianite were found in the district in the form of a single water-worn crystal of the mineral which I picked out from several pounds of otherwise useless *nambu.* On sinking a pit some yards up-stream at the point where the hill course meets the alluvial flat, I obtained about an ounce of thorianite. It was obvious that the crystals from their water-worn character had come from some locality up-stream, but search both up the main stream and up the Ehelipilianda-dola (see map) failed to show any further traces, although many likely looking flats were examined. Finally the small tributary, the Kunwillehenatenna-dola, investigated, and *nambu* obtained from a small stretch of gemming land. This yielded no thorianite, but on working up the gemming flat to the foot of the hill course a small deposit of thorianite was found, a matter of only a few pounds, and the mineral could not be traced above this point.

The place above where the amuhenatenna-dola is joined by the Ehelipilianda-dola, had only yielded zircon and corundum; proceeding to the corrie where the stream has its source, only the lightest *nambu* of tourmaline and spinel was found ; however, half-way between these points in a small *deniya* a deposit of about 2 cwt. of thorianite was found. Within three days the native owners of the land working day and night had exhausted this deposit.

Ma-dola:- This dola rises in the flats and deniyas on the top of the ridge, being at its source formed of three dolas the place where these join being in consequence known as *Tun-dola.* From this place a large sample of thorianite, which I was shown, was stated to have come. It was alsmot impossible to ascend the scarp face of the hill, that is, from the south; working from Pelmadulla, however, I reached the locality after some arduous climbing only to find that there was no time to make excavations. Supposing that this is actually the place the thorianite shown me came from, it might well be wroth working, as there is abundance of alluvium left untouched though some gemming has been done. The inaccessibility of the spot would render work here difficult.

The streams flowing into the We-ganga from the south yielded in almost all cases minerals allied to thorite and annerodite in small quantities, and in two cases thorianite was foud. One was in a *mala-dola* at Migahahena, a water course draining into the Nuge-dola, where a pocket of thorianite was found ; the locality is closed to the top of the hill ridge and should have proved a likely place to find the parent rock, but careful search failed to reveal it.

Pila-dola:- In a small deniya at the bottom of this stream a little thorianite was found, but the place has been gemmed out. Search was made for the mineral in the upper part of the stream, but no further traces were found.

A dredging carried on in the We-ganga itself showed that small quantities of thorianite were to be obtained in the concentrated sands, as well as some monazite.

No doubt the whole district would repay further search. I have had ssmples brought me from Panagama and from Doloswala, but was unable to find time for work in these villages. The country should also be investigated towards Pelmadulla, where thorianite is known to occur.

In prospecting for thorianite or other thorium-bearing minerals the following points should be borne in mind:-

(1) Wherever a single fragment only of such a mineral is found in a sample of *nambu*, the stream is worth further investigation.

(2) Though there is no doubt a tendency for thorianite to remain near the head of a stream, this cannot be relied on, and it may collect in pockets anywhere along a stream course.

(3) Even if no portion of the upper courses of a stream contain more than trace of a thoria-bearing mineral, it may yet be worth while to dredge its sands near its junction with a main river.

(7) THORITE

By the Director and Assistant Director

The occurrence of monazite as fine grains in the sands of many streams and rivers renders it probable that the mineral will be found in larger fragments not reduced by abrasion ; such being the case it is not unlikely that some of the minerals which have been provisionally classed as thorite will appear on analysis to be monazite. It was pointed out in the last Administration Report that the form of many of the crystals obtained çould with difficult be reconciled with that of a tetragonal mineral such as thorite, and further, the analysis there quoted was that of a single specimen*. [1] We have also observed that several specimens among those examined by simple blowpipe processes give a marked reaction for phosphate when brought into solution and treated with ammonium molybdate. The full analysis of a number of specimens is therefore desirable. until this is done, however, we must continue to class certain superficially similar minerals as "thorite", or, preferably, *orangite.*

The following new localities for the mineral may be noted:-

Massena (Balangoda):- Rounded pebbles similar to these previously described with orange-coloured, flattened crystals.

Nagahahena, near Avisawella:- Here a few fragments only were found.

Narunkandure-dola, Dombagammana ; Rila dola, Dela ; Pusse-dola, Dela, in the Ratnapura District (see map):- A few fragents only were found in the gravels of these streams, and no deposits of economic value.

The Narunkandure-dola yielded a few of the flattened crystals. The material from the Pusse-dola was different to the typical "thorite" and strongly suggests monazite.

(8) MONAZITE

By the Director

[1]* Two further analyses have since been received, but they are of specimens from adjacent localities

This mineral, until recently almost the sole commercial source of thorium, but not known to occur in Ceylon, was first detected at the Imperial Institute in thorianite-bearing river sand from the Ņiriella-ganga at Niriella (Sabaragamuwa), which had been forwarded by the Director for further examination.

It was subsequently discovered by the Director in the Avisawella district, occurring in, sufficient abundant in the Getaheta-oya to give a distinctly yellowish tinge to the concentrated heavy sand, otherwise consisting mainly of limonite and magnetite. It occurs in small, well-rounded grains of a waxy honey-yellow colour, with a marked greenish tinge when viewed by transmitted light. It is probably of widespread occurrence in river sands in Ceylon, as it is present in all the river sands examined near Avisawella, viz., the Sitawaka-ganga, Getaheta-oya, Kiriwangalaela, and the Gonagala-oya and Lawpitiya-oya at Mapitigama, as well as the original Nirella locality, and also in sands from the Walawe-ganga at Balangoda and the We-ganga near Dela, and the Narunkandure-dola at Dela, where it is fairly abundant and is associated with thorianite.

(9) NIOBATES

Various minerals, which may be roughly classed provisionally as niobates and tantalates, have been collected by the officers of the Survey, but in no great quantity. Niobium is of no commercial value at the present time, though there is a limited demand for tantalum and its compounds. The mineral referred to in page 10 of the last Administration Report as probably annerodite is the most common. No analysis or reports on this mineral have yet been received from the Imperial Institute, though the specimens have been in the hands of the authorities there for over a year. A paper has been read before the Royal Society (June, 1905) on Annerodite from Ceylon, by Mr. Strutt. It is not however known what specimens were determined or whence obtained. It not annerodite the mineral is probably fergusonite, though the strong re-action for uranium which we observe in blowpipe tests would not be expected in the latter mineral.

It is however possible that at least two distinct though superficially similar minerals may be associated, as there is considerble variation of specific gravity and lustre to be observed in individual specimens.

(10) CASSITERITE

By the Director and Assistance Director

The first record of the occurrence of cassiterite in Ceylon will be found on page 10 of the Mineralogical Survey's Administration Report for 1904. The material there referred to was brought in the latter part of 1904, and it was stated that it had been obtained at Niriella, Palle pattuwa, Nawadun korale, Sabaragamuwa.

Two weeks were accordingly spent by the Director at Niriella in February, 1905. The cassiterite (alluvial tin stone) was found to occur in the bed of a very small stream called the Etunkahena-dola, at a point where gemming had been carried on ; no work is being done at present owing to a legal dispute. The deposits of *illam* are exceedingly limited ; the proportion of *nambu* obtainable is also small ; it consists of ilmenite, cassiterite, & c. The cassiterite is in the form of rather small, well-rounded, or sub-angular fragments,

Sketch Map of District North of Avissawella

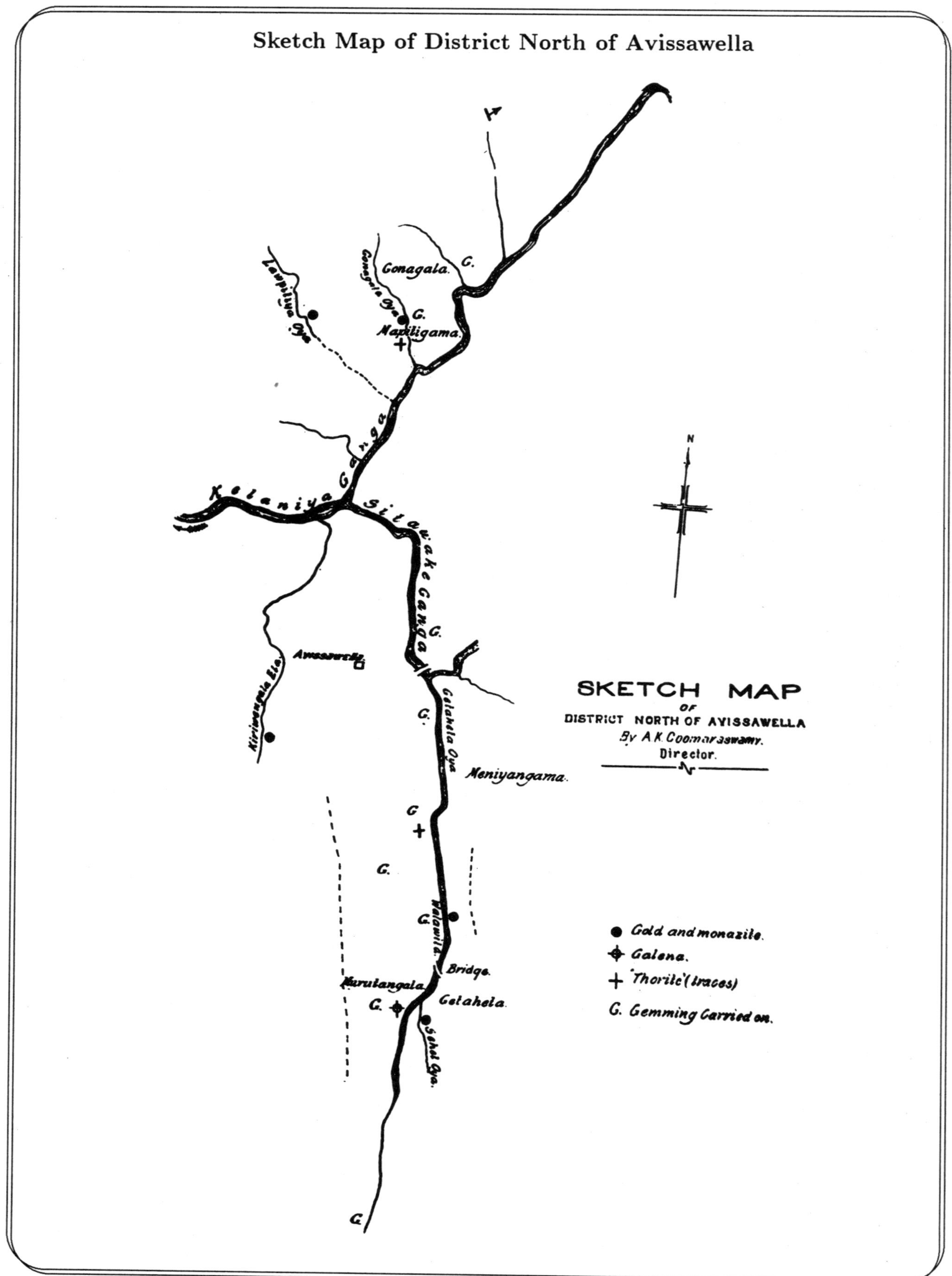

quite black in colour and showing no trace of crystalline form. It is distinguished from ilmenite and rutile, to which it bears a considerable superficial resemblance, by its specific gravity of 6.91 - 7.0, its pale brown streak, and by yielding globules of tin when fused on charcoalwith sodium carbonate in the usual way. Traces of cassiterite could not be discovered elsewhere in Niriella ; but one of two

specimens have been received that must have come from other localities; one of these was stated to have been picked up on Gabbela estate. It was also reported in a newspaper last year that cassiterite had been detected at Hatton, but no specimens reached the Mineralogical Survey.

It is likely that in all cases the cassiterite is derived from isolated exposures of rocks of granitic character belonging to the Balangoda Group. Such rocks have been observed at several localities near Ratnapura, although the varieties exposed were not those yielding any valuable minerals.

Shortly after this the Assitant Director obtained samples of cassiterite from the neighbourhood of Naragala, Navadum korale, Sabaragamuwa.

The exact locality was found to be Induwehena (Crown land) in the extreme south of Marapona, close to the Noragala boundary.

The occurrence is in the alluvium of a small *deniya* about 250 yards long and 5 yeards wide in its widest part, but the *illam* has not probably a greater width that 25 yeards. The lower third of the *deniya* has unfortunately been gemmed out illicitly, and pits have quite recently been sunk. Much *nambu*, however, does not appear to have been found in this part, which has been worked fro gems, chiefly sapphire and topaz. Nine pits were sunk in various parts of the deniya, and it was found that roughly the upper third yielded no *nambu*. From the middle third, however, four pits yielded abundant zircon and 1 or 2 ounces of cassiterite in all. The distribution of the mineral is somewhat irregular, but it is not believed that more than 2 or 3 lb. could be obtained from the place. The deposit is therefore small and economically unimportant. Search was made in surrounding *deniya* without discovering any further indications of cassiterite. Careful examination of the rocks exposed in the catchment of the *deniya* failed to reveal any deposit of the mineral *in situ*.

Traces of cassiterite have also been found in *nambu* from a pit by the We-ganga near Dela.

Later in the year the Assistant Director found that small quantities of cassiterite occurred at Dalawiladeniya in Pohorabawa (Sabaragamuwa). The occurrence is in an *owita*, which has been and still is, extensively gemmed. Practically the whole of the *illam* has been excavated and washed, and as only about an ounce could be collected it is not anticipated that there is anything approaching a deposit which could be worked commercially.

Further investigation, however, in Kuruwita showed that a somewhat more important deposit occurred in a stretch of paddy land including Minuwan and Hapugaha deniyas, which are continuous, the Hapugahadeniya forming the head of the valley, and is Crown land. The lower part of the Minuwandeniya were found to be completely gemmed out, the higher part yielded traces of cassiterite in two out of three small pits that were sunk. In the Hapugahadeniya, of these small pits sunk progressively up the valley, the lowest yielded no cassiterite, but the other two together, almost half a pound including pieces which are the largest hitherto found, one being about 3 ounces in wieght. The time it was possible to spend in the place was limited, but it will be seen that there is a marked increase in the quantity of the mineral towards the head of the valley, and it will probably be advisable at some future time to prospect further up the valley into the jungle at its head. it is not anticipated, however, that the deposit will be of any considerable extent.

Numerous reports of other localities where cassiterite is alleged to occur have been received, but have not so far been substantiated. The peculiar appearance of Ceylon cassiterite renders it extremely liable to be mistaken for zircon and rutile, especially the latter, by superficial observers.

An analysis was made of the Niriella cassiterite at the Imperial Institute, and the following report forwarded to the Director, Mineral Survey:-

"The specimen of cassiterite from Niriella was too small to subject to dry assay, but the following analysis made by the wet method shows a marked basence of other reducible metals, and na ore represented by such a specimen should produce a very pure metal on smelting:-

	Analysis
Stannic oxide	94.00
Ferric oxide	0.86
Manganous oxide	0.03
Lime	0.50
Insoluble residue, chiefly niobic and tantalic oxides	4.46
Equivalent to metallic tin 74.09 per cent	99.83

"The value of such an ore (calculated on the basis of Ł135 per ton for the present price of metal in ingot) would be a approximately Ł91 per ton".

"From this value would have to be deducted the cost of freight, transport, insurance, and other charges necessary to place the ore on the market".

(11) RUTILE

By the Director

A specimen of the rutile from Hatton, mentioned in the last Administration Report (p.10), has been analysed with the following results:-

Titanium dioxide	..	TiO_2	..	99.30
Ferrous oxide	..	FeO	..	0.60

There is some demand for rutile as a source of titanium for use in the hardening of steel, in the presparation of mordants for leather dyeing, and for the manufacture of filiaments for incandescent electric light.

A company in Virginia, U.S.A., is engaged in concentrating rutile from a deposit, and the resulting material is said to command a ready sale. Rutile in Ceylon, though widely distributed, has not as yet been found in any considerable deposits. The Hatton sample is valued at about Ł30 a ton.

(12) GALENA

By the Director

Small cubic crystals and rounded fragments of galena occur abundantly in the bed of the Getahetaoya at Murutangala, half a mile above the bridge at Getaheta. Well-washed *nambu* consists of more than

half its weight of galena. The *nambu* is mainly derived from the *illam* of the left bank river ; this *illam* is worked by dredging in the river with *garan-udella*, under-cutting the bank. The remainder of the *nambu* consists of pyrite, garnet, tourmaline, zircon, corundum, spinel. A good many fragments of galena are intergrown with pyrite. At Walawita, half a smile below the bridge at Getaheta, very extensive gemming is also carried on, but scarcely a trace of galena occurs in the *nambu* - an illustration of the fact that heavy minerals rarely travel fro from their source. The galena is probably derived from some pegmatitic or granitic intrusion exposed at or above Murutangala, but the propect of discovering this actual matrix would be very small. It seems unlikely that the occurrence of galena is of sufficient extent to warrant commercial exploitation ; this to some extent depends on the proportion of silver which the ore may be found to contain.

Some excellent samples of galena, larger than the above, were brought to the Assistant Director at Dela, the supposed place of origin being Pussells, but an attempt to trace the actual locality was unsuccessful.

(13) MOLYBDENITE

By the Director

The occurrence of molybdenite was early recorded by Dr. Gygax, and it is also quoted as a Ceylon mineral from "Colombo and Adam's Pik, in gneiss", in Leonhard's "Handworterbuch der Topographischen Mineralogie", 1843. It has now been re-discovered in the Kegalla District.

The molybdenite occurs in a vein of pegmatite, intrusive in the charnockite series and exposed at Hettimulla, four mile from Kegalla on the Bulatkohopitiya minor road and one or two hundred yards from the fourth milepost. The exposure is a good one, a few blasts having been put in, and some 20 lb. of molybdenite obtained, under the impression that it was graphite. The pegmatite averages five or six feet wide, and cuts the granulites, which have a general north-west strike. The pegamatite consists mainly of orthoclase (often in graphic intergrowth with subordinate quartz), quartz and accessory pyroxene (decomposed), biotite, molybdenite, and pyrite. The molybdenite forms laminar scales and lenticular flakes, which reach 2 inches or more in length, but tends also to form six-sided crystals under favourable circumstances. It occurs mainly in a central quartzose zone of the pegmatite, but also in the most felspathic parts, and even at the very edge in contact with granulite. The pyrite occurs in small cubes, irregular grains, and thin films coating other minerals.

(14) SHORT ACCOUNT OF GEMMING DISTRICTS VISITED

(a) By the Director

Niriella:- This is not one of the chief gemming districts, but some gemming is done. The river itself (niriella-ganga) is regularly worked. The forest area between Niriella and Karawita, along the bank, is reputed good and has been largely worked illicitly. The *illam* here consists of a ferruginous laterite containing rounded pebbles. Gemming is carried on in several deniyas, but not in rice fields or owitas, where the *illam* lies at a great depth. Kotamulla, two miles from Ratnapura on the Niriella path, is the centre of a gemming district ; some *deniyas* are quite worked out. The *owitas* have not been much worked ; a pit was being sunk near the vihara at the time of my visit, and here the *illam* consisted of laterite, being quite typical *"cabook"*, only distinguishable from the usual kinds by the presence of large rounded quartz pebbles, *illam-gal.* This observation proves the existence in Ceylon of a "low-leverl", secondary, or alluvial laterite, not easily distinguishable from that which forms *in situ* as a decomposition product, except by its mode of occurrence and the presence of the rounded pebbles.

Ratnapura:- Gemming has been carried on with unusual activity during the present year. The deep pits at Veralupe have been much worked. Deep *illam* is also obtained at several places in the neighbourhood of Marapona. In such cases, where extensive deposits of gravel are tapped, it is very unusual to meet with heavy minerals containing rare elements. These are chiefly found in smaller deniyas and streams, in which they have not travelled very far from their source.

Getaheta:- The whole length of the Getaheta-oya is very extensively worked, yielding an abundance of good sapphires. The deposits on either bank are also worked extensively, both above the bridge at Getaheta (*i.e.,* in Murutangala) and between this bridge and Avisawella. These localities are the scene of persistent and active work although the *illam* lies deep and the water is very troublesome. A system of payment by contract for keeping the pits free of water is adopted.

Avisawella:- Gemming is carried on here and there between Getaheta and Avisawella, and also at Maniyangama east of Avisawella, and in the Sitawaka-ganga itself. The high-level gravels by the railway bridge are well spoken of. Gemming, however, is totally unknown to the west of Avisawella.

The following general observations on gemming may be set down here. It appears that gemming is usually confined to gem-bearing districts, *i.e.,* the absence of any gemming indicates that a district is bare of gems. Not only have my own observations in non-gemming districts usually failed to reveal even the presence of a piece of corundum or zirrcon, but one hears reports of previous unsuccessful trials having been made. This is not to say that gems will never be found where not at present known, but indicates that the country has been to some extent prospected already and no very startling discoveries need be expected.

In the good gemming districts themselves the good places are pretty well known, and usually more or less worked out, the deeper *illam* often alone remaining. In such places as Getaheta the gemming of the river bed is carried on continously again and again, and the deposits on the banks are not yet exhausted. Gemming operations in the stream itself are so carried on as to undermine the bank and get at the deposits of *illam.* This process may greatly widen and even alter the course of a stream.

Gemming, I believe, can never be profitably undertaken on any scale by Europeans. Even for Ceylonese it is more a lottery than a regular source of income. The gemmers fall into at least three classes, viz., illicit gemmers and men who occasinally work on share with other; fairly prosperous men who gem their own lands, also employing other men on various terms; and rich men who have their pits worked for them or rent out the land. Mr. Brampy Pieris's pits at Haldola (four miles from Ratnapura) are examples of the latter class. Here the pits are very remunerative. The *illam* occurs some six

fathoms below the surface ; oridnary village gemmers ar employed to cut down to it, but it is only washed in the presence of a responsible person, who attends the pit when the *illam* is reached ; three-fourths of the gems are given to the owner, the remaining fourth being shared amongst the men, who get also their food, but now ages. Mr. D.J. Jayatileke works his pits at Potgulkanda on a half-share system.

There have already been several unsuccessful attempts at gemming companies made, but it is my opinion that the scattered ownership of the gemming lands, the careful supervision required, and the inevitable uncertainities make it unlikely that such companies should succeed. It might be otherwise with a company confining its attention to river beds, and working these for gems, gold, and thorium bearing minerals.

(b) By the Assistant Director

In the course of prospecting for heavy thorium-bearing minerals the following gemming districts have been visited:-

Pasdun korale East:- Gemming has been extensively carried on in the villages of Pelawatta, Hewessa, Gurulubedda, Hedigala, Dehipitiya, and other villages in the valley of the Magura-ganga, but the *illam* in these places seems to be now practically exhausted, though work is still carried out at Pelawatta. The district was famous for varieties of chrysoberyl, including alexandrite ; and in the *nambu* which may now be obtained fragments of chrysoberyl take the place of the zircon which is so common in other gemming districts. It has been stated that rubies have been found in Pelawatta, but I saw no variety of corundum, and am inclined to believe that the stones found were garnet, the villagers being for the most part very ignorant of gems and gemming as compared to the natives of the Ratnapura District.

Kukula korale:- The gemming of the south-eastern part of this korale has been described in the last Administration Report. A considerable amount of gemming has been carried on in Kukulugama and in Kalawane, and work is still carried on in this last village, but in Kukulugama only occasionally. As is usually the case, the places where the *illam* lies within a few feet of the surface have been entirely exhausted, while in places where it is at a depth of two or three fathoms it is often untouched. A model section of apit sunk at Kukulugama showing the various *tattuwas* to scale has been prepared and placed in the Colombo Museum. Varieties of spinel, zircon and corundum are found. The gemming of these villages cannot, however, be described as good. It is not until the village of Karawita is reached that the gemming assumes the importance and value which has made the Ratnapura District famous. Here the land on each side of the Karawita-gang has been honeycombed with gemming excavations, the ground in places having been dug over three times, on the chance that some fragments of *dalam* may have been left by former gemmers. Where the *illam* lies at a depth of two or more fathoms in the paddy fields there is still untouched ground, though this was being worked at the time of my visit and will doubtless be soon exhausted. Gemming is carried on in the next valley to the north, viz., that of the Homgamuwa-ganga, but I did not investigate this thoroughly.

In the valley of the We-ganga are extensive gemming lands in the villages of Horagala, Nivitigala, Dela, Dombagammana, and Erabadda. The hill dolas are naturally exhausted, but near the river deep pits are sunk and valuable gems often found. Gemming is also carried on in the river itself, the *illam* being dredged with *garan-udella*.

The gemming at Pussella was examined, also that in the village of Pohorobawa, where good gems were obtained in the Dalawiladeniya, and at the time of my visit the *illam*, which was left in small patches, was being energetically worked over for the third or fourth time by a small army of villagers including women and children, the first time that I have seen women engaged in the work.

Nuwara Eliya:- The district was gemmed by the Sinhalese before British occupation ; gems were also found during the search for gold under the direction of Sir Samuel Baker in 1854, and again by Mr. Le Mesurier, late C.C.S., about ten or fifteen years ago. The swampy land now converted into Nuwara Eliya lake was known as the Ruby Plains and reputed to be good gemming land. Though no systematic gemming has been done of late years, most of the suitable land appears to have been worked at some time, and traces of old pits are abundant round the borders of the lake and on the Moon and Elk Plains, especially the former, and even where there are no obvious traces of old pits search with the long *illam kuri* shows that many spots now long grown over have been dug.

The nature of the superficial deposits on the Moon Plains (which name, by the way, has no relation to the supposed occurrence of moonstones there) has been described in the section on Gold. Under the peaty soil is stiff white clay, which overlies an *illam* of rounded quartz gravel, there is more clay, and then a second layer of *illam*. These, however, would be described as top *illams*, as the bed rock on the centre of plains has never been reached, and consequently the bottom *illam* resting on *malawa* has not been exposed. Sir S. Baker found rounded quartz gravel at a depth of 18 feet, but did not at that depth reach the bed rock, which appears to lie at a much greater depth. I sank several pits in various places on the Moon Plains to a depth of 10 feet through the two top *illams*, which I extracted and washed. Pale blue corundum (*dalam*) was fairly abundant, also blue spinel ; tourmaline was common, zircon not so much so and in small crystals unsuitable for cutting. A number of small pieces of cat's-eye were found, and tiny rubies were abundant. A piece of *dalam* was found on the path on the rifle range, which is in a long narrow valley in which flows a small stream, along which rock *in situ* is often exposed. Pits were sunk in the valley below this spot close to the 900 yards stand. A top *illam* was found at a depth of about 3 feet which yielded fragments of corundum and cat's-eye and abottom *illam* resting on *malawa* at a depth of about 4 feet, consisting of gravel with sand ferruginous cement. This yielded fairly abundant corundum and an exceptional amount of tiny fragments of cat's-eyes (with some slightly larger pieces), zircon in small brown crystals, and fairly transparent tourmalines.

There is gemming land in the Kandapola forest where I sank several pits, only obtaining tourmaline, some of which was quite pale green in colour. No doubt there is still good gemming land left at Nuwara Eliya, which it will repay some prospector in the future to investigate.

(15) THE SINHALESE METHOD OF HAND DREDGING

By the Director

In the gem-bearing parts of Sabaragamuwa, Sinhalese are accustomed to work the beds of reviers fro gems, and less often for gold, by a process of dredging known as ganga-adinawa. It is said that the work (for gold) was also carried on by Muhammadans in parts of the Sitawaka-ganga.

At least six or eight men are required to to the work properly. Each man is provided with a special long-handled mamoty called *garana-udella* or *poru-udella*. These mamoties measure about 7 × 17 and weigh about 12 lb. The handles (for which a light spirngy wood is used (usually *riti*), are twenty feet or more in length ; sometimes, however, the work is started with shorter handels and they are lengthened from day to day as required. When handles are made of more than one piece of wood they are very firmly and skilfully bound with cane in the manner shown below:-

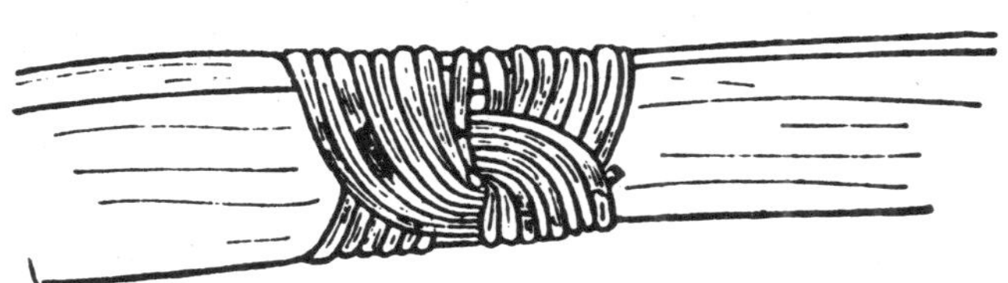

A convenient place in a river is selected, where there is a good current of water not too deep, and where, if possible, no large boulders will interfere with the work. If boulders are subsequently met with they are, if not too heavy, removed with the aid of stout poles used as levers.

It is usual to make a low fence partly acorss the stream from either side, in order to increase the flow of water in the middle where the workers stand. The men stand in a row across the stream, facing up stream, and being to rake up the river bed as far as the mamoties will reach, ultimately scooping out a large hole under water in front of them. If gems are required, the overlying layers of sand and clay are thus removed until the *illam* is reached, when this is scooped up in the same way and washed in baskets as usual. If gold is required the material forming the uppermost layer of the *malawa* and immediately underlying the *illam* must be worked. This sand and gravel composing the *illam* and immediately overlying the malawa (which must be reached as a sign that the lowest superficial deposits are being worked) are scooped up and allowed to drop into the water around the men's feet. The current then acts as anatural sluicing machine, carrying the lighter materials down stream and allowing the gems, heavy minerals, and gold to accumulate on the bank of dredged material on which the dredgers stand. A s the work proceeds, and a larger area of *malawa* is laid bare, the position of this bank of dredged material naturally shifts down stream, and the quantity of material continually increases. It is desirable to carry on the work for at least a week and sometimes for a month if weather permits. If the water is clear the gold is then obtained as follows:- the black sand (mixed with gravel, & c.) is scooped up by hand and placed in coconut shellf (*pol-katu*) and *nembilis* and therein concerntrated by hand washing. A *teliya* or shallow dish, if often used to complete the process. The gold occurs in very small thin flakes and is troublesome to collect ; it is said that in some districts mercury is used. The experiments made did not yield gold in sufficient quantities to even nearly repay the cost of the experiment, but I understand that gold is sometimes obtained in sufficient quantities to remunerate the workers at the rate of 50 cents to Re.1.50 per day. The work is very laborious, and can only be done by experienced men, who are rarely willing to accept less than 75 cents or Re.1 per day as wages when working for others.

(16) RECENT DEPOSITS AND SUBAERIAL DENUDATION

(a) By the Director

It does not appear necessary to give any further detailed account of gemming in the present report, with the exception of gemming in rivers by dredging (see separate description above). A few good sections seen in gem pits will, however, be worth describing. Extensive gemming is carried on in and near the Getaheta-oya, above and below Getaheta [see also under Galena and Monazite]. The best sections were exposed at Walawita, in fields on the left bank of the stream, half a mile below Getaheta bridge. Here the section was as follows:-

1. *Pasa* (alluvium)	.. ..	4	10
2. *Kabuk* - (mottled clay with feruginous concretionary nodules, resulting from alteration of [1] ; a gradual transition from [1] to [2])	.. ..	4	8
3. Clayey sand with small pebbles	.. ..	3	2
4. Light blue clay, 3 in. to	.. ..	0	7
5. Gray clay full of leaves and wood, 2 in. to	.. ..	1	9
6. *illam*, quartzose gravel with gems, 4 in. to	.. ..	0	6
7. Sand (yields gold and monazite) about	.. ..	0	6
8. *Malawa*	.. ..	1	4
		16	10

The leaf-bed has an extended distribution along the Getaheta-oya valley ; it is found at Murutangala, half a mile above the Getaheta bridge, and again at Nagahahena half a mile or more below the Walawita locality above described. At Nagahahena well-preserved seeds and resin (*dummala*) as well as leaves and wood are found in the leaf-bed. The leaves are difficult to preserve as they are apt to crack and peel away from the clay on drying. The appearance of the leaf-bed and overlying mottled clay strongly recall exposures of the Woolwich and Reading beds in England, but are of course much nmore recent, and probably contain none but recent species. It is remarkable that gravels of various agres are rarely met with in Ceylon, the majority of *illams* occurring not far from the present water level. Gravel terraces some 20 feet above present water level occur at *Welimada* and elsewhere in the Badulla District, and there is a large gravel terrace high above the present river a tPeradeniya and probably elsewhere along the mahaweli-ganga valley ; but a regular system of river terraces of different ages, such as we are familiar with in Europe, is not to be found. The reason for this is not altogether clear, but it may be partly explained by the torrential character of the rivers (in the wet seasons) and their narrow valleys, which are determined in direction, and narrow in form in consequence of the influence of the foliation planes of granulites.

The leaf-bed[*2] evidently corresponds to clayey deposits yielding wood, & c., which are often met elsewhere underlying *illam*; see for example, the section at Kukulugama described below by the Assistant Director, and this section at Walawita, Mapitigama, Kegalla District :-

[2*] *Cf.* also Carl Scherzer, "Voyage of the Frigate Novara," 1861, p.390, where an ccount of some gem pits near Ratnapura is found. The book speaks of a "bituminous clay" with "organic remains such as leaves, trunks of trees converted into a substance resembling lignite, and not infrequently elephants' tusks and bones of animals". As a matter of act, however, animal remains are exceedingly rare, the only one that I have heard of bring a piece of elephant's tooth given to me by Mr. G.S. Saxton and stated to have come from the *illam* of the Veralupe dola, Ratnapura.

	ft.	in
Alluvium	7	0
Clay with tree trunks	1	0
illam above *malawa*	1	0

Resin (*dummala*) was found rather abundantly in the Kanuketiyadeniya at Mapitigama ; it is said to be sometimes regularly worked in such deposits, being required for use in devil-dancing ceremonies and for the preparation of oil for olas.

A tour was taken by road from Dambulla to Trincomalee and thence to Anuradhapura. From the hills some miles south of Dambulla there is a gradual slope of gently undulating country extending northwards to the sea of Jaffna, east to Trincomalee, and west to Chilaw. The greater part of this sloping plain is exceedingly dry. Hills of various sizes rise abruptly from the plain, and their bare smooth sides have a remarkable aspect, leading to their being spoken of by Walther as consisting of "domoid" gneiss. The rounded ofrms of these hills are the result of exfoliation produced of by intense heating during the day and considerable radiation at night. Exfoliation takes place along curved surfaces parallel to the exterior surface, and independent of original structural planes. The prevailing strike over a large area is nearly north and south (along the Anuradhapura road), but this strike bends gradually eastwards as one approaches Trincomalee, until it is then nearly N.E. and S.W., thus differing greatly from the direction usually observed in Ceylon.

Of the isolated hills, one of the best known and most striking is Sigiriya, which probably owes its form and preservation to the horizontal disposition of the massive bedded granulites composing it, contrasted with the more usually vertically or steeply inclined foliation to be observed in the surrounding country. The conspicuous hill of Batalegala in the Kegalla District (the "Bible rock") has similar form and owes it to similar causes, though in the latter case the hill forms the definite crest of an anticline (well seen from Danagirigala), and only a very small area can be occupied by granulite with really horizontal foliation plains. While on the subject of hill forms, another and commoner type may be referred to well seen in Hantana near Kandy. In his type one side of the hill is formed by a long dip slope, the other by a steep carp face running along the line of strike. It is often supposed that the forms of mountains in Ceylon are volcanic, or result from violent upheavals and upthrusts.[†3] These popular beliefs are quite erroneous, and result from ignorance of the affects of the ordinary process of denudation. The rocks of Ceylon must have consolidated far below the surface of the earth, and they are now exposed to view as the result of long and continuous denudation which they owe their present form and outline. To return to the North-Central Province the plain referred to is a plain of subaerial denudation and may fairly be called a penepla central part of the district rainfall is very small, and even the largest river beds are dry during greater part of the year. Some areas are practically areas of internal drainage, and falls often does not reach the sea. The consequent retention of salts which would ordinally carried out to sea in solution by rivers goes far to explain the remarkable hardness of the water which has given so much trouble in connection with water supply for the Northern railway.

[3†] Sir Emerson Tennet's account of Sigiri is absurd in this respect

The lime salts are doubtless mainly derived from exposures of crystalline limestone occurring abundantly throughout the district, and in part from the ordinary charnockite and leptynites of the charnockite series. They are so abundant that a coating of lime carbonate is often deposited on stones, roots of trees, & c., in the beds of streams, as may be very well seen in the Sigiriya-oya near the resthouse at Sigiriya, where also the section in the stream bank shows two or three feet of calcareous concretions (*Kankar*) forming the subsoil. With the exception of such beds of calcareous concretions occasionally met with, no river deposits were observed between Dambulla and Trincomalee or Anuradhapura, nor are they met with along any part of the North road to Anuradhapura. The occurrence of *kankar* in Ceylon geology.

Trincomalee:- Recent deposits of a very different nature were examined at and near Trincomalee. The undulating peneplain above described extends to the sea at Trincomalee, where the same rather abrupt hills are still found. The Swami rock is one of these, and is remarkable as one of the very few and probably the highest marine cliff in Ceylon. The general absence of good sections on the coast is a serious hindrance to the study of Ceylon geology.

South of Dutch Point, Trincomalee, we find a raised beach consisting of pebbles and boulders of quartz and granulite and pieces of coral cemented into a tough conglomerate. This is found attached to the rocks just above, and extending some way below, high water mark. The conglomerate is thus attached to the rock platform forming the present beech plane ; it occurs also filling up cracks and joints. The cognlomerate has a maximum thickness of about two feet.

At *Challi*, five miles north of Trincomalee, beds of coral rock crop out in the shore in a corresponding position, and this material has been burnt for lime. Several well sections in the district expose similar beds ; one at Mankanay, three-quarter mile inland from challi, gave the following section:-

		ft.	m.
Soil	.. ..	2	0
Calcareous rubble	.. ..	2	0
Calcareous sand, with many pebbles in lower part	.. ..	2	0
Pebbly calcareous sand, well bedded	.. ..	2	6

At the mouth of the *Salape-aru*, twenty miles north of Trincomalee, there are shelly sandstones exposed on the right bank of the river. These sandstones contain abundant marine shells including oysters. The bed is from six inches to two feet in thickness, and is exposed about one foot above high water mark and stretches down the sloping bank between tide marks.

More interesting, however, are the *Matti-aru crab beds*, and these have a slight economic interest. Fossil crabs are round washed out on the banks of the *Matti-aru* and are collected and sold for from twenty-five cents to a rupee for use as medicine, *e.g.,* for diabetes ; the crabs are gound to a fine power and taken with milk or water.

The *Matti-aru* is one of several rivers which have the *Salape-aru* as their common estuary. Proceeding inland up the river the banks rise to a height of some 6 or 10 feet above the water. Alternating with these vertical banks are stretches and pits of sand, which are literally covered with marine shells, particularly "window-pane oysters" (*Placuna*) as well as nodules containing various fossils, including the crabs; it is said that fish and turtle remains are also sometimes found. The bed from which these fossils are derived was found *insitu* some one and half mile from the cọast, near to a conspicuous hill, Nachchiyarmalai. Here the section observed on the right bank was as follows:-

		ft.	in
"Crab bed-stiff dark clay full of marine shells especially *Placuma* and containing nodules. One crab nodule was found *in situ.* Surface of this bed about one foot above high water mark	..	9	0
More sandy beds underlying clay		2	6

The crab bed probably covers a very considerable area. Crabs are also reported from Mullaittivu far up the coast. There are good specimens in the Museum labelled Palanti-aar, th exact situation of which place cannot be traced.

All the shells probably belong to recent species still living in the neighbouring seas. They are beautifully preserved, traces of the original colour often remaining. Many are evidently identical with shells that can be picked up on the beach less than two miles away.

Beside the shells, fragments of wood occur in the crab bed, and also occasional crystals of selenite, a mineral not hitherto recoreded from Ceylon (but see above p.2)

The fossils have been identified by Dr. Henry Woodward and Mr. R. Buller Newton (see Geol. Mag. Dec. V., Vol. II., No.497, November, 1905, pp-508-510); their names are given in the following list:-

CRUSTACEA

Macropthalmus Latreillei (Desmarest); *Scylla Serrata* (De Haan).

GASTROPODA

* *Telescopium telescopium* (Linnaeus).
* *Potamides fluviatilis* (Potiez & Michaud)
* *Purpura carinifera* (Lamarck)

Nassa ornata (Kiener)

* *Melogena pugilina* (Born)

Natica sp.

LAMELLIBRANCHIA

* *Scapharca rhobea* (Born)

Arca sp.

* *Placuna placenta* (Linnaeus)

Tapes undulata (Born)

Tapes testrix (Chemnitz)

Diplodonta of oblonga (Hanley)

Tellina sp.

Dosinia salebrosa (Roemer)

Chione allied to *imbricata* (Sowerby)

Those marked with an asterisk occur also in corresponding on the coast of the Madras Presidency.

The crab determined by Dr. H. Woodward as *Macropthalmus Latreillei* (Desmarest), is the "Medicine Crab" of the Chinese pharmacopoeia (see Daniel Hanbury", "Notes on Chinese Material Medica", Pharamaceutical Journal and Trans. July and August, 1860, and February, 1862, published separately by theauthor, February, 1862, Plough Court, Lombard street, E.C.); and a chelate pincer belongs to *Scylla serrata* (De Haan), the great Indian swimming crab. *M. Latreiller* is also found in the post tertiary clays of southern China (Island of Hainan), and *S. serrata* has been reported from similar beds in the Philippine Island.

(b) By the Assistant Director

1. *Beach deposits near Bentota:-* These are representative of a series of deposits about seal level (*i.e.*, exposed between low and high water levels), which appear to occur at intervals along the western coast of Ceylon. They are exposed near Bentota, about a mile to the south of Barberyn lighthouse, about three-quarters of a mile to the south of Benotata resthouse, and again about three miles south of the resthouse. The deposit nearest the resthouse consists of a conglomerate with feruginous cement, and full of shells, corals, and comminuted fragments of the same. The other two consist of horizontally bedded stretches of sandstones and conglomerates, of which the upper surface is just lapped by the waves at low water and covered at high water. The upper beds are chiefly sandstones with small pebbles, cemented by culcareous material, but perfect shells are rare. The lower beds, at least in the northern deposit, are of a dark ashy colour and almost entirely made up pieces of coral and more or less perfectly preserved shells, all of which appear to belong to species ot be found on the present beach. These fossiliferous beds occur at a depth of about three feet ; there was no means of ascertaining the depth to which they extended.

It does not appear possible that these deposits could have been formed in their present position-indeed they are now being rapidly eroded by the waves. We have therefore good evidence, of the subsidence of an old beach or sand pit, and since there is no evidence in the present beach of any cementing oaction of its materials being in progress, at all events above water level, it is [probable that this was preceded by an earlier upheaval.

These deposits very closely resemble the "litoral concrete" described by Oldham (Geology of India, 2nd ed., pp.406-409) which he regards as formed about sea-level, and cemented by re-deposited carbonate of lime after being upheaved. There is strong evidence of depression in many places on the east and west coasts of India, *e.g.*, at Madras, where the thickness of the alluvium is 55 feet, and at Pondicherry, where the thickness is over 550 feet, practically all below the sea level in both cases.

At Tinnevelli there is a submerged forest, also at Bombay, where it is near the raised litoral concrete indicating an oscillation such as must have happened on the west coast of Ceylon.

It is interesting to note that there is a tradition that in the reign of Tissa, about 200 B.C., in consequence of the torture of a priest by the kind, the sea at Kelaniya enroached on the land for "seven leagues", deluging towns and villages (see "The Tamils Eighteen Hundred Years ago", by V. Kankasabhai, 1904, p.21, and *Mahawansa*, Chap.XXII).

Tennent asserted that there was lack of evidence to corroborate this tradition, but this, the occurrence of submerged litoral concrete now supplies. J. Lomas (Brit. Assoc. Report, Southport, 1903) attributes the formation of Adam's Bridge to the emergence of sand spits or paars, and considers that the account of the building of the "bridge" in the *Ramayana* may

be a tradition founded on events witnessed by man. The original formation of the litoral concrete may well be due to such an upheaval. He, however, holds that there is no reason to suppose that the islands now forming the "bridge" were even united.

It would, however, now appear probable that there has been submergence of these islands, and that they may well have been at on time united, though, so far as is known, there is no direct evidence of this in the neighbourhood of Mannar.

2. *Subaerial denudation in the north-west of the Island*:- Kurunegala is at the north-western extremity of the hill country, and the hill there take the form of rounded rock masses, scantily covered with jungle and shaped by prologed denudation into forms bearing a fancied resemblance to certain animals, *em.g.,* the Lion rock, the Elephant rock, Eel rock, & c. As one progresses towards Puttalam the hills become lower and are finally represented by smooth, rounded, rock surfaces, either bare or covered with a thin layer of soil, and often level with or only a few feet above, the surrounding country. Such rock surfaces are characteristic of the whole north-western dry zone, and bear a superficial resemblance to rock surfaces in a glaciated region, which illusion is heightened by the frequent occurrence of masses of huge boulders. Their formation seems largely due to exfoliation caused by sun heat combined with the heavy periodical rains which prevent any accumulation of soil. The rocks are also more obscurely foliated than in the southern portions of the Island and have therefore little tendency to split in one direction rather than another. This, however, is not so much the case near Anuradhapura, where a well-defined north and south strike of the foliation gives rise to elongated rock masses, which formed suitable sites for temples and *gal-gewal.*

The extreme slowness of denudation is well shown by an ancient incription on an exposed rock face at Tonigala, which, except for a very slight sunflaking in one place, is practically as fresh as when cut over 2,000 years ago.

3. An interesting series of exposures of sedimentary rocks occurs on the Puttalam-Anuradhapura road between the 8th and 10th milestones. The rocks, which have an ancient appearance, reminding one in places of *Millstone grit* and in places of *Triassic conglomerate*, consist of sandstones, conglomerates, and grits. The rock forms a good building stone and is used as such in the bridges and culverts about this part. No fossils are to be found in the grits, but a block of calcareous sandstone by the roadside was full of corals and shells. Though not *in situ* it probably belonged to the series as no evidence could be obtained of such a block being brought to the locality. The deposit is probably marine, and its height above sea level is probably between 57 and 70 feet, according to bench marks on the milestones.

4. *Alluvial deposits*:- Among the alluvial deposits investigated in the course of gemming operations the following call for special notice:-

Pasdun korale:- Pasdun Korale East lies on the borderland between the low country and the hill country and is characterized by broken hill ridges, rising abruptly from flat, swampy land. The alluvium is characterized by the absence of ferruginous material, and there is a sub-soil of angular quartz fragments meixed with pieces of quartz felspar rock. The sub-soil is, as a rule, only two or three feet deep and rests on a white clayey *malawa*. The angular gravels often lie at some height above the existing river courses.

Kukulugama:- The village lies in the broad strike valley of the Kukulu-ganga, on a fine stretch of *owita* (meadow land). The river meanders through this, but keeps much nearer to the north side of the valley. The alluvium appears to be chiefly brought down by the lateral streams flowing from the southerly range of hills, and near the points where these streams enter the main valley the gravels are gem-bearing. Below the surface soil is a well-marked sandy bed containing vegetable matter, of which the lower portion is full of well-preserved leaves, and corresponds to the leaf-bed (*Kola-tattuwa*) described above by the Director.

The section is as follows:-

	ft.	in.
Top soil-yellow clay	5	0
Black clay with old wood	3	0
Sand with vegetable matter	1	0
Leaf bed (*kola-tattuwa*)	1	0
Gray clay	1	0
Gravel (top *illam*)	0	2
Gray sandy bed with ilmenite and garnet ..	1	0
Bottom *illam*, with garnet, spinel, zircon, ilmenite, and abundant pyrites (about) ..	1	0
Malawa (clay)		

On the north side of the river is a curious little patch of high gravel, which lies about 20 feet above the present level of the river. The surface soil is about 1 1/2 foot deep, below which comes the *illam*, which consists of well-rounded pebbles cemented by ferruginous material. It yields abundant spinel on washing. Below this comes red earth mixed with pebbles, and known as *pas illam*. The lowering of the beds of the lateral streams as that of the main river was eroded, has resulted in the isolation of the gravel patch as a small plateau, a result which has no doubt been helped by the resisting character of the cemented gravel.

Moon and Elk Plains, Nuwara Eliya:- These consist of undulating land and low hills separated by shallow depressions. The shallow valleys and lower slopes of the hills are boggy and covered with coarse grass, the only trees being scattered rhododendrons. The hills are covered with jungle which ends in curiously abrupt lines against the grass land. The plains are for the most part gemming lands, and many pits have been sunk in them in the search for gold and gems. In this connection they are referred to above. Here it is sufficient to say that their formation seems due to prolonged denudation, lowering the hills and silting up the valleys, which, judging by the depth of the alluvium, must have been of considerable depth. The late Sir Samuel Baker, in search for gold, sank many shafts (see under Gold). He found at the top peaty soil about 18 inches in depth then a layer of brown clay, below this quartz gravel mixed with "pipe clay", below this again "pipe clay" to a depth of about 18 feet, which was succeeded by more quartz gravel with rounded stones of considerable size. Another pit was sunk to a depth of 40 feet without reaching the bed rock. It is hard, however, to distinguish the alluvial kaolin from that formed *in situ* by the decomposition of underlying rock. Judging by the rapid decrease in depth of the alluvium towards the hillsides, the sides of the old valleys must have had a steep slope. The alluvium deposited by the small streams on the hill slopes has only a depth of 3 or 4 feet, and usually has one or two beds of well-wounded gravel.

E16 *Ceylon Administration Reports for 1905.* *[IV.-Education.*

5. Reference may be made here to the manufacture of salt at Puttalam, which was examined in June. Naturally-formed deposits of salts do occur in the neighbourhood of Puttalam, but the deposits are periodically destroyed, and the output of salt is solely derived from the salterns where it is prepared under Government supervision. There are seven great salterns stretching along the sandy and muddy flat shores of the Puttalam lake, actually an arm of the sea, extending north and south and separated from the outer sea by a long sandy spit and by a number of low sandy islands. The salterns are divided into a number of rectangular pans or beds, each roughly ten yards by eight, a group of 20 or 30 such beds forming a "waykal", of which in the western saltern there are 110. Each "Waykal" is owned by one or more Muhammadan properietors, who manufacture the salt under Government supervision. No doubt the water in the "lake" has a higher salinity than in the outer sea, but in the Palavi saltern a portion of the lake is enclosed as a reservoir which allows a preliminary concentration. In the case of the other salterns the water is led direct from the lake into a channel running through the saltern.

From this it is baled into certain of the pans in each "waykal" set apart as "warmers". Here the water concentrates for about six weeks, until it is sufficiently saline to crystallize on the finger when dipped into the water and withdrawn. Meanwhile the remaining beds in each "waykal" have been prepared by hammering down the floor and spreading a thin layer of white sand to act as 'crystallizers", and into these the brine from the "warmers" is run. Crystallization takes place rapidly, and more water is run in from the "warmers", the process being repeated until each pan is full. Care has to be taken not to allow the crystallization to go too far, or there forms "needle salt", which appears to be mainly magnesium sulphate. This is stated to take place much more readily in a slatern fomred on sandy soil, as Palavi, than on clay as in the case of the western saltern. The salt is drawn from the pans and piled up on the "waykal" boundary, when the quantity is estimated and a moiety of its value is paid to the manufacturer. When all the salt is drawn it is removed to the *kottu* or huts, where it is temporarily stored and the remaining moiety of its value paid. The salt as drawn is full of mud and other impurity, and no attempt is made to purify it by recrystallization. It is curious to note, however, that the salt when stored in the *kottu* appears to effect some recrystallization spontaneously, with the result that much of the dirt is expelled and forms a black crust on the surface of the heap, the interior being comparatively white-indeed as pure as it will ever be.

Samples of the salt and the needle salt have been forwarded to the Imperial Institute for analysis. A more detailed account of the manufacture of salt is given in the 12th Interim Report of the Survey (MS.), and an account of the Salt Department from a financial point of view in the "Ceylon Manual", 1905, pp.395-398.

(17) CRYSTALLINE LIMESTONES

By the Director

Very few exposures of crystalline limestones have been studied during the present year. Those passed on the road between Matale and Dambulla have been examined some years ago and are referred to in a paper "On the Crystalline Limestones of Ceylon", Quart. Journ. Geol. Soc.

The white stone used at Sigiriya for stairways consists of the typical crystalline limestone of Ceylon, often showing characteristic accessory minerals such as blue apatite, & c. I do not acutally know of any exposures of limestone *in situ* very near Sigiriya, but it is probable that these do occur. On the Trincomalee road, beyond Habarana, there are roadside exposures ; and at Trincomalee itself, near the dockyard, are some small obscure exposures, accompanied by mixture rocks, recalling the Galle group.

The most important exposures examined, however, are at Mahakandarawa, four miles from Mihintale. Here the limestone is quarried and burnt for lime, but is still better exposed in the nearly dry bed of the Weli-oya. These exposures illustrate exceedingly well the remarkable mixtures of granulite and limestone which are found, and the occurrence of streaks and lumps of granulite, more or less peripherally modified, in the limestone itself. The streaks of accessory minerals marking the foliation of the limestone weather out slightly above the general surface, but are distinguishable from streaks of modified granulite which stand out a good inch above the general surface. It is impossible to avoid the conclusion that, whatever the origin of the limestone, the two rocks have existed together in a state akin to fusion, and owe their present foliation and disposition to general causes. If the limestones are to be regarded as original sediments, they must have been intruded upon and melted by the invading charnockite series, and at a depth so great that the pressure sufficed to prevent the escape of carbon dioride and the formation of *wollastonite*; movements took place or continued in the whole mass which ultimately consolidated in its present form as a result of slow cooling, the carbonates consolidating last of all and thus being able to behave, under pressure, as an intrusive rock, as was suggested by Cotta 1853 (when his original view of the eruptive origin of crystalline limestones was abandoned).

The Weli-oya exposures show, beside the interestreaking of limestone and granulite, the inclusion of large isolated masses of acid granulite in the limestone. The other localities in Ceylon where these phenomena can most easily be studied are the Maha-oya below the Hanguranketa bridge (Central Province) Ella (Uva), and Gongala (Sabaragamuwa).

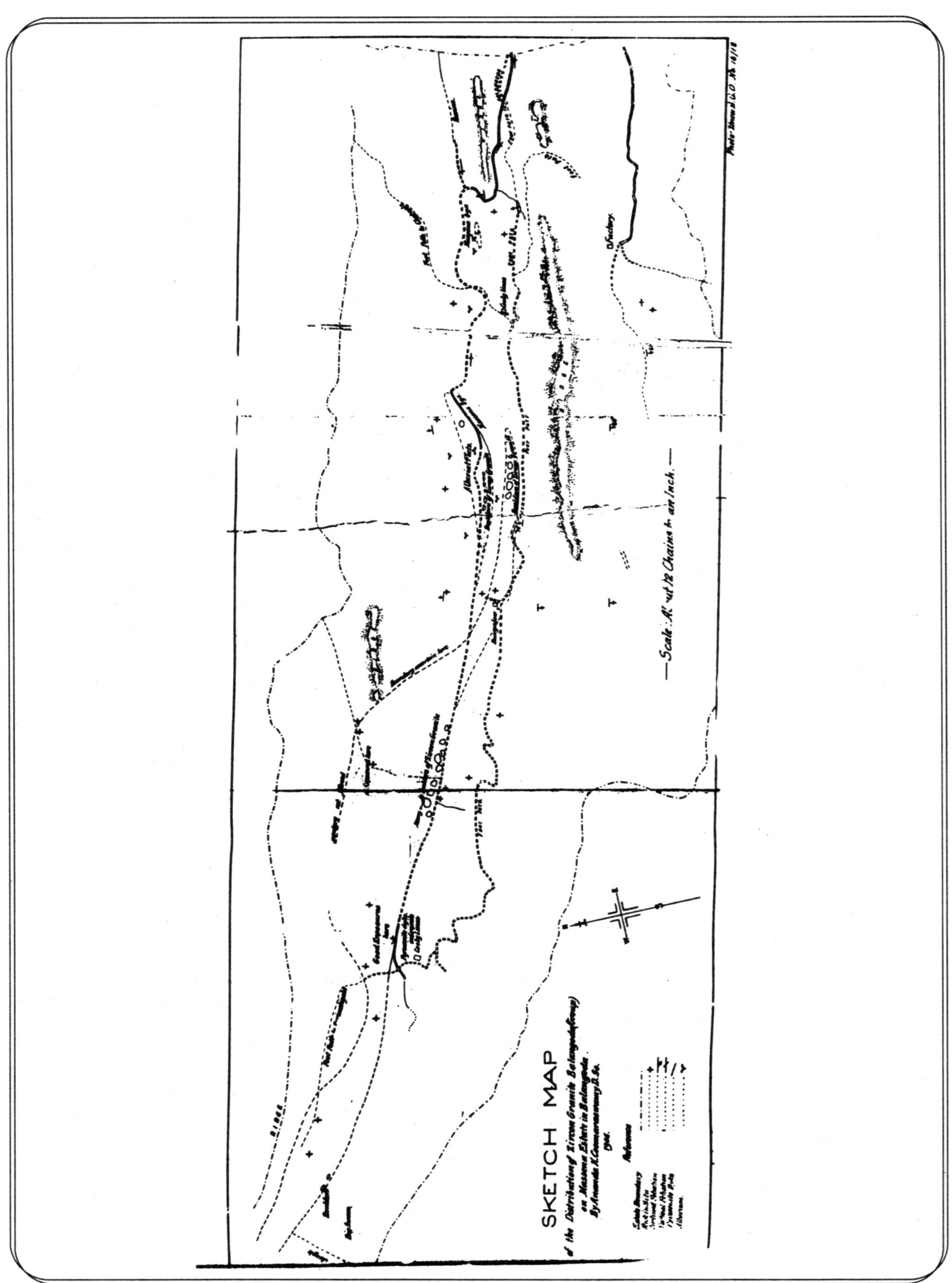
SKETCH MAP
of the Distribution of Zircon Granite Balangoda (group)
on Massena Estate in Balangoda
By Ananda K. Coomaraswamy B.Sc.
References
—Scale: 1" = 12 Chains to an Inch.—

(18) BALANGODA GROUP

By the Director

During the year no new exposures of rocks containing characteristic accessory minerals suggesting the "Balangoda Group" have been met with, but opporutnity was taken while dredging operations were being conducted in the Walawe-ganga, to spend two days at Massena (by the kindness of Mr.C.J. Smale) and make a rough sketch map of the disposition of the exposures of zircon granite there. The matter is of some economic interest, because it is to be expected that if the rarer minerals, such as thorite, thorianite, monazite, and annerodite, are ever found *in situ*, it will be in granitic rocks of this type and occurring in a similar way.

The main part of the Massena estate consists of a valley running a little north of west and south of east and determined by the foliation of the granulites, whose strike is constant in the same direction. The zircon granite seems to form an irregularly lenticular mass intruded along the granulite foliation plains, in a mass of very considerable thickness and extent.

No other equally extensive exposure of rocks classed as belonging to the Balangoda Group is known. Some previous observations will be found in an article on the Balangoda Group in the Geological Magazine, No.482 (1904). It is evident that these rocks occur most typically in lenticular sills or laccolites intruded

between the granulite foliation planes ; and also in dykes which cut these planes at various angles. These observations will be of value should any of the rarer valuable minerals be discovered *in situ* in similar rocks.

(19) INTRUSIVE PYROXENITES, MICA-PYROXENITES, AND MICA-ROCKS IN THE GRANULITES

By the Director

During the three years of the Mineralogical Survey's work a number of scattered observations have been made on the various rocks above referred to; these are embodied in a paper appearing in the Geological Magazine, No.494, August, 1905. The following is a summary of these observations. A group of pyroxenites, in which the minerals diopside, phlogopite, hornblende, and scapolite predominate, while sphene, plagioclase, pyrite, apatite, spinel, and carbonates are often accessory, occurs in small sills and dykes intrusive in the granulites. (One of these small dykes appears also to be intrusive in the Massena zircon granite.) These pyroxenic intrusions have usually a zoned or laterally symmetrical structure analogous to that of mineral veins. The intrusive material never shows a chilled edge, but there is a rapid transition from the pyroxenite to the granulite. The course of the sills and dykes, which are rarely of any size, is usually determined by the foliation planes and joints of the granulites.

These rocks, or at least the mica-rocks or mica veins, which I have classed with them, are of some economic interest. The mode of occurrence of mica in veins is described in previous Administration Reports.

It is interesting to observe that the rocks described closely resemble the pyroxenic rocks associated with the apatite deposits of Canada described by A. Osann ("Notes on certain Archaean Rocks of the Ottawa Valley", Geol. Surv. Canada, Part 0, Ann. Rep., Vol. XII., 1902), and by C.H. Gordon ("Pyroxenites of the Grenville Series in Ottawa County, Canada", Journal of Geology, Vol. XII., No.4, 1903, pp.316-325) who states that the evidence there too "clearly warrants the conclusion that the pyroxenic rock is intrusive". Detailed descriptions and further references will be found in the papers mentioned.

(20) LAND SUBSIDENCE AT MANIPPAY

A good deal of attention has been attracted by a subsidence of paddy land which took place at Manippay on the 20th of May. At dawn on that day the villagers living near the Kirai fields heard a strange sound which was found to have been caused by the collapse of the surface soil over an area about ten feet in diameter. The pit thus formed contained a pool of salt water which gradually widened till about 4 p.m., when the pool was about 40 feet in diameter ; soil from the sides continued to fall in for a time, but the hole is not now increasing in width. The depth of water was about 30 feet ; and its surface about one foot below the surface of the field. (It is now much lower).

The explanation given, that salt water passing through underground passages gradually undermined a place where the subsoil (recent coral sand, & c.) was moe than usually friable, appears likely to be correct. Such subsidences in limestone districts in England are known as "swallow-holes". It is said that previous examples of similar occurrences have been known in the Jaffna District.

(21) GLOSSARY OF TECHNICAL TERMS

(Supplementary to the Glossary given in the Administration Report of the Mineralogical Survey of 1904.)

Borella	..	Sandy gravel without gems; sometimes applied to upper barren part of *illam*
Maladola	..	Small stream, only running in wet weather. *Cf.* English "winterbourne".
Pattan	..	Facetted, naturally or artificially.
Yakun-kapapu	..	"Devil-cut", applied to idiomorphic crystals with well-preserved facets, which are regarded as artificial.
Madulu	..	Variety of graphite breaking readily into smooth lenticular pieces resembling the segments of ripe jak fruit ; a good quality.

Yabura .. Hard graphite of bad quality, of little value owing to the presence of pyrites, which in a decomposed state gives it an ashy appearance, whence the name.

Guru-pachcha, .. Name applied to certain brown or liver coloured minerals found in some *illams*, and which probably belong to annerodite, fergusonite & c.

Gona-piti .. Green chert of opal.

Kahondangala .. Brown chert (formerly used for striking fire).

ERRATA IN FORMER GLOSSARY

For read — *For "garanidama" read "garana-idama"*

For "kuri" read "kura" — *For read*

(22) MUSEUM WORK

List of Gems, & c., purchased for the Museum in 1905 from balances available on the Mineralogical Survey Vote

			Rs.	c.				Rs.	c.
651	..	Matale garnets	1	50	645	..	Beryl ("large uncut acquamarine")	75	75
709	..	Zircon	12	0	618	..	Cat's-eye	1	0
644	..	Topaz (clear, rough colourless piece	4	0	661	..	Gold	1	0
771	..	Sapphire	1	50	832	..	Topaz (Green)	100	0
772	..	Chrysoberyl (twin crystal)	5	50	834	..	Sapphire (42 carat)	350	0
782	..	Tourmaline (green)	1	0	835	..	Chrysoberyl (12 3/4 carat)	100	0
783	..	Thorianite	1	0	837	..	Topaz	30	0
666	..	Gold	1	75	838	..	Sapphire (32 carat)	126	0
769	..	Chrysoberyls ("katta")	3	0	839	..	Ruby	7	50
654	..	Topaz (green)	15	0	840	..	Zircon	5	0
655	..	Sapphire (rough)	5	0	841	..	Tourmaline	5	0
669	..	Zircon	1	0	842	..	Tourmaline	5	0

In addition to the foregoing purchases, a small quantity of thorianite (20 oz.,) and a few zircons were exchanged with a mineral dealer in Germany for a collection or wooden crystal models and a set of foreign thorium-bearing minerals for comparative purposes, costing altogether 110 marks. The thorium-bearing minerals and a selectgion of the crystal models are exhibited.

The following books and instruments have been purchased during the year:-

			Rs.	c.
Weinschenk, E.	..	"Grundzuge der Gesteinskunde"	2	62
Grubemann	..	"Die crystallinen Schiefer"	11	15
Rosenbusch	..	"Mikroskopische Physiographie der Mineralien", Part I	15	0
Dana	..	Mineralogy	40	75
Suess	..	"The face of the Earth", Vol.I	19	81
Westphal balance and accessories			59	6
Herbert Smith's Refractometer			75	75
Jolly pring balance, also a contact goniometer			58	50

A photograph of the new sections exposed in cutting the new tunnels on the Kadugannawa incline has been added to those already exhibited. A sketch map of exposures of zircon granite (Balangoda Group) at Massena, near Balangoda, has also been hung on the walls.

The arrangement of the mineral gallery is now fairly complete, so far as the present collections are concerned; the accommodation, however, will hardly admit of many more additions, and it will be necessary to provide for one additional wall-case and one table-case with drawers during the present year. The total number of specimens catalogued is 843.

The number of applications for the identification of minerals has greatly increased during the present year, mainly in connection with the boom in thorianite, which was at its height during the middle of the year. The majority of the minerals supposed to be thorianite, however, belonged to magnetite, ilmenite, & c. Specimens of mica and iron pyrite continue to be sent in with an inquiry as to whether they contain gold. Cinnabar is also reported from time to time, but as many of the specimens sent in are artificial, and as the one locality supposed to have been discovered by a prospector during the present year was proved to have been salted, the evidence for the occurrence of this mineral in Ceylon remains insufficient.

It was hoped that information would have been received from the Imperial Institute during the year regarding various rare minerals which cannot be satisfactorily determined without a complete chemical analysis ; as, however, no information has been received even relating to minerals sent home in the early part of 1904, it has been impossible to complete the classification and arrangement of the rare minerals on hand. If the required information is received during the current year the minerals can be arranged in August or at Christmas.

Amongst the additions to the collection there must be mentioned the gems which have been purchased to fill up gaps, partly out of the Mineralogical Survey Vote for contingencies and partly out of the balances available at the end of the year on all votes for the Mineralogical Survey.

Amongst the minerals submitted for examination was a small green stone, cut *en cabochan*, sent by Mr. W.C.Wild. This stone was remarkable for showing a broad ray something like that of a cat's-eye or moonstone, but very highly coloured, with the spectrum colour from deep red to deep violet. As the specific gravity was found to be about 4.2., the stone must be a variety of zircon, and it is probable that the stripe of prismatic colours is an effect analogous to that seen in opals, but here much more regular and produced by the zoned strucutre of the zircon, which is composed of successive very thin plates of slightly differing refractive index. A zoned structure is however very usual in zircon, and is not particularly conspicuous in this specimen, and if it be the actual cases of the peculiar appearance it is strange that it should not have been observed before.

Amongst some *nambu* obtained in the Ratnapura District was a small twin-crystal of chrysoberyl, consisting of three individuals twinned on p.(031). The forms present were a (100), b (010), s (120), x (101), o (111), n (121). This is the first twinned chrysoberyl to be added to the collection ; the absence of others is unfortunate, as Ceylon is a noted locality for good chrysoberyl twins ; but it often happens that these living crystals are also good cat's-eye and are worth too much to find their way to the Museum.

(23) MINERALS EXPORTED IN 1905

Statement*[4] showing the Exports of Minerals (the Produce of Ceylon) from the Island during the year 1905.

Country to which Exported	Plumbago		Thorianite		Thorite		Mica	
	Qty.	Value	Qty.	Value	Qty.	Value	Qty.	Value
	Cwt.	Rs.	Cwt.	Rs.	Cwt.	Rs.	Cwt.	Rs.
United Kingdom	164,915	1,929,505	172	70,397	1	300	-	-
British Colonies								
British India	1,065	12,461	-	-	-	-	-	-
Burma	42	491	-	-	-	-	-	-
Victoria	1,761	20,604	-	-	-	-	-	-
Foreign Countries								
Austria	222	2,597	-	-	-	-	-	-
Belgium	54,286	635,146	2	1,665	-	-	-	-
China	128	1,498	-	-	-	-	-	1
France	3,158	36,948	-	-	-	-	-	-
Germany	110,629	1,294,360	5	2,118	-	-	-	-
Holland	1,496	17,503	-	-	-	-	-	-
India, excluding British	5	58	-	-	-	-	-	-
Japan	23,111	270,399	-	-	-	-	-	-
Russia in	2,863	33,497	-	-	-	-	-	-

4*Information kindly supplied by the Principal Collector of Customs

Country to which Exported	Plumbago		Thorianite		Thorite		Mica	
	Qty.	Value	Qty.	Value	Qty.	Value	Qty.	Value
	Cwt.	Rs.	Cwt.	Rs.	Cwt.	Rs.	Cwt.	Rs.
Europe								
Sweden	201	2,352	-	-	-	-	-	-
United State of America	248,966	2,912,902	-	-	-	-	-	-
Total	612,848	7,170,321	179	74,315	1	300	-	1

N.B.- In this return in some cases the value is given, although the quantity of the article exported is omitted, because this is less than the unit of measurement.

A.K. COOMARASWAMY, Director

CEYLON.

ADMINISTRATION REPORTS, 1906.

PART IV. — EDUCATION, SCIENCE, AND ART.

MINERAL SURVEY.

REPORT of Dr. A.K. COOMARASWAMY, F.L.S., F.G.S., Director.

Geological Survey & Mines Bureau
1995

MINERALOGICAL SURVEY

Director : ANANDA K. COOMARASWAMY, D.Sc., F.L.S., F.G.S., M.R.A.S

Assistant Director : JAMES PARSONS, B.Sc., F.G.S.

REPORT OF THE DIRECTOR OF THE MINERALOGICAL SURVEY FOR 1906

(1) DISTRICTS EXAMINED BY THE DIRECTOR

The first part of January was occupied in the completion of Administration Reports ; during the second part the Director was laid up with influenza. The first part of February was spent at Kitulgala (Kegalla District) in the examination of alluvial deposits there. From Kitulgala a move was made to the village of Deraniyagala. Nothing of special importance was found in either of these districts. The beginning of March was spent in the Maskeliya district, with headquarters at Laxapana, for a fortnight. An occurrence of corundum at the head of the Maskeliya valley was next investigated, after which a short stay was made at Hatton to examine the occurrence of andalusite and sillimanite on Dunbar estate. Easter was spent in Kandy.

The greater part of May was spent at Pattipola in the investigation of alluvial deposits, and a short visit was also paid to the Harton Plains. Amonth's leave was then taken. In July an expedition was made to Madugoda and Mimure in Uda Dumbara. The beginning of August was spent in Colombo. On 8th August mr. Parsons proceeded to Europe on six months' leave on half pay, it having been decided that the Mineralogical Survey should be continued for a further period of three years; and that Mr. Parsons should be Director, the present Director not desiring to carry on the work.

During the latter part of August, whole os September, and first portion of October the Director was engaged on other work connected with the Ceylon Rubber Exhibition at Peradeniya, having been seconded to organize an exhibition of the Arts and Crafts of Ceylon.

After a week's leave in the middle of October the Director proceeded to Pelmadulla and Balangoda to examine occurrences of copper, mica, and graphite. The latter part of October was so extremely wet as to make field work quite impossible. In the beginning of November a short visit was made to Mirupe to complete a sketch map of part of the Kandy District for the present Administration Report. The remainder of the year, the time selected for recess, was occupied as usual with reports and with work in Colombo.

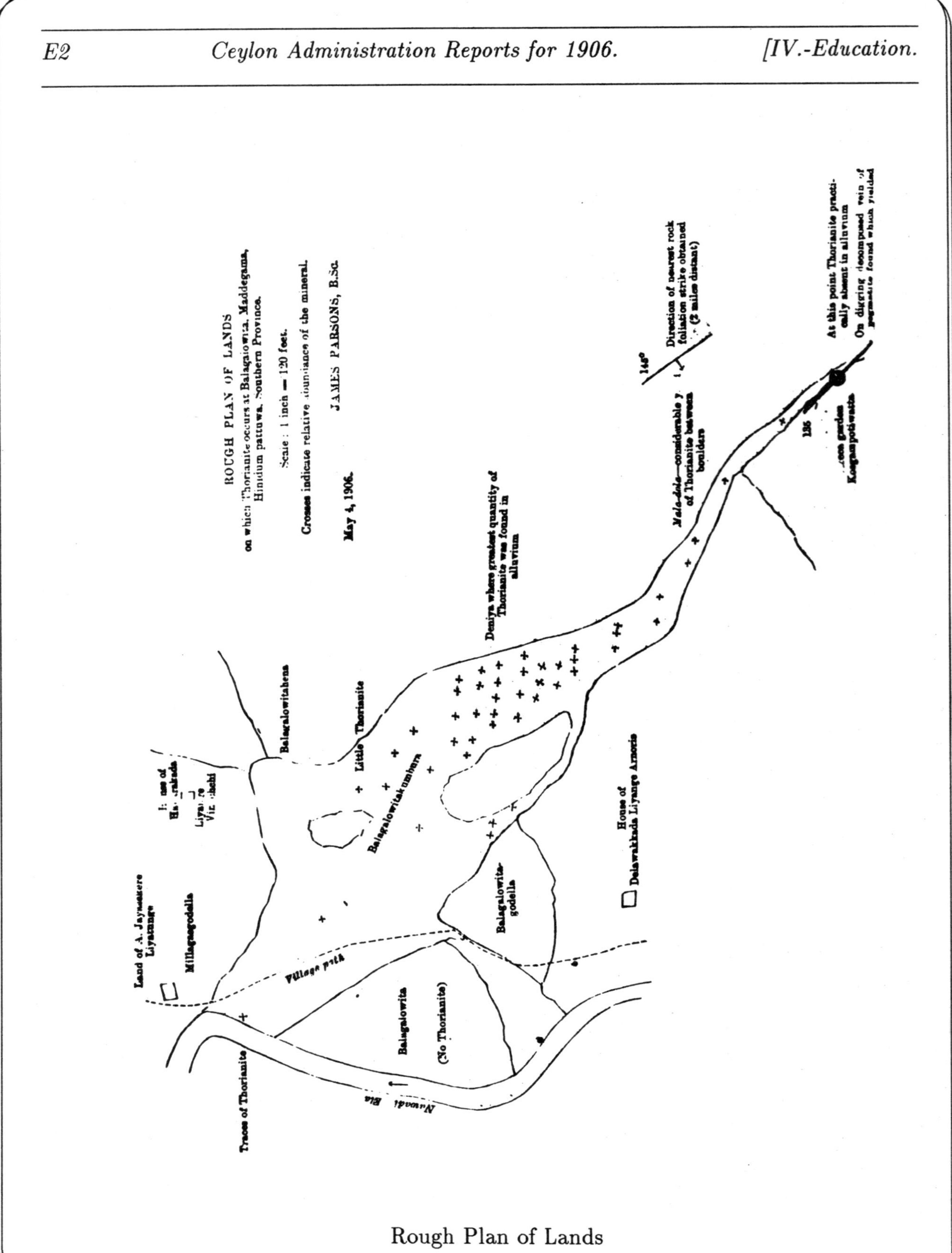

Rough Plan of Lands

(2) DISTRICTS EXAMINED BY THE ASSISTANT DIRECTOR

Work was commenced in January in the Galle District, when an occurrence of asbestos near Hikkaduwa was investigated as far as possible. Seventeen days were spent near Dodanduwa, after which I travelled to Udugama *via* Baddegama and Wanduramba, staying a few days at threse places. From Udugama I proceeded up the Gin-ganga to Hiniduma, after which I went on to Neluwa, a village about ten miles to the north of Hiniduma by Gansabhawa path. A period of about three weeks was spent at Neluwa investigation the occurrence of thorianite in the neighbourhood, the results of which investigation fully justified the somewhat long stay in one place. The Hinidum pattuwa had been partially prospected over by me in 1904, but it was necessary to visit the district again to investigate the reported occurrence of thorianite. On my return journey from Neluwa I stayed a few days at Hiniduma, whence I returned *vai* Udugama and Galle to Kandy at the beginning of April, when the first twelve days of the month were occupied in general office work, determination of minerals, and preparation of the interim report. The remainder of the month and the first three days in May were covered by the Easter vacation and by leave. I then proceeded to Ragalla in the Central province, and after a stay of a week proceeded to Welimada in Uva, where it was my intention to stay only one or two days, as the district had been traversed in 1903, but he unexpected discovery of thorianite in the locality led me to prolong my stay to eighteen days. After this I established my camp for a week at Dehiwinna, and then proceeded to Bogoda, where eleven days were spent and another occurrence of thorianite noted. At the close of my work at Bogala I retunred to kandy, the month of July being spent in further examination of the occurrences of mica in the Kandy District, in preparing my interim report and my portion of the Administration Report, necessitated by six months' leave being granted me from the beginning of August.

Science and Ard.] *Mineralogical Survey.* *E3*

(3) THORIANITE AND OTHER RARE MINERALS

The most important observations on thorianite have been made by Mr. Parsons in the Galle District. Perhaps a ton of thorianite has been obtained at Maddegama, Hinidum pattuwa, in the Galle District. The deposit appears to have been first discovered about August, 1905, by Delawakkada Liyanage Arnoris one of the owners of the land, to whom the occurrence is said to have been revealed in a dream. The land is owned by three families, two of which parted with their shares to a Mr. Kotalawala of Galle, by whom the deposit has been worked in conjunction with K.L.Arnoris, owner of the third share.

The deposit is found in the course of a *maladola, i.e.,* a stream in which water runs only in wet weather. This stream rises in an areca garden near its boundary against Balagalowitahena, on the slope of a hill range running approximately parallel to the strike of rock foliation in the district, and whose crest forms the boundary between Maddegama and Batuwangala villages. The stream for about 190 yeards has a rocky bed, now excavated and reexcavated for the thorianite; the fall is about 1 in 7; in it then reaches and irrigates a paddy field (Balagalowitakumbuta) in the valley of the Nawadi-ela, which here forms the boundary between Maddegama and Neluwa. The Nawadi joins the Gin-ganga a quarter of a mile below.

Thorianite was first found in the dneiya, where the greatest amount was obtained; the paddy field right down to the ela was dug over, but he yield was much less. Working then from the deniya up the *maladola* considerable, though diminishing, quantities were found , and the stream course was dug up over a width of 46 feet. As the top of the stream was reached the thorianite-bearing gravel became thinner, until in the areca garden it was represented by a layer of sandy soil only, in which the thorianite was so scarce as hardly to pay washing. The deposit seemed to be completely exhausted. The rocks seen in the stream course are quartz-orthoclase and mica-quartz-orthoclase pegmatites, and ordinary types of acid granulite ; above the point at which the thorianite ceased and as far as the top of the hill only acid granulites are seen. However, at the point where the last traces of thorianite were found, an excavation was made by mr. Parsons' director, and at a depth of 2 feet the thorianite began to become more abundant. Following the maximum yield of thorianite a trench was now made, evidently along the outcrop of a decomposed pegmatite vein. Kaolin from the decomposed felspar and hydrated mica are the conspicuous constituents of the pegmatite in its present condition. About 600 lb. of the wet decomposed rock yielded 1 3/4 lb. of thorianite, or about 6 lb. to the ton ; the 1 3/4 lb. represented one man's washing for five hours. The thorianite obtained from the vein has an outer skin of ferric oxide, which is absent from the alluvial material. The individual crystals are very sharp and often intergrown in groups. Twinning is also common. No zircon occurs in the washings from the pegmatite, or in the alluvium of the *maladola* or the deniya ; but large crystals are common in the alluvium of the paddy field. The gravels in the paddy fields are therefore probably chiefly deposited by the Nawadi-ela.

Apart from its scientific interest, this discovery of thorianite *in situ* should be of considerable economic value. There seems no reason why the pegamtite vein should not be mined to a considerable extent. It appears that the more acid portions of the pegmatite, which withstand weathering best and are therefore exposed at the surface, are just those in which thorianite is absent or scarce, which should be a guide to sinkers of trenches or pits in other localities, such as Bambarabotuwa, where thorianite has been traced to the head of a stream, but where no exposure of thorianite-bearing rock has hitherto been found.

It is interesting to note that the thorianite does not travel very far from its source ; traces were found as far away as the bank of the Gin-ganga opposite the mouth of the Nawadi-ela, but not in the Gin-ganga below that point. Traces of thorianite were also observed in a deniya above Balagalowita on the hill range known as Galkinissa on the south-east, where indeed any continuation of the pegmatite vein might be looked for, but the trace was of the slightest.

The deposits are very isolated, being twenty miles distant from those at Erabdda and thirty miles distant from those at Bambarabotuwa. The only other occurrence noted in the Southern Province was at Monrovia estate, Dodanduwa, where traces were obtained by panning the *malawa* from on or two pits, though it was certainly absent in the *illam*. Monazite was also present.

In Sabaragamuwa work has been continued by Messrs. Finlay, Muir & Co., and by Mr. Holland in the Bambarabotuwa district, and at least one or two tons of thorianite obtained. It has been found worth while to wash the soil from the banks of the Kuda Pandi-oya, and it is also found that the bed of the stream itself continues to yield thorianite after every fall of rain. No trace has been found of the parent rock from which the thorianite is derived. The best way to prospect for this rock would be to drive a series of trenches about the head of the stream in which the thorianite has been traced. The rock is doubtless decomposed and concealed by soil, vegetation, and perhaps large boulders. Specimens of thorainite found on Ferndale estate have been shown to the officers of the Survey.

In the Kegalla District traces of thorianite were met with in the sands from the bed of the Deraniyagala-ganga, but none in the *illam* of its banks. The price of thorianite has now declined to Rs.4 per lb. as a result of the action of the German ring, who control the world's supply fo monazite. The samples of thorianite have been found at the Imperial Institute to vary considerably in composition. A variety "from galle" (? Hiniduma), which formed the subject of a paper communicated by Professor Dunstan and Mr. B.M. Jones to the Royal Society in 1906, contains less thoria and more uranium oxide than those previously analyzed. Samples of monazite, on the other hand, have been found to contain nearly 10 per cent. of thoria in place of the 5 or 6 per cent. usually found in monazites from other sources. This of course increases the value of the Ceylon material.

Traces of thorianite have been found by Mr. Parsons near Welimada, on the Uva-patanas. About half a mile from the resthouse, close to the Bandarawela bridle path; in an exposure of spinel-mica rock associated with an orthoclase-quartz pegmatite. In places the rock is almost entirely composed of pinkish blue spinel; elsewhere it psses into soft mica veins and pockets. The rock, which is very friable, is exposed at the side of a small stream irrigation a paddy field (*Hatdowala*), which after joining a similar stream flows into the Welimada-oya near the resthouse. The resemblance of the soft micaceous veins to the thorianite matrix at Maddegama, Hinidum pattuwa, suggested further investigation. A small pit was sunk in the stream close to the exposure of rock. About 2 feet of gravel was found resting on one of the soft decomposed mica veins. The washings of the gravel yielded a number of small fragments of thorianite, a few crystals, but mostly irregular pieces. None of the fragments were appreciably

E4 *Ceylon Administration Reports for 1906.* *[IV.-Education.*

waterowrn, indicating their derivation from a source near by. Washings from the top of the *malawa* (here the decomposed top of the mica vein) yielded traces of thorianite and some crystals of thorite. Subsequent further investigation revealed only the presence of the thorite in the *malawa*, so that perhaps the thorianite at first obtained was not really *in situ* in the mica vein, but derived an embedded in its soft surface. This cannot be regarded as proved, however, for even at Maddegama the proportion of thorianite from a single washing is very small, and the thorianite is unevenly distributed in the vein. Pits were sunk above and below that just referred to; of these, the upper pit yielded neither mineral from the gravel, and only a trace of thorite from the *malawa* ; the lower pit yielded both minerals from the gravel, and the *malawa* could not be reached on account of the presence of large boulders. A pit was also sunk in the paddy field about 200 yards below, in a spot where heavy minerals might be supposed to collect, but only black spinel and a little zircon could be found. Four other pits sunk in selected spots yielded only traces of ilmenite, rutile, zircon, and spinel, but no thorium-bearing minerals. Two pits sunk on the bank of the Welimada-oya yielded a trace of gold, but no other heavy minerals. Washings from terrace gravels about 20 feet above the present stream and exposed about a mile above the footbridge yielded only traces of ilmenite and zircon.

Investigations at Bogoda, about five miles from Badulla, were also carried on. There is exposed at Landewala a spinel-mica rock associated with orthoclase-quartz pegmatite as at Welimada, and also with mica in fair-sized crystals, which had been worked commercially. A small pit was sunk by the side of a stream close to the exposures of psinel-mica rock in a paddy field. Washings from gravel yielded a few fragments of thorianite. Another pit sunk a few yards lower down yielded a few fragments of thorianite and some crystals of thorite. Other pits sunk above and below these points yielded only the merest traces of thorite.

The mica-spinel rock with the associated pegmatite apparently forms a leticular boss, the extent of which is difficult to estimate ; the exposure would indicate that it is about 150 yards long by 50 yards wide. It is cut by the stream referred to above approximately at right angles to its longer axis. At one point in its course across the exposure the stream expands into a pool known as the *Balawala* ; in the sides of this pit a vein or pocket of soft micaceous rock is exposed ; the presence of this soft rock has no doubt been the cuase of the formation of the pool. A considerable quantity of the rock was crushed and washed ; it yielded a few fragments of thorite, but no thorianite ; considering, however, the extreme scarcity of thorianite in the gravel, it is sufficiently likely that the mica rock is really the source of both minerals in this locality. Though the deposits at Welimada and Landewela are so small, they are nevertheless of the greatest interest, as showing that thorianite is associated with, if it does not actually occur in, a peculiar spinel-mica rock, itself intimately and probably genetically associated with an orthoclase-quartz pegmatite ; and it has been proved that thorite occurs in micaceous veins in the same rock. It should not be forgotten that, as described above, the thorianite at Maddegama, Hiniduma, occurs in similar micaceous veins, and is worked commercially.

These observations show that it may be well worth while prospecting for thorianite in other than reputed gemming districts, and further afford some guide to prospecting, which has hitherto been rather a matter of chance. Careful prospecting amongst micaceous rocks may lead to the discovery of other valuable deposits similar to that at Maddegama. It should be noted that the occurrence of thorianite in the micaceous rocks at Welimada and Landewela could not be actually demonstrated ; it does not necessarily follow that crushing and wshing the rock on a large scale might not be remunerative, as a washing of 12 lb. of wet material at Maddegama yielded only .079 gram of thorianite, representing .032 lb. to the ton ; yet the average washing on a large scale yielded as much as 6 lb. to the ton, and was highly remunerative.

Of other localities, small quantities (microscopic grains) of thorianite were detected by the Director in pannings from sand of the Sitawaka-ganga at Deraniyagala (Province of Sabaragamuwa), and by the Assistant Director near Dodanduwa (South Province) in the superficial *malawa* in pits sunk in paddy fields (especially Ratgama and Gonagaladeniya on Monrovia estate).

The balck mineral from Bambarabotuwa, referred to in Administration Report fro 1904, p.10, has proved to be the rare mineral zirkelite, containing 20 per cent of thoia.

Another rare mineral from Bambarabotuwa is tscheffkinie ; the amount of thoria in this mineral is about 5 per cent.

The mineral known to the Sinhalese as *guru-pachcha*, and which has been found in many localities is Sabaragamuwa, proves to be fergusonite, containing about 2 per cent. of thoria and over 4 per cent of uranous oxide.

The mineral hitherto known as thorite continues to be met with in small quantities in various localities. Of these, the most interesting is an occurrence *in situ*, near Nugatenna gap, of a yellow mineral rich in thoria and containing silica, and which appears to be thorite. Near and about Nugatenna pass (Uda Dumbara, Central Province) are considerable masses of coarse augen gneiss, consisting mainly of quartz, orthoclase, and biotite. The orthoclase occurs in large augen, sometimes as much as 6 or 8 inches in length and almost as wide; the interspaces are occupied by a finer-grained mixture of quartz, felspar, and streakily disposed biotite. The rock often appears to have been affected by pressur, the orthoclase cleavage faces being usually more or less bent. Several exposures (below Nugatenna on the Kandy side, on Woodside estate, and on the Madugoda road) show that the granitic rock is intrusive in the granulites ; earth movements have not obscured this intrusive character. The gneiss or granite tends to occur in masses interbanded with or gradually transgressing the granulites, which have a general N.N.E. strike. The junctions of granulite and gneiss are never perfectly sharp, and though quite distinguishable in the field are vague in hand specimens, as is generally the case in Ceylon.

The rock often forms large weathered boulders externally decomposed and not well adapted for study, but freshly blasted specimens are to be found on the Madugoda road a little beyond the 26th milepost. Here the supposed thorite occurs in the form of waxy honey-yellow grains, generally closely associated with the biotite ; the largest grains have a diameter of about 1/8 inch. Only a very little of the material could be collected. Under the microscope some cleavage flakes show what appears to be a biaxial interference figure, the axial angle being very small ; other cleavage fragments not showing any figure have faces inclined at about 78 degrees to each other, extinction coinciding with one of these cleavage faces. The mineral has a specific gravity of about 4.47. A small specimen has been forwarded to the Imperial Institute for analysis.

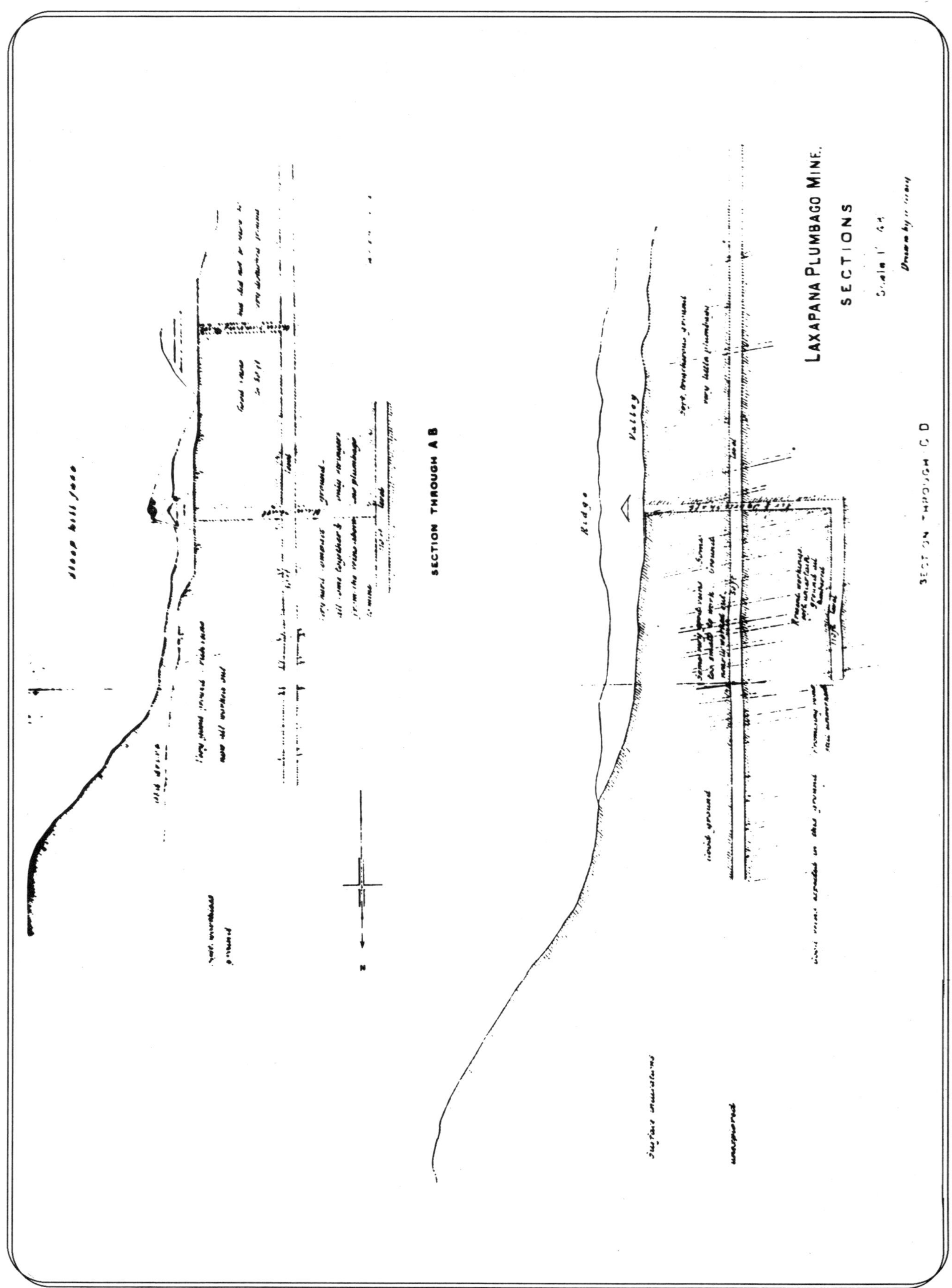
LAXAPANA PLUMBAGO MINE.
SECTIONS
SECTION THROUGH A B
SECTION THROUGH C D
Steep hill face
Ridge
Valley

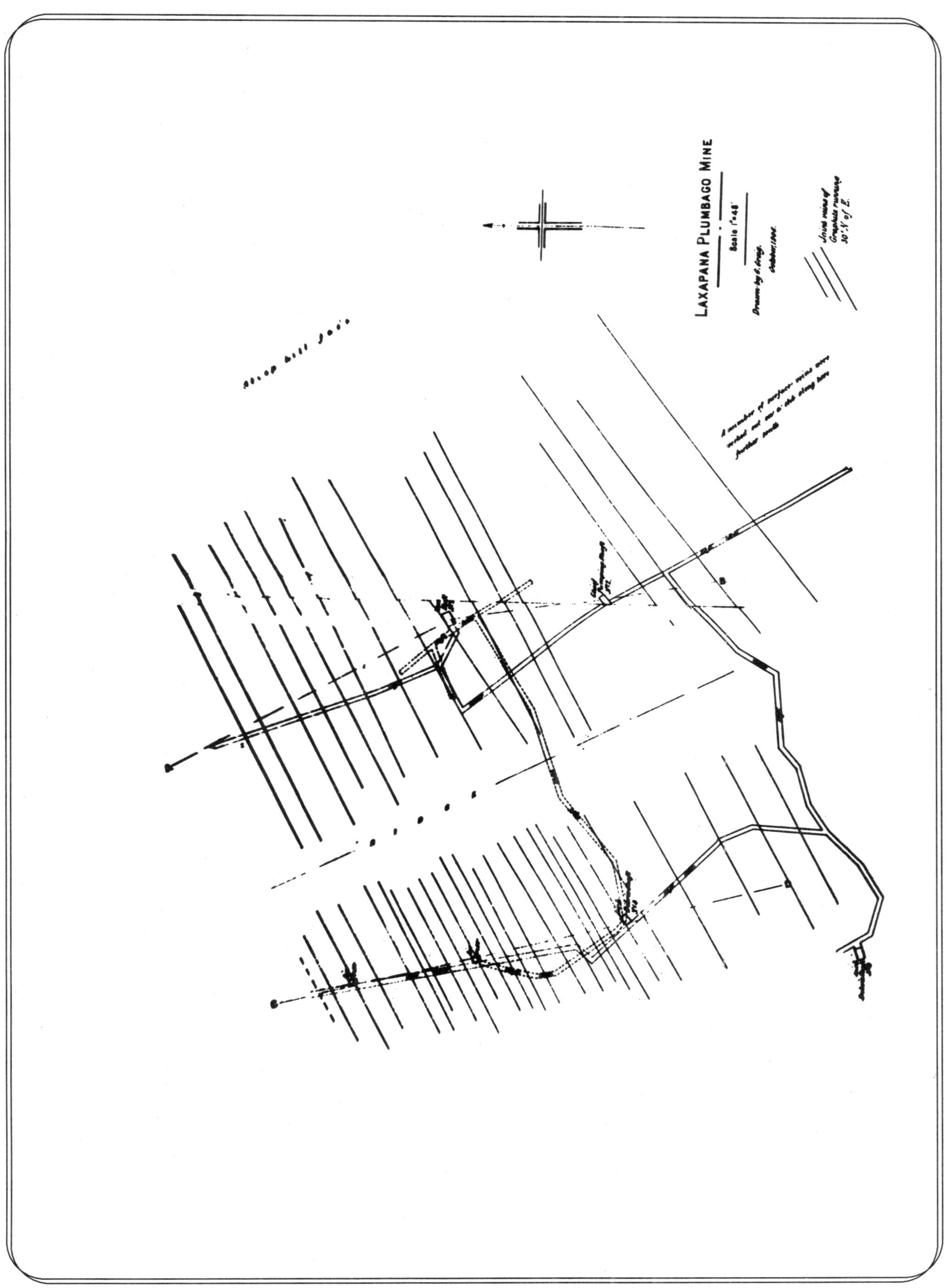
LAXAPANA PLUMBAGO MINE
Scale 1"=48'
Joint veins of Graphite running 30°N of E.

(4) COPPER

Native copper has been found in the neighbourhood of Pelmadulla, and is a new record for Ceylon. The exact locality is Higgastenna, Denawakandakada, Nawadun korale ; the land is bounded on the north by Dewatagahahena. One P.T. Doresamy has taken out a plumbago license (No.92) with a view to further search for copper. It seems that when the Gansabhawa path was last cleared pieces of copper were picked up in the soil, which consists of friable laterite with ferruginous concretions. No solid rock is exposed near to the spot where copper has been found. On the occasion of my visit only one piece of copper could be discovered after a careful search ; the place had already been thoroughly well searched over, though no excavation of any importance had been made. I recommended the lessee of the mining rights to make shallow excavations and to wash the material so obtained and to continue the work for a few weeks till some idea of the quantity of copper obtainable could be gathered. The amount is probably small. Unfortunately the nearest stream is a quarter of a mile away, so that a small tank would have to be prepared to hold water for the washing.

The nuggest of copper are very irregular in form; some are very small, but other pieces range up to half a pound or more in weight. The nuggest have a very rough surface, and can only with the greatest difficulty be detected amongst the ferruginous concretions with which they occur and which they closely resemble.

There is nothing to suggest an artificial origin for the nuggest. The copper is probably derived from an oxide or carbonate by reduction by organic substances, as suggested by Knop and others (Hintze, vol. 1., p.200). Native copper occurs thus in Namaqua and Damaraland, especially where alternation process have been extensive. The occurrence of copper carbonate has been recorded in Ceylon.

(5) ASBESTOS

Asbestos has been found at Tiranagama, a village near the coast about 12 miles north of Galle. Small veins of crystalline limestone containing asbestos have been met with at a depth of 180 feet in a trial shaft, from which no graphite has yet been obtained. The asbestos appears to be serpentine (chrysotile), and is probably developed from forsterite, a common mineral in the crystalline limestones of Ceylon. In places the limestone is also rich in small flakes of phlogopite mica. Unfortunately there seems to be only a very small quantity of the asbestos obtainable. If more extensive, the occurrence would probably be of commercial value. The mineral as brought from the pit is in a wet and "soppy" state, but when dry appears in the characteristic fibrous state.

Another occurrence is at Nawadiwila, a village about three miles distant. The asbestos seems to have been found in some quantity in graphite pits now abandoned and filled in, at a depth of about 80 feet. The pieces were stated to reach a yard in length, but this no doubt was an exaggeration. When brought to the surface is dried to a friable earthy material. The only specimens obtainable by Mr. Parsons were of this character. Such material would obviously only be suitable for lagging for steam pipes, but for this would be eminently fitted, especially if mixed with finely divided mica such as could be obtained almost anywhere. As there is considerable use of lagging on steam pipes and boilers in tea factories, for which imported material is now used, this asbestos might command a local sale.

(6) GRAPHITE

The well-known plumbago mine at Laxapana was visited and examined; it is of special interest, as it has been systematically worked and plans dept. A plan of the mine drawn by Mr. George Greig, Superintendent of Laxapana estate, is through his kindness reproduced in the present report. This plan indicates very well the mode of occurrence of the graphite in parallel veins, which do not extend for great distances laterally. The veins are parallel joint veins, nearly vertical, and striking 55□E, of N. The largest are nearly a foot thick. They can rarely be followed with profit to any great distance, hardly even extending more that 200 feet horizontally or 50 feet vertically. Their form is therefore that of very thin lenticles. Sometimes they almost completely die out, continuing only as a thin line and then again thickening. Branching veins have not been met with, but the veins occasionally leave joint planes to follow the foliation planes. A remarkable cavernous hollow linedwith graphite was once met with, called a "vug" by the Cornish miner in charge ; this cavity was about 15 feet high, oval in section, and more than a yard wide ; its walls were lined with flaky graphite, with which the floor was also covered to a considerable depth. The graphite is of exceedingly good quality, and the output about 300 tons per annum.

No other occurrences of importance are known in the Maskeliya district. The only other graphite of importance examined was in the neighbourhood of Hikkaduwa, Southern Province, where there are several groups of pits, of which Mr. E. de Soya's at Trianagama is an especially large one ; none others call for special mention.

(7) NITRE CAVE

The most noted nitre cave in Ceylon is found in the rather inaccessible Mimure district. it is a large cave in a limestone cliff, and owes its existence to the weathering out of the limestone above its junction with the seemingly sheared decomposed micaceous granulites underlying it. The crystalline limestone is very impure, but does not otherwise call for remark. A hundred years ago it is said that 60 lb. per annum of nitre were obtained here. Very little nitre can be seen now. The nitre is derived solely from the dung of the bats, which are abundant in the cave ; the earth on the floor, or rather amongst the interstices of the huge limestone boulders with which the floor is covered, was collected and washed, and the product crystallized for use in the preparation of gunpowder. In modern times the bats' dung has been used for manure by a neighbouring planter. The quantity to be had, however, is not such as to be of commercial importance. There is no real cave earth or any trace of a stalagmite floor such as might suggest the possibility of the occurrence of animal remains ; this appears to be unfortunately the case with most or all Ceylon caves. There is however in the Museum a specimen of the floor of 'a Vedda cave" in which animals occur, presented by the late H. Nevill, and the matter is referred to by him in the *Taprobanian.* it may be noted in passing that there are said to have been as many as

twenty-two "nitre caves" in Ceylon, from which was obtained nitre for the gunpowder expended in the wars with, the Portuguese, Dutch, and British. The only other caves which have so far been visited by the officers of the Mineral Survey are the cave in crystalline limestone on Lower Albion estate, Hakgala ; the travertine cave at Hangala-ela near Welimada ; a cave in Ravanakanda, near Ella ; and a small cave in crystalline limestone and with fairly good stalactites in the Ratganga valley, Sabaragamuwa.

(8) CORUNDUM

Corundum occurs in prismatic blue weathered crystals in surface soil near the south-east boundary of Anandale estate (Upper Maskeliya), extending thence across the estate boundary on to Meeriyakota ; and the corundum is again found rather more sparingly near the middle of Meeriyakota estate. These occurrences lie roughly along the direction of the foliation strike (50 degrees W.of N.). The amount obtainable on Anandale could hardly exceed a few hundred weight, and there would be little chance of reaching the matrix without considerable excavation.

(9) ANDALUSITE

The mineral andalusite, not of commercial value except in rare cases as a gem, occurs in Ceylon, but has never been found *in situ* here. Aggregates of andalusite and spinel, not waterworn, sent by Mr. H.T. Armitage from Dunbar, Hatton, led to the hope that the mineral might there be found *in situ.* The aggregates are obtained in a bed of whitish kaolin exposed at several points along one side of a small valley or deniya near the middle of the estate. Several excavations proved that this andalusite-bearing clay did not continue. Indeed, in one case it was found to rest immediately on *illam* (about a foot), and that on *malawa.* As all the rocks were here decomposed to a great depth, and the clay was not even *in situ*, no further investigation was made.

(10) MOONSTONE

The moonstone pits on Rajawela estate near the main road are now being re-opened, as the result of an increased demand for moonstone. The quarries form a trench or series of trenches and extend for some distance along the outcrop of a band of granulite, which crops out in the middle of the limestone band which runs through the estate. The granulite and limestone possess, as usual, a common foliation.

The moonstone-bearing zone is a band of rock about a yard wide, characterized by the presence of large augen orthoclase. A marked feature of this rock is the elongation and flattering of the quartz in the rock (a white quart-felspar leptynite). The quartz is spread out over planes like slickensides, and seems to have been rolled out, perhaps while in a viscous state. But deforming movements have certainly taken place subsequent to the complete consolidation of the rock. The ends of the augen of orthoclase are sheared and rounded, the cleavage faces are bent, and the mineral itself traversed by fissures often occupied with a fine-grained quartz felspar mosaic, evidently a product of mechanical degradation of the original material.

This moonstone *illam* is associated with other acid granulites less obviously sheared, which have an outcrop about 12 yards wide, and dip steeply westwards, the whole forming a band, within the wider band of crystalline limestone, traceable either way along the strike to Attaragala and Gangapitiya, where moonstone is also mined.

(11) KHONDALITES

Some observations on these sillimanite-bearing rocks were made at and near Pattipola. Exposures of decomposed rock on the Horton Plains path, not far from Pattipola, show that these rocks occur interfoliated with the other types, forming bands or lenticles of no great thickness. The large round garnets decompose with the formation of limonite, but do not break up, so that the rotten rock has the appearance of a coarse pisolite.

It is usually impossible to meet with these rocks in anything like a fresh state, and even exposures of rock *in situ* are rare. An exposure in the middle of the third railway tunnel below Pattipola was therefore of special interest. An apparently lenticular band of coarsely garnetiferous granulite occurs amongst other granulites which could not be well examined. The typical varieties consist almost entirely of pink garnet, sillimanite, and kaolinized felspar. In another variety biotite becomes abundant. The coarsely garnetiferous rock is a most handsome rock, more conspicuous that any other Ceylon rock known to me. The rock decomposes with the greatest readiness, and it is difficult to obtain even from the blasted material perfectly fresh specimens. The rock as exposed in the tunnel wall is soft and friable, showing that considerable decomposition has taken place even since the tunnel was made.

(12) SERPENTINE

An interesting occurrence of serpentine near Ragalla was examined by mr. Parsons. The rock is exposed in the bed of a stream flowing into a tributary of the Halgran-oya in the village of Rupaha. The serpentine forms a band of rock about 10 feet thick, overlying a band of limestone interbedded (as usual) with the granulites. Between the limestone and the serpentine is an intrusive rock composed mainly of plagioclase, with some pyroxene and a little quartz. Below this the limestone is in places rich in pyroxene and the limestone passes into the usual contact types at the junction with the granulites. The serpentine is traversed by dykes striking approximately E. and W. (the general strike of the limestone being N. and S.), which show a dark margins of (?) serpentinized pyroxene and a core of mica. In small pockets the mica is well developed and pieces 2 or 3 inches square can be obtained, but even if the size were sufficient the laminae are too buckled for the occurrence to have commercial value. The serpentine appears to pass by insensible gradations into the limestone, losing its green translucent waxy look and becoming white, opaque, and earthy. The upper part of the bed is a misture of serpentine and calcite, and may be called a cipolino. The rock appears to resemble the cipollinos of the Chindwara district, Central Provinces, India, recently described by Mr. L. Leigh Fermor

(Rec. Geol. Surv., India, XXIII., pt. 3, 1906). His suggestion regarding the origin of the Indian rock by the partial carbonization of a rock originally entirely composed of pyroxene can hardly be applied to the present instance, for we should then have to suppose that all the crystalline limestone of Ceylon originated in this way ; and there is no evidence of that. In any case the exposure is of interest, as, although serpentine is not rare in slightly

altered crystalline limestones, no such large mass is known to occur elsewhere. The rock is thought to have a medicinal value (for gasteric complaints), and is often seen in the possession of villagers and for sale in the boutiques.

(13) ALLUVIAL DEPOSITS

Dodanduwa:- Mr. Parsons examined deposits in the Dodanduwa district, especially on an newar Monrovia estate. This district is a fairly important gemming centre, although large and valuable gems are seldom found. Zircon is exceedingly common, though transparent and well-coloured specimens are rare. Chrysoberyl is common, usually in small fragments, sometimes with enough "silk" to be cut as cut's-eye. Alexandrites are also found, but it is a always stated that varieties of chrysoberyl from the low-country lack the fire and clolur of the Sabaragamuwa, and this appears to be actually the case. Corundum, chiefly in the form of *dalam*, is not uncommon, though good sapphires are rare, and rubies still rarer. The sapphires are of a good blue, something like Rakwana stones.

The gemming lands are typical of the low-country, being situated in wide-spreading flat paddy fields and owitas ramifying between low hills caused by the outcrop of some more acid and harder rock. These low hills are often entirely surrounded by paddy fields forming low "islands" (*duwa)*, which are utilized as sites for one or more dwelling-houses or coconut or areca gardens. The soki in the paddy fields is as a rule from 1 to 2 feet deep, and is succeded by about a foot of sandy angular-quartz *illam*, which rests on *malawa* of white or yellow kaolin. In such quartz *illam* chrysoberyl is generally found ; the *illam* and *malawa* bear a close resemblance to that of Pasdun Korale East, where gold is almost invariably yielded by the *malawa.* In many cases the top inch or so of *malawa* was washed here too, and specks of gold were not infrequently met with. In several pits thorianite and monazite were also obtained by panning the black sand, though these minerals were absent in the *illam.*

Hitherto, in panning only the top inch or two of *malawa*, it has been assumed that the gold, thorianite, and thorite if and when present are derived, *i.e.* in no way genetically connected with the local *malawa.* I do not feel quite sure that this is always the case ; at any rate it will be advisable next year to make some trials of decomposed rock where derived material is absent, or from a depth of a goot or more below the *malawa* surface. Another point to be investigated in the course of future work is the proportion by weight of heavy sands occurring in the different deposits and the proportion of rare minerals, if any, in the heavy sand. For this purpose Mr. Parsons is bringing out a magnetic separator, and statistics will be as far as possible collected.

Neluwa:- Mr. Parsons also investigated the gemming at Neluwa, where he was stationed for some weeks examining the occurrence of thorianite at Maddegama. Traces of gold, thorianite, and thorite were met with, but hardly in promising quantities. A dredging in the bed of the Gin-ganga, about 1/4 mile below the mouth of the Nawadi-ela, showed about 3 feet of sand overlying a bed full of leaves and seeds - the *kola tattuwa* of the 1905 report (p.16) - a deposit which has not been previously observed in the river beds, and here probably shows that the river is flowing over the site of an old backwater. Below the leaf bed was the old gravel or *illam*, which yielded only a little zircon. The *malawa* was reached at a depth of 5 feet. A trace of gold was obtained. There is little to remark about other pits investigated. Zircon is very common in the district ; of other gem minerals, chrysoberyl (including alexandrite and cat's-eye) is not uncommon.

Kutulgala:- Alluvial deposits at and near Kitulgala were examined by the Director, ut little or nothing of importance was discovered. All the sands examined yielded gold and monazite, but no thorianite. The Kelani-ganga at Kitulgala is not adapted for dredging owing to the rocky bottom.

Deraniyagala:- Here, too, alluvial deposits are examined, though nothing of special interest was found ; a series of specimens of *nambu* was, as usual, sent to the Imperial Institute for further examination. As mentioned above, trace of thorianite were seen in pannings from dredgings in the Sitawaka-ganga at Deraniyagala. On the whole, Kitulgala and Deraniyagala appear to be unpromising districts for gems, and no occurrences of mica or graphite are known or worked.

Maskeliya:- It is well known that gems occur in the Maskeliya district ; but as superintendents are not encouraged to, and often not allowed to, make investigations, not much is actually done ; in some cases Sinhalese gemmers paying a monthly rent (Rs.50 is an ordinary payment for amonth) are allowed to work on the estates ; and some planters make investigations for themselves. There can be no doubt that valuable gems occur ; a noted example of this was the well-known case of a cooly who picked up in gravel of the Maskeliya-(Kelani-)ganga, near Maskeliya church, a valuable cat's-eye, which afterwards became the subject of a law-suit. The examination of a series of washings from Dotala, Moray, and Gartmore estates showed that most of the ordinary gem minerals occur. Old gem pits are to be seen on the high patana near Garawilatenna ; the section here shows about 1 foot of *illam* underlying 4 feet of mud and soil. The only gem mineral observed in washings of *illam* was corundum, but not of any value ; pannings from the *malawa* surface yield traces of gold and monazite. The deposits here are not more than three miles in a straight line from the original thorianite locality at Bambarabotuwa ; but no trace of that mineral was detected.

The country east of Ormidale and Bargrove estates yields no gems ; on News Brunswick, however, and estates further not very good gems are said to have been found. Most of these estates have been partially explored at one time or another, either by the proprietor or by Sinhalese. The same has been the case at Norwood, where gem pits are occasionally sunk on flats near the Hambantota-oya. Between Norwood and Hatton there is an estate with the suggestive name of Menikwata. At Hatton itself, Dunbar estate has been systematically explored by the proprietor, Mr. H.T. Armitage

; all kinds of gems have been obtained, some being of very fair value ; but gems do not occur in such abundance as to repay the trouble taken, commercially speaking. This may be taken as a general indication of the usual state of affairs in the Hatton and Maskeliya districts.

Pattipola and Horton Plains:- A considerable number of the gravel deposits of the streams between Pattipola and Ambawela were examined by the Director. Many streams show only a deposit of coarse sand (*Borella*), resting upon *malawa*, and have no deposit of gravel. On the other hand, two high level gravels of the Pattipola-ela were observed. On estimated at 15 feet above the present stream is exposed on the path from Pattipola to Horton Plains, on the slope just before the

E8 *Ceylon Administration Reports for 1906.* *[IV.-Education.*

ela is crossed, about 1/4 mile from Pattipola resthouse. The coarse waterworn gravel attains a thickness of about a foot, but dies away upwards and downwards on the slope. Washing produced a fragment of chrysoberyl, but little else of interest.

The latest gravel is found at the present stream level, overlain in the banks by 6 or 8 feet of alluvium. The gravel is 2 feet or more in thickness, but yields little of interest.

Another high level gravel is exposed on the same path a mile or more from Pattipola, just before the path leaves the main valley. This gravel is about a foot thick, and may be as much as 40 feet above the present stream. There have thus been observed in this neighbourhood three grvels at different levels, approximately 40 feet, and 15 feet above, and one at present water level.

The deposits vary much within short distances ; thus, pit sunk at the junction of two streams about one-third mile south-west of Ambawela station produced no *nambu* at all, while following up the southern branch for half a mile another pit yielded *nambu* with abundant zircon, a few pieces being almost worth cutting.

A short visit to the Horton Plains was also made. Here the same conditions prevailed, and it was difficult to meet with satisfactory *illam*. A pit sunk near the old trout-breeding trenches not far from the resthouse was however promising. The *illam* here is of varying thickness (0 feet to 1 foot), resting on *malawa* and overlain by 18 inches or 2 feet of peaty earth. The proportion of *nambu* is unusually large, consisting chiefly of zircon, with ilmenite, rutile, corundum, & c. The number of fragments of clear pale blue corundum (good *dalam*) was unusually large. Other localities examined showed little of interest.

Uda Dumbara:- Streams examined by teh Director in this district proved very unpromising ; the deniya at Nugatenna gap yielded a couple of fragments of corundum, and isolated zircon crystals are generally to be met with in washings from the neighbourhood of Madugoda and in the Mimure valley. Garnet of worthless quality is particularly common in the Mimure valley. From the Kahatagolla-oya in Haliyala, near Madugoda, fragments of "thorite" were obtained, as well as corundum and zircon.

Garnet of a better quality is rather abundant in the river deposits at Teldeniya. Dredging is unsatisfactory, as the bottom of the river consists of sand resting on *parugala* (decomposed rock *in situ*, but not so soft as *malawa*), and not true *malawa* such as usually underlies deposits of any value.

The following section appears in the river bank:-

Alluvium (about)	..	20 ft.
Bed of coarse concretionary "race" (about)	..	6 ft.
illam	..	1 ft. 6 in.

resting on sandy decomposed rock of the dark noritic type. Garnet is specially abundant in the *illam*.

(14) MAP OF PART OF THE KANDY DISTRICT; EARTH MOVEMENTS IN CEYLON

The officers of the Mineralogical Survey have not been asked or expected in the course of a mineralogical reconnaissance to make geological maps of the country. To do this adquately would require the co-operation of several persons working for very many years, for there are no clearly marked and very distinct formation the limits of which could be easily and rapidly sketched. Anything less than detailed work corrsponding to that of long-established geological surveys would therefore be more or less waste of time ; and there do not at present exist sufficiently detailed large-scale maps, except in a few limited areas. A portion of the Kandy District, however, had been studied in some detail by the Director [*1] previous to the establishment of the present survey, and opportunities have been taken from time to time of carrying on the work thus begun. A small area near Hakgala had also been previously examined by the Director ; a small portion of the Badulla District was examined by the two officers of the Survey in 1903 ; and isolated observations have been made in all the districts inevestigated. The structure of the part of the Kandy District shown in the accompanying map by the Directo ris now fairly clear ; and as there is considerable uniformity of structure thoughout Ceylon, the strucutre of this typical district will prove a sueful guide in the examination of others.

Mention has already been made [†2]of the series of N.N.W. and S.S.E. folds into which the crystalline rocks of Ceylon have been thrown, and the name *Taprobanian* has been given to the earth movements that produced the folds. The present map exhibits the structures in greater detail. speaking now with special reference to the area shown in the map, it will be seen that the rocks are here thrown into two main folds (a syncline and an anticline) with axes trending N.N.W. and S.S.E. The axes themselves plunge steeply to the N.N.W. The main or Dumbara syncline, as it may be named, covers most of the broad cultivated valley so called ; the broad end of the outcrop stretches away between Wattegama and Katugastota ; the narrow end runs up the Haragama valley ; the roa dcutting at the top of the hill coincides with the acute apex of the narrowing outcrop of the syncline. At this point the apparently single band of limestone must consist of an isoclinal overfold dipping E.N.E. The outcrop continues down the hill on the further side towards Hanguranketa.

[1*]Q.J.G.S.LVIII., 1902,

[2†] Splia Zeylanica, I., p III, 1904

Nearer to the centre of the syncline, at Haragama, the structure is still better illustrated by the main band of limestone, which following the strike sweeps sharply round, the trend of the outcrop changing from N.N.W. to W.N.W. Within this again there are no bands of limestone to display the strucutre, which can, however, be clearly made out when the strikes of the granulite are mapped.

The main anticline in the area referred to may be called the Uduwela anticline from the Uduwela valley, across which its axis pases. The broad end opens out over the Uduwela valley towards Deltota ; the narrow end crosses the mahaweli-ganga valley at Peradeniya, and passes up the narrow valley beyond Gangaruwa towards Kobbekaduwa. A slight fault probably exists between Hantane and Matinapatana, at a point where the change of strike is most abrupt. The anticline is not so well shown on the map as the syncline, owing to the fact that the bands of limestone die out towards Kandy and Hantane ; nevertheless, there can be recognized outcrops at Kandy (near the Hindu temple), Getambe, and Peradeniya (first road bridge south of the railway station), which represent the continuation of the band of limestone which splits up and tails out at Ampitiya.

The existence of minor folds can often be demonstrated. For example, in the Hanguranketa valley, between the Dolagala and Diyatalawa hills, there at first appears to be merely a succession of beds dipping in one direction, the two hills mentioned having dip slopes to the south-west. Within the valley however minor folds can be detected, not interfering with the main structure ; the best example is an anticline with axis almost horizontal and directed N.N.W. and S.S.E., exposed about a mile above Hanguranketa at a bend in the main road.

Between Padiyapalella and Munwatta there occurs a clearly recognizable, though hardly a minor, anticline (this and the last-mentioned are both outside the limits of the map). This anticline can be studied from the slopes of Diyatalawa, and the view in this direction is remarkable for this peculiarity, that the contour of the surface of the country closely follows its internal strucutre. The outcrops of granulite above the limestone form a series of clearly defined small cliffs, following round the crest of the anticline which passes into the hill, and immediately on the west of this anticline the stream which falls over the cliff at Munwatta bungalow can be seen flowing in a valley, which is actually the most perfect and diagrammatic syncline imaginable. The foldings have almost everywhere determined the direction of streams, roads, and railways ; it sometimes happens too, as in the case just described, ridges however are practically never coincident with anticlinal axes, but are crests (like Hantane) with escarpments on the one hand and dip slopes on the other. Valleys are usually either the troughs between two successive crests of this kind or are determined by joints ; these two valley systems are naturally disposed at right angles to each other, and there are many examples illustrating this feature with diagrammatic precision.

It may now be remarked that the N.N.W. and S.S.E. trend, though very usual in

Ceylon, is not universal. Towards Trincomalee, for instance, the strike swings round to the N.E. ; between Bandarawla and Pelamadulla it is often nearly W. and E., but towards Ratnapura bends again to the north. Around Badulla the general strike is almost exactly N. and S., a disposition which in its northward continuation has determined the straight N. and S. course of the mahaweli-ganga below it junction with the badulla-oya. The typical strucutre and its effects on the contour of the country are probably best seen around Hatton, where the arrangement of strike elements in the strucutre thus described : (1) the Taprobanian folding, and (2) the N.N.W. inclination of the axes of the folds. Sufficient work has not been done to show whether the axes of the folds themselves indulate, so as to pitch now to the N.N.W. and now to the S.S.E., as is suspected to be the case. If so, there must have been foldings in a direction at right angles to the Taprobanian movements. To these secondary movements, whether merely a tilt in one direction or a series of actual folds, the name of Vijayan may be given, commemoration the name of the leader of the first Aryan settlers in Ceylon. How far these or other movements have affected the internal strucutre of the crystalline rocks may be a matter of dispute. I am inclined to think that original structures are often preserved ; there are nevertheless frequent evidences of strain and sometimes of fracture. The moonstone-bearing leptynite at Rajawela, for instance, shows strain and fracture of the large felspars, which are degraded to the form of augen and traversed by cracks and strings of mosaic. Large felspars, both in pegmatites and in the granulites, usually show strain. A peculiar feature often seen is the displacement of foliation planes in opposite directions on the opposite sides feature often seen is the displacement of foliation planes in opposite direction on the opposite sides of small pegmatite veins, as shown in the accompanying diagram ; the disturbance extends at the most

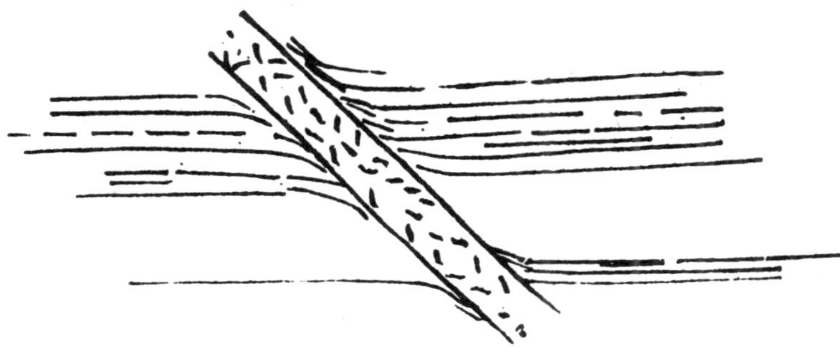

for a few inches on either side of the pegmatite, and is only observed when the pegmatite crosses the foliation at an oblique angle. In extreme cases, as at an exposure about 5 1/2 mile south-east of Balangoda, on a low rounded partly bare hill on the south-west side of the Balangoda-mahawalatenna road, not far from the point (Weligepola) where the steep descent commences, the results of more conspicuous shearing movement can be recognized. In this place the garnetiferous pyroxene granulites form smooth barebacked hills, scantily covered with xerophytic plants. There are to be seen very abundant granitic sills or pegmatites running parallel to the direction of the granulite foliation planes, as well as oblique thereto ; of the oblique veins, some are branches of the parallel ones, others cross these and are thus slightly alter. The parallel sills often fail to follow the foliation exactly, but cross it very gradually ; they are mostly narrow, say 6 inches to 8 inches, the largest 27 inches. It can be easily seen even in the parallel sills that the cleavage faces of the larger orthoclase individuals are strained and bent. In most of the oblique sills clearer evidence of movement is recognizable. Most of them coincide with zones of shearing extending for two or three inches on each side of the intrusion, accompanied by faulting on a small scale producing displacements of a few feet only. The shearing and displacement are liable to die out and come in very suddenly and capriciously. In extreme cases the foliation planes are bent aside as already described, and the pegmatite itself is mylonized and foliated in a idrection parallel with its length. Phenomena of this kind are very rare in Ceylon. Foliation planes of the granulites in Ceylon not infrequently show the presence of a good deal of brown mica, which appears to be of secondary origin developed by movements (exposures on emile below Wilson's Bungalow, exposures on Chetnole road at Balangoda, & c.)

E10 *Ceylon Administration Reports for 1906.* *[IV.-Education.*

On the other hand, delicate centric strucutes and graphite intergrowths, which are common features in microscopic sections, are almost certainly original structures. Perhaps the granulites were in a more or less viscous condition at the time of movement, some more, some less, so, and thus have suffered unequally. The limestones probably suffered deformation without disruption. With regard to the subject of flow, it may be remarked that the flow strucutre of the granulites should be distinguished from any foliation afterwards impressed. The flow strucutre is an original feature which is generally preserved. The direction of flow is generally to be determined from the streaks on the foliation planes ; if the direction of this flow be mapped, it may be learnt whether the flow strucutre is uniformly oriented independently of the disposition of the structural planes ; I am inclined to suppose that this may be the case, but have not done sufficient work to prove it.

Returning now to the map, reference may be made to the crystalline limestone. These have been the principal guide in unravelling the strucutre of the country. The most important band can be traced from beyond Wattegama to Haragama and round to Ampitiya, and, as suggested by means of isolated exposures, on the Peradeniya, a total distance of over twenty miles. The strucutre is here and there confused by the varying bulk of the masses of granulite separating the bands of limestone. There are also the minor folds, but these are not toublesome. It will be observed that practically no evidence of fauliting has been detected. There may be a small fault between Hantane and matinapatana, but I do not know of any visible or even proved fault in Ceylon with a throw of more than a few feet. The limestone bands do not die out suddenly, or taper away, but split up into narrow bands separated by wider and wider masses of granulite, and are thus divided up till they can be traced no further. There are met with further away long the line of strike lenticular masses, which may be regarded as the continuation of the bands which could no longer be followed.

(15) Bibliography of Ceylon Geology

Arsandaux, H.

"Analysen einiger Mineralien" (Forsterite from kandy). Bull. Soc. France. Min., XXIV., p. 475 (1901).

it Baker, Sir S.:

"Eight Years in Ceylon", London, 1855.

Bauer, Max.:

(1) "Durchsichtiger blauer Spinell von Ceylon", Neues Jahrb. 1895, p.281.
(2) "Chrysoberyllkrystall von Ceylon", *Ibid.,* 1897, p.106

Bellamy, C.V.:

"Geology of the District of Jaffna". Monthly Literary Register and Notes and Queries for Ceylon, III. (1895); and Vanni District manual, p.8.

Bournon, Count de:

(1) "Observations sur quelques-uns des Mineraux, soit de I'Isle de Ceylan, soit de la Cote de Coromandel, rapportes part M. Leschenault de la Tour", Paris, 1823.

(2) "Description of certain gangues of Spinelle brought from the Island of Ceylon". Phil. Mag. LXIII., pp. 30-36 (1824).

Brodie, A.O.:

"Topographical Account of the District of Nuwarakalawiya", Journ. roy, Asiatic Soc., Ceylon, III., pp. 150-180.

Busz, K.:

"Mittheilungen aus dem Mineralogischen Museum der Universitat Bonn". Zeit. fur Kryst., XV., p 616 (1882).

Ceylon:

By an Officer of the late Ceylon Rifles. London, 1876.

Church, A.H.:

(1) (Notice of) "Analyses of Ceylon Apatite". Min. Mag., 1899, p.14.
(2) "Precious Stones", London, 1905.
(3) "Nature", LXIII., p.464 (1901) (Letter).

Coomaraswamy, A.K.

1. "Ceylon Rocks and Graphite", Quart. Journ. Geol. Soc., LVI., pp.590-614 (1900).
2. "The Crystalline Limestones of Ceylon". *Ibid.*, LVIII., pp.399-422 (1902).

3. "Origin of the Crystalline Limestones of Ceylon". Geol. Mag., 4th decade, IX., pp.
4. "The point de Galle Group (Ceylon); Wollastonite-Scapolite Gneisses". Quart. Journ. Geol. Soc. LVIII., pp.680-689 (1902).
5. "Note on the Scenery of Ceylon". Rep. Brit. Ass. Belfast, pp.613, 614 (1903).
6. "Occurrence of Corundum *in situ* near Kandy". Geol. Mag. 4th decade, X., pp.348-350 (1903).
7. "The Crystalline Rocks of Ceylon", spolia Zeylanica, L. pp. 105-111 (1904).
8. "Uraninite" (Thorianite). *Ibid.*, pp.112, 113 (1904).
9. "Silicification of Crystalline Limestones". Geol. Mag., 5th decade.
10. "Mineral Resources of Ceylon". Official Handbook, Ceylon Court, St. Louis World"s Fair, pp.146-152, Colombo (1904).
11. "The Balangoda Group", Geol. Mag., 5th decade, I., pp.418-422 (1904).
12. "Administration Report, Mineralogical Survey of Ceylon, 1903". Colombo (1904).
13. "Report on Thorianite and Thorite". Colombo (1904).

Science and Ard.] *Mineralogical Survey.* *E11*

14. "The Rocks and Minerals of Ceylon". Spolia Zeylanica, III., pp.50-56, and in "Guide to the Colombo Museum". Colombo (1905).
15. "Report on the Occurrence of Cassiterite in Ceylon". Colombo (1905). Includes reports by James Parsons and W.R. Dunstan.
16. "Recent Marine Clays at Kuchavelli, Ceylon". Geol. Mag., 5th decade, pp.508, 509 (1905).
17. "Minerals New of Rare in Ceylon". Spolia Zeylanica, III., pp.198,199.
18. "Administration Report, Mineralogical Survey of Ceylon, 1904". colombo (1905). (Portions by James Parsons.)
19. "Administration Report, Mineralogical Survey of Ceylon, 1905". Colombo (1906). (Portions by James Parsons.)

Cordiner, J.:

"A Description of Ceylon, containing an Account of the Country, Inhabitants, and Natural Productions, with Narratives of a Tour round the Island in 1800", London, 1807.

Crook, T., and Jone, B.M.:

"Geikielite and the Ferro-magnesian Titanates", Min. Mag., XIV., pp.160-166 (1906).

Davy, John:

(1) "Description of certain Rocks in the South of Ceylon", Quart. Journ. Sci., V., pp.233-235 (1818).
(2) "On the Geology and Mineralogy of Ceylon". Geol. Trans., V., pp.311-327 (1821).
(3) "Travels in Ceylon", London (1821).
(4) "Sur les Sources Chaudes de Ceylan". Ann. de Chemie, XXIII., pp.269-272 (1823).

Dick, A.:

"On Geikielite, a New Mineral from Ceylon", Min. Mag., X., pp.148-160 (1892).

Diersche, Max.:

"Beitrag zur Kenntniss der Gesteine und Graphite-vorkomnisse Ceylons", Jahrb. d.k.-k. Reichanstalt, XLVIII., pp.231-288 (1898).

Dixon G.G.:

"Report on Gold Prospecting in Ceylon", Colombo (1903).

Donath, E.:

"Der Graphit". Leipzig, 1904.

Dunstan; W.R.:

(1) "The Occurrence of Thorium in Ceylon", Naure, March, 1904, pp.510,511.
(2) "Occurrence of Thorium in Ceylon", Bull. Imperial Institute, II., p.13 (1904).
(3) "Report on the Occurrence of thorium-bearing Minerals in Ceylon". Colombo (1904).
(4) "Report on the Results of the Mineral Survey (Ceylon) in 1903-1904." colonial Reports. London (1905).
(5) "Report on a Consignment of Mica forwarded by the Government of Ceylon" (with Note by A.K. Coomaraswamy), *Ceylon Government Gazette* No.6,057, pp.457-459 (1905).

Dunstan, W.R., and Blake, G.S.:

"Thorianite, a New Mineral from Ceylon". Proc. Roy. Soc., LXXVI., pp.253-265 (1905).

Dunstan, W.R., and Jones, B.M.

Proc. Roy. Soc., 1906

Ferguson, A.M.:

(1) "Plumbago, with Special Reference to the Position occupied by the Mineral in the Commerce of Ceylon, and the Question discussed of the alleged existence in the Island of the Allied Substance Anthracite". Journ. Roy. Asiatic Soc., Ceylon, X.,pp.171-266 (1885).
(2) "gold, Gems, and Pearls in Ceylon". Colombo, 1888.

Fletcher, L.:

"On Baddeleyite (Native Zirconia), a New Mineral from Rakwana, Ceylon". Min. Mag., X., pp.148-160 (1892).

Foote, R.B.:

"Translation of Walther, I. Rec. Geol. Surv., India, XXIV., pp.42-45 (1891).

Gardner, George:

"Sketch of the Geology of Ceylon in Lee's Ribevro's History of Ceylon", pp.203-209. Colombo, 1847.

Gmelin, C.G.:

(1) "Analysis of Cinnamon Stone from Ceylon". Edin. Phil. Journ., XI., pp.129-132 (1824).
(2) "Analysis of a Black Mineral from Candy in Ceylon, named Candite", *Ibid.*, IX., pp.384-387(1823); and Ann. de Chemie, XXV., pp.208,209 (1824).

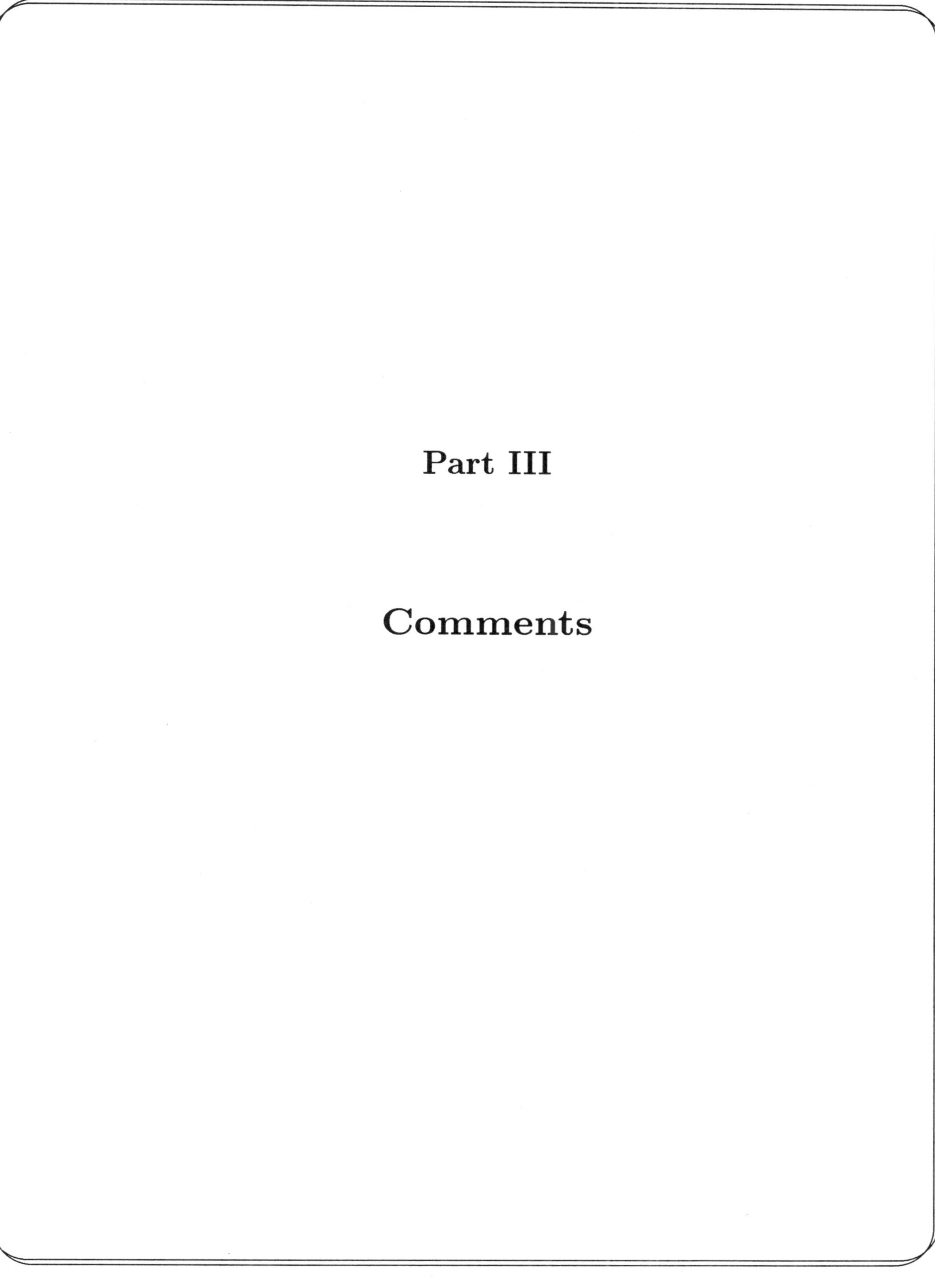

Part III

Comments

510 Nature [March 30, 1904

The Occurrence of Thorium in Ceylon

Letters to the Editor, Nature 69, March 31, 1904, p.510.

The Government of Ceylon determined last year to carry out, with the cooperation of the scientific and technical department of the Imperial Institute, a systematic survey of the economic minerals of Ceylon. Mr. A. K. Coomaraswamy and Mr. H. G. Pearsons were selected to conduct the survey in Ceylon, and to despatch the specimens of the minerals found to the Imperial Institute for chemical examination and commercial valuation. Among the specimens thus received were those of a mineral existing in smalll black cubical crystals found in the refuse washings near Balangoda, in the Sabaragamuwa Province, which had been identified by Mr. Holland, a resident in Ceylon, as probably uraninite or pitchblende. The same mineral has been since observed by Mr. Coomaraswamy in a vein of pegmatite at Gampola, in the Central Province of Ceylon.

The specific gravity of the mineral was found to be 9.32, and an analysis by Mr. G. S. Blake, of the scientific staff of the imperial Institute, furnished the following results :-

		Per cent.
Thorium oxide	ThO_2	76.22
Cerium oxide	CeO_2	8.04
Lanthanum and didymium oxide ...	La_2O_3 Di_2O_3	
Zirconium oxide	ZrO_2	trace
Uranium oxide	UO_3	12.33
Ferric oxide	Fe_2O_3	0.35
Lead oxide	PbO	2.87
Silica	SiO_2	0.12
		99.93

The mineral is clearly not pitchblende, since the percentage of oxide of uranium is only about 12 per cent., whilst the principal constituent is oxide of thorium (thoria), which is present to the extent of more than 75 per cent., an amount far higher than that contained in any mineral hitherto examined. This mineral appears to be new, and I suggest for it the name of *thorianite*. Since it is radioactive, it will no doubt be found to be an important source of radium or radio-active earths, and will probably furnish helium, points which will be investigated as soon as more material has been obtained.

A second part of the same specimen furnished the following results on analysis :–

		Per cent.
Thorium oxide	ThO_2	76.22
Cerium oxide	CeO_2	6.39
Lanthanum and diymium oxide ...	La_2O_3 Di_2O_3	0.51
Zirconium oxide	ZrO_2	3.68
Uranium oxide	UO_3	11.19
Ferric oxide	Fe_2O_3	1.92
Lead oxide	PbO	2.25
Silica	SiO_2	1.34
Insoluble residue		0.41
		99.93
Specific gravity		8.98

The two sets of analytical data prove that the material has essentially a uniform composition, the differences observed being apparently due to inclusions of zircon in the second portion analysed.

In the meantime Sir Wiliam Crookes has received a specimen of the supposed pitchblende from Ceylon, and has found it to be radio-active to about the same extent as Cornish pitchblende.

The second mineral examined was found by Mr. Holland in the same gen washings at Balangoda, and was identified as probably monazite. This mineral was pale brown, and when fractured exhibited a purple brown interior with a

resinous lustre. The specific gravity was 4.98. An anlysis by Mr. Blake furnished the following results :–

		Per cent.
Thorium oxide	ThO_2	66.26
Cerium oxide (and Cerium earths)...	CeO_2	7.18
Zirconium oxide	ZrO_2	2.23
Uranium oxide	UO_3	0.46
Ferric oxide	Fe_2O_3	1.71
Lead oxide	PbO	0.35
Phosphoric oxide	P_2O_5	1.20
Silica	SiO_2	14.10
Water	H_2O	6.40
		99.89

This mineral is therefore thorite, consisting chiefly of thorium silicate. Both these minerals are under further investigation at the Imperial Institute. Careful explorations are now being made as to the extent of their occurrence in Ceylon.

It is obvious that apart from the scientific interest attaching to the determination of their composition, the discovery in Ceylon of two minerals rich in thoria, now so largely employed for the manufacture of incandescent gas mantles, may be of considerable commercial importance.

Imperial Institute, March 29. **WYNDHAM DUNSTAN.**

A New Mineral from Ceylon

Letters to the Editor, Nature 69, April 7, 1904, p.533.

In the beginning of February I bought from Mr. Holland 5 cwt. of the mineral described by Prof. Dunstan in last week's NATURE (p. 510). It crystallises in cubes, and the density is substantially that found by him. Mr. Tyrer, of the Stirling Chemical Works, Stratford, was so kind as to promise to work it up for me, and the process is still being carried on.

I had hoped to have positive and definite results to communicate before describing its constituents, but the publication by Prof. Dunstan of an analysis, and his statement that he is still engaged in its investigation, makes it necessary to write this letter.

The mineral, when heated alone, gives of 3.5 cubic centimetres of helium per gram; fused with hydrogen potassium sulphate, the amount is increased to 9.5 cubic centimetres. From this source I have already stored about 12 cubic feet of pure helium extracted in Mr. Tyrer's works.

It was at first believed that the mineral was rich in uranium, but different specimens contain only from 8 to 12 per cent. of that element, agreeing in this respect with the analyses published by Prof. Dunstan. Next, the other main constituent was believed to be zirconium, but the high density of the mineral rendered this improbable. An analyst of high standing, whose daily business is to analyse minerals of this kind, returned 82 per cent. of zirconia as a constituent; the percentage of thorium was trifling-under 1 per cent. The mineral contains particularly no thorium; this has been repeatedly confirmed in my laboratory. Nor does it contain any appreciable amount of cerium, lanthanum and didymium. The oxalate is almost completely soluble in excess of ammonium oxalate-a reaction which excludes thorium and the cerium group, but which points to zirconium. The equivalent of the elements of the oxalate group, which I at first took for zirconium, excludes the presence of any large quantity of zirconium, although that element is undoubtedly present. Fractionation shows that the oxalate precipitate (the portion soluble in ammonium oxalate) gives equivalents between 250 (the most insoluble portion of the double sulphate) and 447 (the most soluble portion); by far the major part of the element has the last mentioned equivalent. The separation of this portion is now being carried out with large quantities of material; several hundredweights are being worked up.

Assuming that the element is a tetrad, which is probable from its behaviour, it undoubtedly possesses an equivalent approaching the highest number(44.7), and for this there is a gap in the periodic table between cerium and thorium; one at least of the elements present (supposing that there is more than one present) will probably have an atomic weight of about 177, preceding tantalum (182.5) in the horizontal row of the periodic table.

I am at present engaged in mapping the spectrum of this new body or bodies.

As for radio-activity, the mineral was bought in the hope that it would have a high content of radium. There is a trace of radium present, due to no doubt, to the spontaneous change of the uranium which the mineral contains. But the radio-activity due to this source is certainly not 5 per cent. of the total.

The period of decay of the emanation appears to point to the presence of a radio-active element closely resembling thorium X. The half value is 50 or 51 seconds, and while this is not quite the time for the decay of thorium emanation, it very nearly approaches it; at present the balance of evidence appears to point to the presence of an element closely resembling thorium, but not identical with it. The total radio-activity, moreover, is much greater than can be accounted for by the supposition that the one consists of pure thoria. Within the limits of a letter I am obliged to omit many more characteristics of this curious ore which have to be ascertained, but I hope soon to be able to publish

more definite results; as it is, I regret to have been obliged to tell an imperfect story.

I should like to conclude by acknowledging the great assistance given me in this work by Mr. Tyrer and by my students, Messrs. Gimmingham and Les Rossignol.

University College, April 4. **WILLIAM RAMSAY.**

A New Mineral from Ceylon

Letters to the Editor, Nature 69, April 14, 1904, p.559.

Since writing last week, I have made further experiments on the cubical mineral, and have myself carefully examined the earth constituents. The statement made last week, that there is only an insignificant amount of Thorium present, must be modified. On re-determining the equivalent of the crude oxalate, prepared after the yttrium metals had been separated by treatment with potassium sulphate, it has come out higher than I expexted; indeed, assuming the metal present to be a tetrad, its atomic weight is even higher than that of thorium-about 240, as the mean of two closely concordant determinations. The lower equivalents mentioned in the previous letter were determined as fractions of the double potassium sulphate, prepared on a large scale. This high atomic weight points to the presence of unknown elements of higher atomic weight than thorium; indeed, the mineral appears to be of very complex composition. It may be incidentally remarked that the crude oxalate mentioned above must have contained all the cerium group, and if any considerable proportion of the elements of this group is present, the amount of the element with higher atomic weight than that of thorium would have to be proportionately increased. The high radio-activity would point to the presence of the elements obtained from thorium residues mentioned by Prof. Baskerville, which he states to be radio-active.

The equivalent was determined by comparing the weight of oxide from a known weight of oxalate with the percentage of oxalic acid, as determined by titration of another sample of the same preparation.

WILLIAM RAMSAY.

The letter dealing with the composition of a new mineral from Ceylon contributed by Sir W. Ramsay to NATURE of April 7 (p. 533) reveals certain disrepancies between the analytical results obtained with this material at University College and those of the Scientific and Technical Department of the Imperial Institute recorded in Prof. Dunstan's letter on this subject (March 31, p. 510). Sir W. Ramsay's results indicate that this mineral is practically free from thoria,whereas those recorded by Prof. Dunstan show that it is particularly rich in this oxide. As Prof. Dunstan is at present abroad, and therefore unable at the moment to comment on Sir W. Ramsay's letter, I may be permitted to direct attention to two observations mentioned by Sir W. Ramsay, which appear to be open to question.

He states that the oxalate obtained from a solution of the mineral is soluble in excess of a solution of ammonium oxalate, and that this reaction excludes the points to the presence of zirconium. This inference is not in harmomy with with the observation recorded by Bahr (Annalen, 1864, 132, 231), that thorium oxalate is soluble in excess of ammonium oxalate, a fact since confirmed by Bunsen and by Brauner (Journ. Chem. Soc., 1898, 73, 951). Further, the solubility of the thorium salt in excess of ammonium oxalate has been used by Hintz and Weber (Zeit. Anal. Chem., 1897, 39, 27) and by Glaser (ibid., p.213) as a method of separating thoria from monazite and similar minerals. It would appear, therefore, that the principal evidence brought forward by Sir W. Ramsay in support of his conclusion that the mineral contains no thoria in reality supports Prof. Dunstan's statement that it is rich in this oxide. It may be added that the solubility of the oxalate obtained from the mineral in ammonium oxalate had already been observed in this department.

Sir W. Ramsay appears to be of the opinion that the principal constituent of the mineral is the oxide of a new tetravalent element with an equivalent weight of 447. If this were the case the specific gravity of the mineral would probably be less than 8.2, whereas the determinations of this constant

560 *Nature* *[April 14, 1904*

made here and at University College indicate that its specific gravity is about 9, and this figure agrees well with that required for a mineral containing 75 per cent of thoria.

T. A. HENRY

Scientific and Technical Department,
Imperial Institute, S.W., April 11.

253

Thorianite, a New Mineral from Ceylon.

By WYDENHAM R. DUNSTAN, M.A.,LL.D., F.R.S.,, and G.S. BLAKE, A.R.S.M., Assistant in the Scientific and Technical Department of the Imperial Institute.

(Received May 1, – Read May 18, 1905.)

The mineral which forms the subject of this paper was collected in Ceylon during the progress of the mineral survey of the island, which was commenced in 1903 under Professor Dunstan's supervision, with the principal object of determining the extent to which economic minerals, such as graphite, mica, etc., occur, and if possible of discovering other minerals of commercial importance. The minerals collected by Mr. A.K. Coomaraswamy and Mr. James Parsons, the officers entrusted with the surveys in Ceylon, are submitted to examination and analysis in the Scientific and Technical Department of the Imperial Institute, and are subsequently subjected to such technical trials as may be necessary in order to ascertain their precise uses and to determine their value.

Among the materials thus received from Ceylon at the Imerial Institute was a small quantity of a heavy black mineral occurring chiefly in small roughly cubical crystals. This mineral was, in the first instance, furnished to the officers of the mineral survey by Mr. W.D. Holland, who believed it to be uraninite or pitchblende. He had previously sent specimens to several persons in this country under this name.

A chemical examination of the small specimen recieved at the Imperical Institute showed, however, that the mineral contained a large proportion of thorium in the form of the dioxide (thoria) and only a small portion of uranium. It was evident that this was a new mineral chiefly composed of uncombined thoria (ThO_2). Pending the arrival of more material to enable a further examination of its properties to be made, a preliminary account of its composition and properties was communicated by one of us to 'Nature'*, and the name *thorianite* was suggested for the substance, which contained at least 75 percent of thoria.

It was also stated that the mineral was radio-active and that it apparently contained helium, points which would be made the subject of further investigation.

The publication of these results was followed by a communication to 'Nature'† from Sir W. Ramsay, who announced that he had purchased 6 cwt.

* March 30, 1904, p.510.

† April 7, 1904, p.533.

of the mineral from Mr. Holland some months previously and had been engaged in its examination. He was unable to confirm the statement that it contained thoria, but announced that it had furnished considerable qunatities of helium. In a second communication to 'Nature'* Sir W. Ramsay modified the statement he had previously made that the mineral contained practically no thorium.

Further small supplies of the mineral having since been received at the Imperial Institute, we have been enabled to confirm the conclusion first arrived at that the substance is chiefly composed of thoria, and we now desire to bring before the Royal Society an account of this remarkable new mineral for which the name thorianite is apporpriate.

A general account of the composition, properties, and uses of thorianite is included in Professor Dunstan's official 'Reports on the Results of the Mineral Survey in Ceylon, 1903-4', which was issued as a Parliamentary Paper (Cd. 2341) in January, 1905.

Occurrence of Thorianite in Ceylon.

The following account of the occurrence of the mineral, at first supposed to be "uraninite," has been given in official reports by Mr. A.K. Coomaraswamy, B.Sc., the Director of the Mineral Survey of Ceylon.

"Mr. Holland of Dikmukulana has for long taken an interest in local mineralogy, amongst other things obtaining specimens of 'nampu' (gem-bearing gravels) from as many localities as possible by offering a small reward to the native 'gemmers' who bring it. It is difficult to persuade the native that gems are not required or to get him to reveal the true source of 'nampu'. These difficulties overcome, the examination of 'nampus' is an ideal method of gaining a knowledge of heavy minerals of any district. Early in 1903 Mr. Holland obtained 'uraninite' amongst his samples and was gradually able to get as much as 6 cwt. brought to him along with other stuff. This material was sent to England and sold to Sir W. Ramsay at the rate of £50 for 6 cwt. Subsequnetly Mr. Holland induced a native to show him the source of the material, and finding that some quantity occurred, he took out a prospecting license in the full belief that the locality was Crown land.

"Thorianite, together with a mineral regarded as thorite, and a number of other heavy minerals, is found near Kondurugla, Bambarabotuwa, Province of Sabaragamuwa. The principal deposit occurs in and near the bed of the upper part of the Kuda Pandi-oya, a small stream, which at first occupying a small north-west and south-east strike valley north of the Hopewell-Hapugastenna bridle path, turns through nearly a right angle and joined the Maha Pandi-oya, and in a small 'deniya' swamp just below the swamp, the thorianite is to be obtained in considerable abundance. It could not be discovered *in situ*, but it can hardly be doubted that it is derived from some rock outcropping not far distant from the highest part of the little

* April 14, 1904, p. 559.

stream; the stream is so small that after a few days of drought (not very usual in this wet district) no running water is met with above the camp. The matrix is no doubt a rock of granitic type similar to those containing zircon, allanite, etc., which have been met with elsewhere in the Balangoda district, intrusive in the Charnockite series, and classed as belonging to the Balangoda group. Of these rocks the largest exposure (a narrow lenticular mass two miles or more in length) is that of zircon granite on Massena estate (six miles);* the rock consists of felspars, quartz, biotite, zircon, and illeminite; smaller exposures of zircon granite are found on Hermitigla (eight miles) and opewell estates (15 miles); at Denegama bridge, on the road between Balangoda and Belilhul-oya, and near the 91st milepost on the same road. There are also similar granites without accessory minerals exposed – *e.g.*, on the main road about a mile below Balangoda. The allanite-granite, or pegamatite, is best seen on Lower Benegama estate, where it occurs as a dyke 3 or 4 feet thick, forming a conspicuous ledge in the bed of the stream which runs through that part of the estate. The rock is composed of felspars (chiefly orthoclase), quartz, biotite, and allanite. The allanite forms thin tubular idiomorphic and larger irregular crystals, the biggest having a greatest diameter of 3 inches. Almost identical rocks occur (1) in the Weweldola on Dikmuklana estate (11 miles), and (2) at Weligepola (9 miles). It is important to observe that a pegamatite rock, composed mainly of pink orthoclase, quartz, and biotite, with accessory apatite, tourmaline, ilmeite, etc., has been observed on Ambalawa estate, Gampola,† containing, in addition to the above-mentioned minerals, a few cubic crystals of a black mineral at first regarded as uraninite, but which is almost certainly thorianite.

"To return to the Kondurugla locality, the thorianite is probably derived from a granitic rock with no very large outcrop, but in which it occurs in considerable abundance; perhaps together with the zircon, thorite and ilmenite, which quite possibly, however, occur separately in other rocks of a similar character. The outcrop of this rock could not be discovered owing to the dense jungle and thick soil and landslips; rock (mainly decomposed granulite) is indeed exposed at several points in the beds of the Kuda Pandi-oya and of the small streams joining it, but any search for the outcrop of a particular rock in the adjoining jungle could only be expected to succeed at the cost of a large expenditure of time and money, and might even then result in failure.

"There is a large area, including at least the whole of Sabaragamuwa and parts of the central, western, and southern provinces, wherein this or other rare heavy minerals may be looked for.

"It is unlikely that any very extensive deposit of any of these rare heavy minerals will be found, but they may be expected to occur at vaious points in moderate amounts. In the Kuda Pandi-oya valley a total of 5 tons might, perhaps, be obtained; from the Alupola-dola or the Kuda-oya I doubt if half a ton could be profitably extracted, but these estimates are quite uncertain. The Gampola occurrence is quite without commercial value. The total amount actually obtained in Bambarabotuwa, so far, does not exceed 15 cwts. In the

* The distances are quoted from Balangoda.
† 'Spolia Zeylancia,' vol.1, part iv, 1904.

immediate neighbourhood of Kondurugala the Walaweduwa jungle seems the most favourable distict for investigation. An examination of the Ratganga valley towards the west showed that thorianite is absent there, and that even zircon is very scarce; but on the north-east one or two heavy minerals not yet determined were actually obtained in the southern part of the Walawaduwa Crown Forest Reserve, and further work on that side of Kondurugala might be useful. It must be mentioned, however, that the district is a very wet one, the jungles swarming with leeches; it is also very inaccessible, provisions and camping effects having to be carried fully 20 miles from Balangoda. Camping in these jungles is abosolutely useless except in fine weather; and in the absence of a detailed map observations cannot be very accurately set down.

"Since August, 1904, a small deposit of the mineral provisionally identified as thorite was discovered at Duruyakanda, South of Gilimale, about six miles from Ratnapura.

"A further amount of 1,200 lb. of thorianite has since been obtained from Bambarabotuwa."

Description of Thorianite.

Thorianite, as it occurs naturally, is often associated with other minerals, and is not easy to obtain in a completely separate condition.

Dr. J.W.Evans has made a preliminary examination of the crystallographic characters, and intends to further study this subject. It may be stated now that the mineral occurs in small cube-like crystals up to nearly a centimetre in diameter, the largest, so far seen, measuring rather more than 8 mm. In most specimens the colour is a dull grey or slightly brownish black, but those crystals which have not suffered from attrition in the streams are a jet black with a bright resinous or pitchy lustre. The difference may be attributed partly to the grinding of the surface, and partly to superficial chemical changes. The mineral thus differs from uraninite or pitchblende which usually occurs massive, and not obviously crystalline. The streak furnished by uraninite is brown with a tinge of green.

By transmitted light the mineral is opaque except in thin sections.

The double refraction is very low. The refractive index probably exceeds 1.8.

The only faces that are ordinarily developed on the crystals of thorianite are apparently those of the cube. These have a very uneven surface, especially in specimens that have not suffered much from attrition or alteration. This charcter of the surface is mainly the result of the development of a number of small vicinal faces. In some cases larger faces of similar character are present, meeting at very obtuse re-entrant or salient angles, and reminding one of those seen in some crystals of flourspar. In other crystals the faces show a more or less irregular curvauture.

The indefinite character of the surface prevents any accurate determination of the angles. The goniometric readings vary 2 or 3 degrees on either side of 90 degres, according to the portions of the surface from which the image is reflected, but there is in most cases no satisfactory evidence that the faces as a whole are not virtually at right angles. Some crystals, however, appear to be distorted and have angles differing from a right anlge by as much as 5 degrees or even more.

Occasionally twin crystals are met with and these are of some interest. They are interpenetrant twins on a face of the octahedron as twinning plane, and one of the diagonals of the cube as twinning axis, exactly simialr to those of flourspar. As in the case of that mineral the coigns of one cube project as pyramids on an isosceles triangular base on the face of the other. Sometimes the compound form is almost completely regular, four edges of each cube meeting at one point or approximately so at both ends of the common diagonal that forms the twinning axis; at other times the composition is more irregular, and a number of coigns project from the faces of a cube in such a manner that though the faces of one coign are parallel to those of another, they are not in the same plane. In these forms the twinning axis and plane are the same for all the coigns, even when there is more that one projecting from the same face, the more acute point of the pyramid pointing to the coign of the cube from which the twinning axis emerges. The diagonal which constitutes this axis has therefore crystallographic characaters different from those of the other three diagonals, about which this twinning cannot apparently take place, and the symmetry of the crystal must be considered as rather rhombohedral than cubic, although the angles are apparently right angles. The same considerations would apply to fluorspar, in which case the essentially rhombohedral character of its symmetry is confirmed by the occurrence of crystals in which only those faces of the four-faced cube or tetrakis hexahedron $\{310\}$ are developed, in which the finite intercepts are of opposite signs. These faces may be represented by the symbol $\{3\bar{1}0\}$, and are those which bevel the edges that do not pass through the coigns at the ends of the unique diagonal. They together form a scalenohedron, which is in fact the scalenohedron $\{13\bar{4}2\}$ of the rhombohedral system, assuming the cube as the fundamental rhomboderon.*

A similar scanlenohedron $\{2\bar{1}0\}$ or $\{12\bar{3}1\}$ is repeatedly met with in crystlas of halite.†

*'Lehrbuch der reinen und angewandten Krystallographie,' 1830, p.178, fig. 572.

† F. von Kobell, "Uber Merkwürdige Krystalle von Steinsalz,"Journal für praktische Chemie', 1861, vol. 84, p. 420; K. Andrée, "Über Steinsalz Krystalle von hexagonal - rhomboëdrischer Pseudosymmetrie aus Sicilien,"'Centralb. für Mineralogie,' etc., 1904, p.88.

It is interesting to notice that chabzite, which is cubic in general appearance, but really rhombohedral, having its angles differing from right angles by nearly 5 degrees, twins in exactly the same manner as flourspar and thorianite.

The inference that thorianite is essentially rhombohedral in character is confirmed by the observation that in sections cut perpendicular to the twinning axis, the substance is practically isotropic.

Thorianite shows no definite clevage, but the mineral is traversed by irregular cracks which appear to follow the direction of the basal plane more frequently than any other. On fracture, it shows an irregular surface, which is more or less conchoidal on a small scale.

The hardness of the mineral is nearly 7, which distinguishes it at once from uraninite with hardness of 5.5.

Under the blow-pipe thorianite is infusible. It decrepitates, and if raised to a sufficienlty high temperature, is highly incandescent.

It is sometimes associated in rolled fragments with a smooth, yellow-brown, apparently amorphous material, of hardness 6, which envelopes it or forms a rounded deposit on its faces. This substance is reddish-yellow in colour when viewed in thin sections by transmitted light. Zircon also sometimes occurs intergrown with thorianite.

The density of different specimens of thorianite varies between 8 to 9.5 and 9.7. The higher numbers probably represent the density of the actual mineral, which in some cases exhibits cavities partially filled with a yellow ocherous material and also inclusions of zircon, one of the minerals generally associated with thorianite, and these associated mineranls, especially those which contain thorium will be investigated if a sufficinet quantity can be obtained separated from thorianite.

The mineral is easily powedered and then dissolves readily in strong nitric acid or in duluted sulphuric acid, with evolution of a gas which is chiefly helium. Thorianite is scarcely attacked by hydrochloric acid.

Thorianite is highly radio-active, and, in fact, may prove to be one of the most radio-active of minerals. The cause of this radio-activity is referred to below.

Composition of Thorianite.

The methods used in determining the composition of thorinaite are founded on those suggested by Glaser*, Meyer and Marckwald†, Fresenius and Hintze‡, and Benz*.

* 'Zeit. Anal. Chem.,' 1897, vol. 36, p. 213.

† 'Ber.,' 1900, vol. 33, p.3003.

** 'Zeits. Anaal. Chem.,' vol. 35, p. 343.

* 'Zeits. Ang. Chem.,' 1902, p. 297.

The results are shown in the table on p.261, which includes the data obtained from three separate specimens, numbered I, II, and III.

In the estimation of the metals, 2 grammes of the finesly powdered mineral were dissolved in about 15 c.c. of nitric acid of specific gravity 1.4, and after decomposition was complete the solution was diluted and filtered. The insoluble residues in specimens II and III nwere very small, and were treated with hydrofluoric acid to estimate silica. In I the residue chiefly consisted of zircon which was fused with potassium hydrogen sulphate to extract zirconia, and the residue treated as before with hydrofluoric acid to estimate silica. Except for this associated zircon, no zirconia was found.

The acid filtrate from the insoluble residue was diluted to about 300 c.c. and 5 c.c. of hydrochloric acid added. Hydrogen sulphide was then passed through the liquid to precipitate lead which was finally weighed as sulphate. The filtrate was boiled to remove hydrogen sulphide, and oxidation of the last traces of this substance effected with bromine water. To the hot acid solution, amounting to about 350 c.c., excess of amonium oxalate was added, and the precipitate of oxalates of thorium and cerium, etc., allowed to settle overnight and then filtered. The filtrate was evaporated to dryness and treated with nitric acid to destroy oxalic acid, and diluted with water. The calcium and magnesium were separated from other metals by precipitating the latter with ammonia and ammonium chloride. This precipitate was dissolved in hydrochloric acid, excess of acid neutralised, and the solution diluted to 500 c.c. A few drops of solution of sodium sulphite were added, and the liquid boiled to precipitate any titanic acid. No titanium was, however, present. The liquid was evaporated, and the iron oxidised with a few drops of nitric acid. The liquid was neutralised with a few drops of ammonia, and finally excess of ammonium carbonate added. The carbonates first precipitated were re-dissolved on further addition of the reagents, showing the absence of alumina. The iron was separated as sulphide. The filtrate was boiled, acidified with hydrochloric acid, and again boiled to remove all the carbon dioxide. The uranium in solution was precipitated with ammonia, and finally weighed as uranoso-uranic oxide U_3O_8.

The precipitate containing the oxalates of thorium and cerium, etc., was dried, and the oxalates were then decomposed by nitric acid, and the metals obtained in solution as nitrates. After diluting to 250 c.c., sodium thiosulphate was added to the slightly acid boiling liquid till no further precipitate of thorium salt was obtained. The liquid was boiled for a short time, the precipitate filtered off and dissolved in hydrochloric acid, and the

operation twice repeated to completely separate the cerium. To the united filtrates ammonia was added, and the precipitate of hydroxides of the cerium, etc., containing a little thorium and any yttrium present, was dissolved in hydrochloric acid, and the trace of thorium precipitated as above described from a small bulb of liquid. The thorium precipitated were dissolved in hydrochloric acid, and the thorium re-precipititated as hydroxide and finally weighed as the dioxide ThO_2. The filtrate from precipitation of the cerium earths with ammonia contains lime if such were present in the original mineral. This was separated as oxalate and weighed as oxide.

Determinations of the equivalent of thorium made with the precipitated dioxide obtained in the course of the analysis of the third specimen (see p. 261) by conversion into the sulphate gave 57.25, corresponding with an atomic weight of 229. The accepted atomic weight of thorium, relative to hydrogen taken as 1, is 230.8.

The solution from which thorium had been thus removed, containing cerium and the associated earths, including any yttrium, was treated with ammonia; the precipitate dissolved in dilute sulphurie acid, and the solution, after neutralisation, saturated with potassium sulphate. No yttrium was found in the filtrate by dilution and addition ammonia.

The double sulphates of potassium with cerium, and of the associated metals lanthanum and didydium, were warmed with ammonia solution, by which means the earths were obtained as hydroxides. These were dissolved in hydrochloric acid and re-precipitated by potash solution. The hydroxides were washed by decantation, a little potash solution added, and chlorine gas passed until the liquid was saturated. The lemon-coloured hydrated ceric oxide was filtered off, re-dissoved, and re-precipitated as hydroxide, and the cerium weighed as the dioxide CeO2. The filtrate containing lanthanum and so-called didymium was acidified with hydrocholoric acid, boiled to remove cholorine, and the earths precipitated with ammonia and weighed as the oxides.

The uranous oxide (UO_2) was separately estimated by Hillebrand's method as follows:— One or 2 grammes of finely powdered mineral were introduced into a stout tube together with 20 to 30 e.c. of dilute sulphuric acid, consisting of one part of acid to five parts of water. The air was then displaced by carbon dioxide, the tube sealed off, and then heated to 180° C. for several hours till decomposition was complete. The solution obtained was diluted with recently boiled water, and the uranous sulphate titrated with potassium permanganate.

The method adopted for the determination of helium and associated gases consisted in decomposing the mineral by means of dilute suphuric acid

consisting of one part of acid to five parts of water.* Ten grammes of the powdered mineral were used, and the gas collected in a gas burette over mercury with the usual precautions. The mixture was heated at 100°. Decomposition at this temperature was almost complete one day, but the experiment was allowed to continue for two days to ensure the liberation of all the gas.

The gas collected amounted to 105 c.c. at standard temperature and pressure, which corresponds to 10.5 c.c. of gas, chiefly helium, from one gramme of thorianite.

Analyses of Thorianite.

	I. Per cent.	II. Per cent.	III. Per cent.
Soluble in nitric acid —			
Thorium dioxide	72.24	76.22	78.86
Uranium ,, ...	11.109	12.33	6.03
Uranium trioxide	—	—	9.07
Cerium dioxide	6.39	8.04	1.02
Lanthanum and didymium oxides...	0.51		
Yttrium oxide	—	—	—
Lead ,, ...	2.25	2.87	2.59
Ferric ,, ...	1.92	0.35	0.46
Calcium ,, ...	—	—	1.13
Helium	—	—	0.39†
Titanium dixide	—	—	—
Phosphoric oxide	—	—	—
Insoluble in nitric acid—-			
Zirconium oxide	3.68	—	0.20
Silica..	1.34	0.12	
Residue form fusion with potassium hydrogen sulphate.....................	0.41	—	—

Of the three specimens of thorianite analysed the first was too small to admit of any treatment to separate associated minerals, and proved to contain associated zircon. Specimens II and III were separated as far as possible from extraneous minerals, and represent a nearer approach to the single mineral. The highest mount of thoria yet found is nearly 79 per cent., which shows that thorianite is the richest mineral in thoria at present known.

* Hilllebrand, *loc. cit.*

† If the whole of the gas is calculated as helium.

Constitution of Thoriante

From the analytical result, it will be seen that the constituents other than thoria show, as was to be expected in a mineral of this character, some variation. In Nos. I and II cerium dioxide occurs up to the extent of 8 per cent., while in No. III it becomes almost insignificant. The uranium appears to occur in the condition of both uranous and uranic exides. It is, however, intended to further investigate this question, since the presence of uranic ochre on the surface considerable superfincial oxidation, and that crystals may be found containing a much smaller proportion of uranic oxide. Of the other oxides present, zirconia in the insoluble residue may be safely classed as an impurity arising from zircons which are invariabley found with thorianite, sometimes included within the crystalline thorianite. Silica and ferrie oxide may be neglected as contaminations. Leaving out of consideration for the present the oxides of lead and calcium (not a constant constitutent) the mineral consists of a large amount of oxide of thorium and a small amount of oxides of uranium, with a smaller and varibale amount of cerium oxide, the precise significance of which is at present doubtful.

In the original condition the uranium may have existed entirely as dioxide. There can be little doubt that the dioxides of thorium and uranium, as well as certain of the salts of these metals, are isomorphous. In nature these oxides have not been found in a pure crystalline state, but crystals of each have been obtained artificially. Crystalline thoria was obtained by Troost and Aouvrard* whilst studying the relation between the double phosphate of potassium and thorium and that of potassium and zirconium. The phosphates were first obtianed by adding to fused potassium orthophosphate, thoria, thorium phosphate or anhydrous thorium chloride. On raising the double phosphate to such a temperature that both alkali and phosphoric acid were volatilised, thoria was obtained in crystals belonging to the cubic system; the cuboctahedron and rhombic dodecahedron being the forms obseved.† Again, uranium oxide was first obtained in an octrahedral form by Wohler by heating a mixture of uranium oxychloride with sodium chloride and ammonium chloride.‡ Later, Hillebrand repeated the experiment, and obtained practically pure uranium dioxide in black octahedral crystals of a specific gravity of about 11.*

The hydrate of thorium sulphate Th $(SO_4)_2 9H_2O$ is also, according to

*'Comptes Rendus,' 1886, vol. 102, p. 1422.

†*Loc. cit.*

‡'Liebig's Annalen,' 1842, vol 41, p. 345.

*' Zeitschrift für Anorg. Chemie,' 1893, vol. 3, p. 243.

Rammelsberg,* monoclinic in crystalline form and isomorphous with the corresponding hydrate of the uranium salt $U(SO_4)_2 9H_2O$.

Our analyses of different specimens of thorianite are hardly sufficiently numerous to enable us to conclude that the oxides of thorium and uranium bear a definite relation to one another in the mineral. It seems probable that the mineral belongs to the class of substances known as isomorphous mixtures of which the simple form would be represented by the formula XO_2 where X represents a tetravalent element the dioxide of which crystallises in the isometric system the extremes of which would be UO_2 and ThO_2.

Thorianite is evidently closely related to uraninite (pitchblende) in constitution. The crystalline form of the two minerals is the same and the constituents of both are similar. Hillebrand's analysis of the Branchville varieties of uraninte, which were stated to be nearly unoxidised, furnished 72.25 per cent of uranium dioxide and only 13.27 of urnium trioxide.† Uranium dioxide very readily oxidises on exposure to air, and it was only by completely excluding the air that Hillebrand succeeded in obtaining the pure oxide artificially. It is, therefore, certain that the natural varieties of the oxide so far examined must have suffered alternation and oxidation, and that originally their principal constituent was uranium dioxide. Most of the massive varieties of uraninite are very impure, and this may account in a large measure for the presence of many oxides in this mineral which appear to have very little in common with the principal constituents. Lead oxide is apparently generally present in thorianite. It occurs in small amount and may be combined as uranate. In the case of thorianite the crystallographic examination has shown that the mineral has crystallised with difficulty, and that it is more or less contaminated with those minerals which crystallised at the same time, as well as with impurities contained in the magma from which crystallisation occurred.

As far as the present invgestigation has gone it appears probable that thorianite is isomorphous with uraninite, and that in the thorianite of Ceylon some of the thoria is replaced by the corresponding uranium oxide. The evidence, however, is not sufficient to show whether this is a case of isomorphous mixture as seems probable, or of true chemical replacement.

It is obvious that the mineral is one of exceptional interest, and that it presents many problems for investigation, among them being the question of the possible occurrence of small quantities of hitherto little known or unknown elements.‡ The material furnishes a satisfactory source of pure

* 'Berl. Acad. Ber.', 1886, p. 603.

† 'Amer. Journ, Sci.', 1890, vol. 40, p. 384.

‡ Since this was written Hahn, in a communication to the Royal Society ('Roy. Soc. (contd. p.264)

264 *Prof. W.R. Dunstan and Mr. G.S. Blake.* *[May 1,*

thoria, a fact which is of commerical importance as well as of scientific interest.

We have shown that thorianite is chiefly composed of thoria, as at present understood. Baskerville believes that he has obtained evidence that the substance at present known as thorium is composed of more than one element, and that he has separated thoria into three oxides differing in density, and one of which is little, if at all, radio-active. These conclusions have, however, been called in question by Meyer and Gompertz,* who assert that thoria shows no evidence of being other than a single substance.

Determination of the rate of decay of the radio-activity of thorianite made in Lord Blythswood's Laboratory by Mr. H.S. Allen have shown that this property of the mineral is probably consistent with the thorium, uranium, and small amount of radium present.

The radio-activity of the mineral was measured in a parallel plate apparatus, using sufficient material to entirely cover the lower plate. The observed rate of leak under these conditions was 6900 divisions per minute equivalent to a current through the apparatus of approximately 5.5×10^{-11} amperes. Thorianite is, therefore, somewhat less active than some of the specimens of pitchblende examined by Madame Curie, whose results with this mineral are as follows:–

Pitchblende from	Johanngeorgenstadt......	8.3×10^{-11}	amperes
,, ,,	Joachimsthal	7.0×10^{-11}	,,
,, ,,	Pzibram........................	6.5×10^{-11}	,,
,, ,,	Cornwall	1.6×10^{-11}	,,

A series of measurements of the rate of decay of activity of the emanation from thorianite have been made; 16.5 grammes of the mineral were heated in a hard-glass tube, and the emanation, previously dried over phosphorus pentoxide, collected in the testing vessel. It was set aside for 6 hours in order to allow the thorium emanation to decay, and then measurements of the activity were made daily.

The results showed that during the first four days the rate of decay was greater than that observed by Rutherford for radium emanation, but that after this period the rate of decay of activity became identical with that of radium emanation. It is probable that the greater rate of decay during the first four days was due to the presence of thorium "activity" which,

Proc.,' 1905) has annonnced the existence in thorianite of small quantity of a new element, which produces the "thorium emanation." This was separated from the 6 cwt of the mineral referred to earlier in this paper. It must not be overlooked that the consignment, thought apparently worked up as a whole, was doubtless a mixture containing several other minerals than thorianite.

* 'Ber.,' 1905, vol.3, p. 187.

according to Rutherford, decays to half value in the course of 11 hours, whereas the radium emanation falls to half value only in about 3.7 days.

These results clearly indicate the presence of radium emanation in the "total emanation" from thorianite, and that consequently this mineral must contain radium.

Commercial Value of Thorianite.

Owing to the increasing employment of thoria for the manufacture of incandescent gas mantles, the demand for minerals containing thorium has largely increased. The demand is chiefly met from the deposits of sand containing a small percentage of monazite (phosphate of the cerium metals and thorium) which occur in Brazil and in North Carolina. Owing to the foreign control of these sands, British manufacturers have experienced difficulty in manufacturing thorium compounds. Thorianite is, we believe, the first deposit of a thorium mineral to be discovered on British territory. Consignments of thorianite from Ceylon, containing about 70 per cent, of thoria, have been recently sold in this country at the rate of £1500 per ton.

For the manufacture of thorium compounds thorianite possesses the advantage, not shared by any known thorium mineral, of containing uncombined thoria, soluble in nitric acid with formation of thorium nitrate.

A Variety of Thorianite from Galle, Ceylon.

By WYNDHAM R. DUNSTAN, M.A., LL.D., F.R.S., and B. MOUAT JONES B.A. (Oxon.), Assistant in the Scientific and Technical Department of the Imperial Institute.

(Received April 11, - Read May 10, 1906.)

In a previous paper* an account has been given of the composition and properties of a new mineral from Ceylon, chiefly composed of thorium dioxide (78 to 79 per cent.) to which the name thorianite was given. The mineral was shown to contain, besides thoria, a certain proportion of oxides of uranium varying in the three specimens analysed between 11 and 15 per cent., the uranium being present partly as dioxide and partly as trioxide, that is to say, in the same condition as in uraninite of pitchblende. It was shown that the evidence strongly supports the conclusion that thorianite and uraninite possess the same crystalline form and that they are isomorphous minerals. It is interesting to note that the new mineral naëgite, from Japan, which is essentially a silicate of uranium containing thorium, appears to be isomorphous with thorite (thorium silicate). It was suggested that the intimate association of thoria with oxides of uranium in thorianite is to be regarded as a case of isomorphous mixture. Such a mixture, or perhaps "solid solution," might result from the crystallisation of thorianite from a fused mixture of magma containing the oxides of both elements.

We have now obtained important confirmation of this view through the analyses of some unusually large crystalline fragments of thorianite received from the Galle District of Ceylon, which show that a still wider variation may occur in the proportions of the two oxides present in the mineral. This variety is composed of black lumps, usually of indefinite crystalline form, some of the pieces being apparently cubic. They are often partially covered with a brownish yellow substance containing a considerable amount of silica and probably derived from the associated rock. Some of the pieces weighed over 50 grammes, and were evidently portions of much larger masses. The fractured surfaces often showed a slightly less lustrous and more pitchy appearance than ordinary thorianite. The appearance of the mineral is thus intermediate between the small separate and cubical crystals of thorianite and the large masses without definite crystalline form, which are usually characteristic of uraninite. A small amount of material from Hinidumpattu in Galle consisted of small cubic crystal almost indis-

* Dunstan and Blake, 'Proc. Roy. Soc.,' A, vol. 76, 1905.

tinguishable in appearance from ordinary thorianite. In hardness, optical properties, density and general physical characters this variety of thorianite closely resembles the ordinary form of the mineral.

Analysis by the methods described in the previous paper* on thorianite shows, however, that it contains a much larger proportion (from 27 to 32 per cent.) of oxides of uranium and less oxide of thorium, whilst the minor constituents of the two forms of the mineral are seen to be similar in nature and quantity. It appeared possible that the massess might consist of uraninite crystallised on a nucleus of thorianite or *vice versa*, or that the two minerals might be separately crystallised within the masses. Analyses of different portions of one and the same piece have, however, given nearly identical results, and the oxides of the two elements appear to be uniformly distributed through the mineral, although there is some variation in the composition of specimens obtained by different collectors.

It may be noticed in the following table that the Analysis I of the small crystals from the Galle District shows rather more uranium and less thorium than that of the larger lumps (II to VI). If the uranium is calculated as UO_2, the molecular ratio of ThO_2 to UO_2 in the small crystals is almost exactly 2 to 1, but there is no evidence that this is more than a coincidence. No. VII was large crystal, 8 mm. cube, of ordinary thorianite from Balangoda.

The fact that thoria in thorianite is naturally associated with quantities of oxides of uranium varying from 11 to over 30 per cent., confirms the conclusions, indicated in the previous paper, that the oxides of thorium and uranium are present in thorianite in that intimate association known as "isomorphous mixture."

Different specimens of thorianite from the Galle district have furnished 58.84, 62.16, 62.30, 63.36 and 66.82 per cent., of thoria, whilst common thorianite from other localities has furnished from 76 to 79 per cent. of thoria.

The relations of thorium and uranium in minerals is a subject of some importance in connection with the present developments of the theory of the chemical elements. Attention has been directed recently to the subject by the Hon. R. J. Strutt.† It may be noticed in this connection that the percentages of thoria recorded by Mr. Strutt in his paper have been calculated from the observed radio-activity of the minerals, and not directly determined by chemical analysis. A comparison of the two series of percentages shows differences which in some cases are considerable. The calculated results are, however, only to be regarded as roughly approximate.

* *Loc. cit.*

† 'Proc. Roy. Soc.,' A, vol. 76, p. 88 1905.

The Galle variety of thorianite, as was to be expected, is radio-active, and it contains helium.

Mr. Strutt has kindly examined the radio-activity of this variety, and compared it with ordinary thorianite. The method of comparison was as follows:–"The quantity taken in each case was about one-tenth of a gramme. It was dissolved in strong nitric acid, and most of the excess of acid driven off by evaporation. The solution was diluted and exactly neutralised with ammonia, to prevent any injurious effect of the acid fumes on the apparatus. The solution was then made up to a standared volume.

The solutions of the two minerals were placed in exactly similar test-tube wash-bottles. These vassels were selected to have eactly the same diameter, and the inner tube in each case dipped to exactly the same depth in the solution. A continuous current of air could be drawn through either of them alternately into an electroscope. A two-way stop-cock made it easy to exchange one wash-bottle for the other. Constant suction was secured by the obvious device of letting in air through a deeper wash-bottle attached to a T-piece on the other side of the electroscope was thus equal to the difference of depth between this bottle and the wash-bottle containing the active solution. The current was regulated once for all by a stop-cock, and was not sufficiently rapid to appreciably disturb the gold leaf of the electroscope.

In making the comparisons, air was first drawn through the solutions long enough to expel all accumulated radium emanation. Three measurements of the rate of leak due to the first solution were taken, then three with the other solution; then three more with the first, and so on. The mean of each of these sets was compared with that of it successor. The mean ratio so obtained was corrected for the slight difference in the quantities of material taken. The correction (or normal leakage of the electroscope) was too small to be worth applying. Two samples of each mineral were weighed out, and compared in the manner above described in every combination. Four values for the ratio of thorium activities in the two minerals were thus obtained.

They were as follows:–

$$B/A = 1.19,\ 1.12,\ 1.10,\ 1.24. \quad \text{Mean, } 1.16.$$

The specimen of thorianite B used in this comparison contained 78.86 per cent. of thoria and 15.1 per cent. of uranium oxides. The specimen of the Galle variety A used was that numbered I in the table of analyses. The results of the radio-activity determinations are in fair accordance with the proportions of uranium and thorium determined by analysis."

In the following table of analyses, I represents the compostion of the small crystals from Hinidumpattu, Galle District; II to VI represent the compositions of the large lumps from the same locality; II, III, and IV are analyses of different parts of the same crystal, III being that of the outer layer; VII is an analysis of a large crystal of ordinary thorianite from Balangoda.

Table of Analyses.

	I.	II.	III.	IV.	V.	VI.	VII.
ThO_2	58.84	62.16 }	66.82 {	–	62.32	63.36	78.98
$(Ce, La, Di)_2O_3$	0.85	1.84 }		–	2.24	1.16	1.47
Y_2O_3	–	–	–	–	–	–	–
UO_2	} 32.74 {	10.32 }	28.24	28.68	27.02	27.99	13.40
UO_3		18.88 }					
PbO	2.56	2.29	2.29	2.50	2.99	2.90	2.54
Fe_2O_3	1.31	1.11	1.22	2.43	2.28	1.27	0.87
CaO	0.19	0.59	0.54	–	0.50	0.85	0.91
H_2O	1.26	1.05	1.00	–	2.16	1.32	1.28
Insoluble in nitric acid	0.45	0.77	0.56	0.54	0.87	0.77	0.47
He, CO_2 }	present	present	present	present	present	present	present

Mineralogists have often assigned a new name to a mineral found in a new locality when it has differed essentially in composition from the previously known mineral. It does not, however, seem desirable, in the light of present knowledge, to regard this variety of thorianite from Galle as a new mineral species, especially when it is remembered that the oxides of thorium and uranium in thorianite are not chemically combined and that from the probable mode in which the mineral has been formed, it is to be expected that considerable variations would naturally occur in the proportions of the two oxides in the mineral found in distinct localities which have therefore crystallised under different conditions. In fact, it is probable that further examination of other specimens from distinet localities may reveal the existence of a series of substances intermediate between the hypothetical pure uraninite, consisting of uranium dioxide, and the pure thorianite, consisting of dioxide of thorium.

Serendibite $Ca_2\ (Mg, Al)_6\ O_2\ [(Si, Al, B)_6\ O_{18}]$

Triclinic (+)

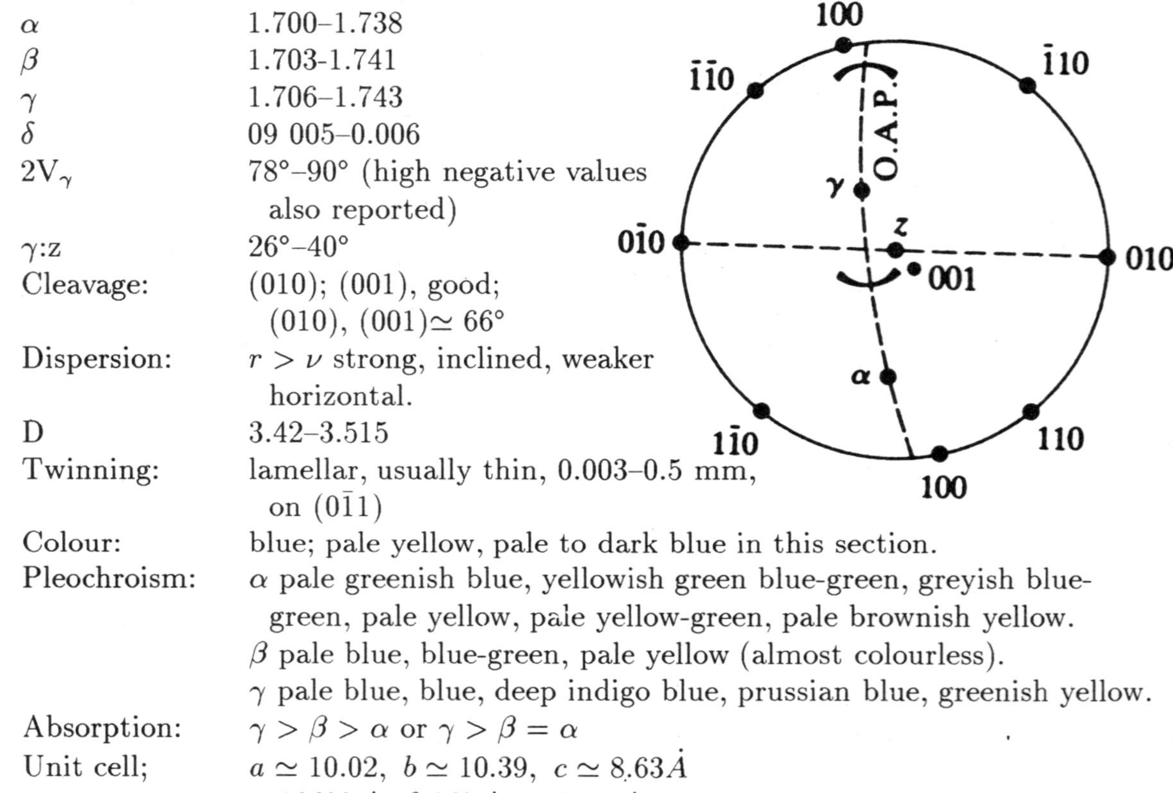

α	1.700–1.738
β	1.703-1.741
γ	1.706–1.743
δ	09 005–0.006
$2V_\gamma$	78°–90° (high negative values also reported)
γ:z	26°–40°
Cleavage:	(010); (001), good; (010), (001)$\simeq$ 66°
Dispersion:	$r > \nu$ strong, inclined, weaker horizontal.
D	3.42–3.515
Twinning:	lamellar, usually thin, 0.003–0.5 mm, on $(0\bar{1}1)$
Colour:	blue; pale yellow, pale to dark blue in this section.
Pleochroism:	α pale greenish blue, yellowish green blue-green, greyish blue-green, pale yellow, pale yellow-green, pale brownish yellow. β pale blue, blue-green, pale yellow (almost colourless). γ pale blue, blue, deep indigo blue, prussian blue, greenish yellow.
Absorption:	$\gamma > \beta > \alpha$ or $\gamma > \beta = \alpha$
Unit cell;	$a \simeq 10.02,\ b \simeq 10.39,\ c \simeq 8.63\dot{A}$ α 106°21′, β 96°4′, γ 124°21′ Z=2. Space group *Pbar*1.

Serendibite is isostructural with aenigmatite (see Table 65). The composition of serendibite (Table 68) varies a little from the ideal formula and there is commonly some replacement of (Mg,Al) by (Fe^{2+},Fe^{3+}) in octahedral sites. Compared with aenigmatite and rhönite, titanium is present in only very small amounts.

The refractive indices of serendibite are considerably lower than those of both aenigmatite and rhönite, and it is easily distinguished from these structurally related minerals by its very low birefringence. The dispersion is strong and the mineral may display anomalous interference colours. Values of $2V_\gamma$ between 78.5° to 82° for λ 460 and 600 nm respectively are reported by Hutcheon *et al.*, (1997). It is also distinguished from aenigmatite and rhönite by its absorption colours, the pleochroism showing mainly blue, green and yellow tints. Twinning is characteristically developed, and is commonly multiple with fine lamellae between 0.003 and 0.5 mm in thickness.

The paragenesis of serendibite is restricted to skarns, in which boron metasomatism has been active, at carbonate with acid rock contacts. Thus at Gangapitiya, Sri Lanka, the locality from which serendibite (Table 68, anal. 3) was first described (Prior and Coomaraswamy, 1903), it occurs in association with diopside, spinel, scapolite, apatite and plagioclase, in a contact zone between limestone and granulite, while at Johnsberg, Warren County, New York (Larsen and Schaller,

Table 68. Serendibite Analyses

	1	2	3	4	5
SiO_2	26.30	20.85	25.33	25.02	22.08
TiO_2	–	0.06	–	0.26	0.17
Al_2O_3	34.05	40.20	34.96	32.46	30.68
B_2O_3	8.37	[5.57]	4.17	6.65	6.95
Fe_2O_3	–	–	–	2.32	5.05
FeO	2.76	3.48	4.17	4.02	6.06
MnO	–	–	–	0.16	0.16
MgO	15.44	12.71	14.91	12.50	12.92
CaO	13.30	17.11	14.56	15.22	14.93
Na_2O	–	0.02	0.51	–	–
K_2O	–	–	0.22	–	–
H_2O^+	–	–	0.69	1.09	0.75
H_2O^-	–	–	–	0.26	0.25
Total	100.22	100.00	100.00	99.96	100.00
α	1.701	1.700	$\simeq$ 1.7	1.719	1.738
β	1.703	1.703	–	–	–
γ	1.706	1.706	–	1.725	1.743
$2V_\gamma$	$83°^a$	80°	–	$\simeq$ 90°	$\simeq$ 90°
$\gamma : z$	–	–	–	–	–
D	3.376	–	3.42	3.453	3.515

	1	2	3	4	5
	Nos. of ions	**on the basis**	**of twenty O**		
Si	3.022	2.489	3.056	3.037	2.716
B	1.660 –6.00	1.148 –6.00	0.872 –6.00	1.393 –6.00	1.475 –6.00
Al	1.318	2.363	2.072	1.570	1.809
Al	3.294	3.297	2.922	3.081	2.641
Fe^{3+}	–	–	–	0.212	0.467
Ti	–	0.005	–	0.024	0.016
Mg	2.644 –6.20	2.261 –6.20	2.693 –6.04	2.262 –6.01	2.368 –6.13
Fe^{2+}	0.265	0.347	0.423	0.409	0.623
Mn	–	–	–	0.017	0.016
Ca	0.000	0.090	–	0.000	0.000
Ca	1.638	2.099	1.891	1.980	1.968
Na	– –1.64	0.004 –2.10	0.120 –2.04	– –1.98	– –1.97
K	–	–	0.024	–	–

$^{a}2V_{\alpha}$

1 Serendibite, grenville limestone-granite contact, Johnsburg, warren county, New York (Larsen and Schaller, 1932). Anal. W.T. Schaller.
2 Serendibite, marble-granite contact, Melville Peninsula, Franklin district, Canada (Hutcheon *et.al.*, 1977),(B_2O_3 by differenc.)
3 Serendibite, contact zone between limestone and granite, Gangapitiya, Kandy, Sri Lanka (Prior and Coomaraswamy, 1903). Anal. G.T. Prior (includes $P_2O_5$0.48).
4 Serendibite. skarn. Tayeznoye, South Yakutia, Siberia (Pertsev and Nikitina, 1959). Anal. I. B. Nikitina.
5 Serendibite. skarn. Tayeznoye, South Yakutia, Siberia (Pertsev and Nikitina, 1959). Anal. I. B. Nikitina.

1932), serendibite (anal. 1) with diopside and phlogopite, is present at the Grenville limestone–granite contact. At Riverside, California, serendibite is developed in thin replacement bands of hydrothermal contact metamorphosed dolomitic marbles. Here the associated minerals include diopside, grossularite, vesuvianite and wollastonite (Richmond, 1939).

A number of serendibite (anals. 4, 5) occurrences in spinel–diopside skarns have been descrihbed from the Tayezhnoye iron ore localities of southern Yakutia (Shabynin and Pertsev, 1956; Pertsev and Nikitina, 1959). The skarn formations are restricted to contacts between carbonate rocks and feldspathic schists, acid gneisses and migmatites, and have involved boron mineralization resulting from post magmatic solutions derived from an alaskite granite. Serendibite is associated with sinhalite, $MgAlBO_4$, and warwickite, $(Mg,Fe,Ti)_2\ BO_4$, in the carbonate rocks, and with tourmaline in the siliceous rocks. In the former the assemblages of serendibite with diopside, pargasite, diopside+phlogopite (all with calcite) appear to be stable. In the siliceous rocks, however, the early formed tournmaline has been replaced by serendibite, which itself has later been partially replaced by an almost colourless second generation tourmaline, presumably at a lower temperature. The replacement of tourmaline by serendibite (anal.2) is also reported from the contact zone between calcite marble and granite, Melvile Peninsula, Canada (Hutcheon *et al.*, 1977).

Serendibite is also found with sinhalite, warwickite and tourmaline, together with chromian tremolite, forsterite, spinel pyrrhotite and graphite in the skarns of the Handemi district, Tanzania (Bowden *et al.*, 1969).

References

Bowden, P., von Knorring, O. and Bartholomew, R. W., 1969. Sinhalite and serendibite from Tanzania. *Min. Mag.*, **37**, 145–146.

Hutcheon, I., Gunter, A. E. and Lecheminant, A. NB., 1977. Serendibite from Penrhyn Group marble, Melville Peninsula, District of Franklin. *Can. Min.*, **15**, 108–112.

Larsen, E. S. and Schaller, W. T., 1932. Serendibite from Warren County, New York, and its paragenesis. *Amer. Min.,* **17** 457–465.

Pertsev, N. N. and Nikitina, I. B., 1959. [New data on serendibite.] *Zap. Vses. Min. Obsch.,* **88**, 169–172 (in Russian).

Prior, G. T. and Coomaraswamy, A. K., 1903. Serendibite, a new borosilicate from Ceylon. *Min. Mag.,* **13**, 224–227.

Richmond, G. M., 1939. Serendibite and associated minerals from the New City Quarry, riverside, California, *Amer. Min.,* **24**, 725–726.

Shabynin, L. I. and Pertsev, N. N., 1956. [Warwickite and serendibite from magnesium skarn of southern Yakutia.] *Zap. Vses. Min. Obsch.,* **85**, 515–528 (in Russian).

Note added in proof: Welshite, a new beryllium-bearing member of the aenigmatite group with composition $Ca_2Mg_4Fe^{3+}Sb^{5+}O_2[Si_4Be_2O_{18}]$, and cell parameters a 9.68, b 14.77, c 5.14Å, β 101°30′, pseudomonoclinic, has been described by Moore (Min. Mag., **41**, 129–132, 1978). The mineral, from the manganese and iron mines at Langban, Sweden, is deep reddish-brown to reddish black in colour, with α = 1.81, γ 1.83, $2E \simeq 45°$, specific gravity 3.77.

Vol. 36 March 1967 No, 277

Anandite a new barium iron silicate from Wilagedera, North Western Province, Ceylon

By D. B. PATTIARATCHI

Geological Survey Department, Ceylon

and EKO SAARI and TH. G. SAHAMA

University of Helsinki, Finland

[Read 9 March 1967]

Summary. Anandite is a new barium iron silicate found in the magnetite ore zone of the Wilagedera iron ore body, in the North Western Province of Ceylon. The mineral is named after the late Dr. Ananda Coomaraswamy, the first director of the Mineral Survey of Ceylon.

Anandite is monoclinic, with a 5.412, b 9.434, c 19.953$A°$, β 94°52′, space group C2/c. D 3.94, hardness 3–4. Optically positive, $b \| \beta$, $\gamma \wedge a$ 12°, $\beta 1.85_5$, $\gamma > 1.88$. Pleochroism: β green, γ brown. Chemical composition corresponds to the simplified formula (Ba,K) $(Fe,Mg,)_3$ $(Si,Al,Fe)_4O_{10}(O,OH)_2$ with $Z = 2$. The data available indicate that the mineral has the trioctahedral structure of the brittle micas.

In 1964 during detailed examination of drill cores of the Wilagedera iron ore prospect in the North Western Province of Ceylon by one of the authors (D.D.P.) monomineralic bands of a block lustrous almost opaque mineral were found within the iron-ore-bearing zones. The optical properties of this mineral could not be determined due to the high body absorption. A rapid chemical analysis carried out by J. P. R. Fonseka in the Geological Survey laboratories gave a BaO content of nearly 19% . A study of the physical properties of the mineral indicated the possiblity that it was a new species. X-ray and further chemical analysis were carried out in the Geological Survey laboratories (Colombo, Ceylon), at Institute of Geology of the University of Helsinki (Finland), and at the Department of Geology of Leeds University (England). The results of this work confirm that this mineral is a new species of brittle micas with the simplified formula $(Ba,K)(Fe,Mg)_3(Si,Al,Fe)_4O_{10}(O,OH)_2$.

The new mineral is named *anandite* in honour of the late Dr. Ananda Coomaraswamy, the first director of the Mineral Survey of Ceylon, in recognition of his many contributions to the geology and mineralogy of Ceylon. The name has been approved by the International Mineralogicl Association Commission on New Minerals and Mineral Names.

Occurrence. The Wilagedera iron ore prospect, which was the first banded magnetite deposit found in Ceylon, occurs within a succession of calc-schists and calc-gneisses of Precambrian age (Fernando, 1958 and 1960, 1961-62).

TABLE 1. Powder pattern of anandite. Filtered iron radiation with internal silicon standard.

hkl	Intensity	$d_{meas.}$	$d_{calc.}$	*hkl*	Intensity	$d_{meas.}$	$d_{calc.}$
*002	60	9.92	9.95	*026	50	2.716	2.712
*004	85	4.99_5	4.98	*131	45	2.681	2.676
022	10	4.27	4.26	$\bar{1}17$	15	2.521	2.513
*113	20	3.700	3.700	*008	80	2.490	2.485
*024	40	3.430	3.430	028	20	2.240	2.238
*006	100	3.320	3.315	0.0.10	35	1.991	1.988
115	25	3.165	3.150	0.0.12	35	1.660	1.657
115	25	2.929	2.928				

* Line used for calculating the unit cell dimensions.

The main ore mineral present is magnetite and is associated with chalcopyrite, pyrite, and pyrrhotine. The magnetite-bearing ore zone is capped by banded magnetite-baryte rock the origin of which is problematical. The ore zone and the interbanded succession of calc-schists and gneisses are of meta-sedimentary origin, the sediments having been subject to high temperature and pressure conditions appropriate to the granulite facies (Pattiaratchi, 1961).

Anandite occurs generally as distinct monomineralic bands $\frac{1}{4}$ in. to 2 in. or more in thickness closely associated with magnetite and suphide minerals. Occasionally the mineral is disseminated as lenses, which are, however, distinct due to the black lustrous property of the mineral.

X-ray crystallography. The mineral was studied with both single crystal and powder X-ray methods. A series of Weissenberg and precession photographs revealed a monoclinic symmetry with the following types of reflections: hkl with $h + k = 2n$, $h0l$ with $l = 2n(h = 2n)$ and $0k0$ with $k = 2n$. Accordingly, the space group is $C2/c$ (or Cc). The uinit-cell dimensions measured from single crystal photographs are: a 5.42/$A°$, b 9.45 $A°$, c 19.97 $A°$, β 95°.

The powder pattern of the mineral recorded with a Philips Norelco

diffractometer is reproduced in table. I. The indexing of the powder lines was made using the IBM 1620 computer of the University of Helsinki. The intensities of the reflections were checked on the single crystal photographs. The unit cell dimensions calculated from the powder pattern are: a 5.412 ± 0.005 $A°$, b 9.434 ± 0.005 $A°$, c 19.953 ± 0.01 $A°$, β 91°52′ ± 10′. These values agree reasonably well with the data obtained from single crystal photographs. The values calculated from the powder pattern are considered to be more accurate and are adopted in this paper.

A complete crystal structure analysis of the mineral has been undertaken by Dr. A. Vorma, of the Geological Survey of Finland. As pointed out by him, the X-ray data show a striking resemblance to the mica minerals representing the polytype $2M_1$.

Physical and optical properties. In hand specimen, the mineral is black in colour. No crystals with measurable faces have been found. The only prominent face is the perfect lustrous cleavage plane parallel to {001}. Occasionally flakes are observed that have hexagonal outlines and probably indicate poorly developed prism faces. Hardness 3–4.

The mineral is transparent only in very thin splinters making the determination of optical properties difficult. Further, due to the perfect basal cleavage, sections thin enough for optical examination could be ground only parallel to the celeavage plane. On the cleavage flake the

following refractive indices were measured in sodium light using high-index immersion liquids: γ' 1.88 ± 0.01 and α' 1.85_5 ± 0.01. The interference figure obtained on the flake indicates that the obtuse bisectrix is nearly perpendicular to {001} and that the optic normal is parallel to α'. Universal stage work showed that the true α direction makes an angle of 12° ± 4° with the crystallographic a-axis.

The optical properties may be summarized as follows: b parallel to β; $\gamma \wedge a$ 12° ± 4°; optically positive; β 1.85_5 ± 0.01; $\gamma > 1.88$; pleochroic with β green and γ brown; strong dispersion. The optic axial angle could not be measured.

The specific gravity of the batch chemically analysed by von Knorring was determined by suspending apparently pure fragments in water. The value 4.00 was obtained. Correcting for the pyrite contamination this value yields 3.94 as the specific gravity of pure anandite.

Chemical composition. Two independent chemical analyses were made, the one by Mr. Fonseka and Mr. de Silva, of the Geological Survey of Ceylon, and the other by Dr. von Knorring, of Leeds University. The results of these analyses are shown in table II. Between some of the cleaveage flakes thin flims of an opaque material were observed. A small amount of this opaque material was isolated by Dr. Vorma, of the Geological Survey of Finland, and was identified as pyrite by a powder pattern. It is extremely probable that all the sulphur shown by the chemical analyses is contained in the admixed pyrite. Table II gives also the results of the analyses corrected for admixed pyrite and for H_2O. The results of the two analyses agree reasonably well with each other although the analyses were carried out independently and on different samples. The atomic ratios (based on $O + OH = 48$) calculable from both analyses are included in Table II.

Acknowledgements. The authors thank Mr. J. P. R. Fonseka, Senior Chemist, and Mr. N. R. de Silva, Chemist of the Geological Survey of Ceylon laboratories, and Dr. Oleg von Knorring, Reader at Leeds University, England, for the chemical analyses; Dr. A. Vorma, of the Geological Survey of Finaland, for his advice and assistance in carrying out this work; and Mr. L. J. D. Fernando, director of the Geological Survey of Ceylon, for his assistance and kind permission to publish this paper.

References

FERNANDO (L. J. D.), 1958 and 1960. Administration Reports of the Government Mineralogist, pp. R. 28 and R. 32, respectively.

—-Administration Report of the Director, Geological Survey, for 1961–62, pp. R. 26 and R. 27.

PATTIARATCHI (D. B.), 1961. Some aspects of the genesis of iron ores in Ceylon (Abstract), indent Proc. 17th Ann. Sess. Ceylon Assn. Adv. Sci.

[*Manuscript received 10 August 1966*]